46945

BIOTECHNOLOGY IN AGRICULTURE SERIES

General Editor: Gabrielle J. Persley, Science Adviser, The Agricultural Research Group, Environmental and Sustainable Development, The World Bank, Washington DC, USA.

For a number of years, biotechnology has held out the prospect for major advances in agricultural production, but only recently have the results of this new revolution started to reach application in the field. The potential for further rapid developments is, however, immense.

The aim of this book series is to review advances and current knowledge in key areas of biotechnology as applied to crop and animal protection, forestry and food science. Some titles focus on individual crop species, others on specific goals such as plant protection or animal health, with yet others addressing particular methodologies such as tissue culture, transformation or immunoassay. In some cases, relevant molecular and cell biology and genetics are also covered. issues of relevance to both industrialized and developing countries are addressed and social, economic and legal implications are also considered. Most titles are written for research workers in the biological sciences and agriculture, but some are also useful as textbooks for senior-level students in these disciplines.

BIOTECHNOLOGY IN AGRICULTURE SERIES

New Diagnostics in Crop Sciences

Edited by

J.H. Skerritt and R. Appels

Division of Plant Industry
CSIRO
GPO Box 1600
Canberra, ACT 2601
Australia

CAB INTERNATIONAL

CAB INTERNATIONAL
Wallingford
Oxon OX10 8DE
UK

Tel: +44 (0)1491 832111
Telex: 847964 (COMAGG G)
E-mail: cabi@cabi.org
Fax: +44 (0)1491 833508

A catalogue entry for this book is available from the British Library.

ISBN 0 85198 934 9

Printed and bound at the University Press, Cambridge

Contents

Contributors

Appels, R. *CSIRO, Division of Plant Industry, PO Box 1600, Canberra, ACT 2601, Australia.*

Bakker, J. *Wageningen Agricultural University, Department of Nematology, PO Box 8123, 6700 ES Wageningen, The Netherlands.*

Chu, P. *CSIRO, Division of Plant Industry, PO Box 1600, Canberra, ACT 2601, Australia.*

Condon, A.G. *CSIRO, Division of Plant Industry, and Cooperative Research Centre for Plant Science, GPO Box 1600, Canberra, ACT 2601, Australia.*

Cooke, R.J. *National Institute of Agricultural Botany, Huntingdon Road, Cambridge CB3 0LE, UK.*

Dewey, F.M. *Department of Plant Sciences, University of Oxford, South Parks Road, Oxford OX1 3RB, UK.*

Eagling, D.R. *Institute of Horticultural Development, Victorian Department of Agriculture, Private Bag 15, South Eastern Mail Centre, Victoria 3176, Australia.*

Gee, S.J. *University of California, Department of Entomology, Davis, CA 95616, USA.*

Gresshoff, P.M. *Plant Molecular Genetics, Institute of Agriculture and Center for Legume Research, The University of Tennessee, Knoxville, Tennessee 37901-1071, USA.*

Griep, R. *Laboratory for Monoclonal Antibodies, PO Box 9060, 6700 GW Wageningen, The Netherlands.*

Hammock, B.D. *University of California, Department of Entomology, Davis, CA 95616, USA.*

Howes, N.K. *Agriculture and Agri-Food Canada Research Station, 195 Dafoe Road, Winnipeg, Manitoba, Canada R3T 2M9.*

Konishi, T. *Faculty of Agriculture, Kyushu University, Fukuoka 812, Japan; Present address: 294 Okada, Mabi, Okayama, 710–13 Japan.*

McCleary, B.V. *Megazyme (Australia) Pty Ltd, 2/11 Ponderosa Parade, Warriewood, NSW 2105, Australia.*

Morgan, M.R.A. *Department of Food Molecular Biochemistry, Institute of Food Research, Norwich Research Park, Colney, Norwich NR4 7UA, UK.*

Reeves, J.C. *Molecular Biology and Diagnosis Section, National Institute for Agricultural Botany, Huntingdon Road, Cambridge CB3 0LE, UK.*

Richards, R.A. *CSIRO, Division of Plant Industry, and Cooperative Research Centre for Plant Science, GPO Box 1600, Canberra, ACT 2601, Australia.*

Schots, A. *Laboratory for Monoclonal Antibodies, PO Box 9060, 6700 GW Wageningen, The Netherlands.*

Skerritt, J.H. *CSIRO, Division of Plant Industry, PO Box 1600, Canberra, ACT 2601, Australia.*

Sward, R.J. *Institute for Horticultural Development, Victorian Department of Agriculture, Private Bag 15, South Eastern Mail Centre, Victoria 3176, Australia.*

Thornton, C.R. *Department of Plant Sciences, University of Oxford, South Parks Road, Oxford OX1 3RB, UK.*

Waterhouse, P. *CSIRO, Division of Plant Industry, PO Box 1600, Canberra, ACT 2601, Australia.*

Preface

New Diagnostics in Crop Sciences reflects a major trend in plant science, where there is a move away from analytical scientists being practitioners in single techniques, to focusing on biological problems. The focus is therefore on the best approach to the solution to the problem. Even if individuals and laboratories specialize in a particular technique, it is becoming clear that multidisciplinary approaches provide the most efficient route to new diagnostic technology. For example, an immunodiagnostics group may utilize many of the methods of protein chemistry and molecular biology in developing recombinant antibodies.

The general principles underlying the development and application of nucleic acid and antibody diagnostic techniques are described to indicate the inherent strengths and weaknesses of the processes involved. Newer approaches in both disciplines (e.g. PCR and non-radioactive reporters for nucleic acid diagnostics and new immunoassay formats for immunoassay) are covered. The book is not intended to be a detailed methods manual but rather to provide practical information to assist plant scientists in considering alternative techniques for their research. We have also defined 'diagnostics' in a broad sense, to include both biochemical techniques based on antibodies and nucleic acid probes and instrumental methods such as carbon isotope discrimination and near-infrared reflectance spectroscopy. In many cases the instrumental methods involve less sample handling and are thus better suited to 'on-line' or 'real-time' analysis of samples by staff with minimal training.

Chapter 1, by the Editors, provides an overview of the principles behind the methods described throughout the volume, as well as discussing methods not covered elsewhere in the text, such as near-infrared spectroscopy and nuclear magnetic resonance. Chapter 1 also addresses experiences in the commercialization of diagnostic methods and kits in the plant sciences.

Correct identification of crop plant varieties (Chapter 2) is important because different varieties of plants will differ in their disease resistance, response to environmental conditions and in suitability for different end-uses. The need to be able to readily identify and distinguish varieties has received added impetus from the introduction of plant variety rights in many countries. This brings an onus to demonstrate that the new variety is distinguishable, uniform and stable in its characteristics. A wide variety of techniques have been employed to do this, ranging from molecular/genetic analyses, protein electrophoresis and HPLC, image analysis and simple chemical or enzymic staining methods.

Chapter 3 includes an overview of the specific challenges and methodology in the preparation of monoclonal antibodies to plant targets, including purification of antigens and the development of appropriate screening assays. Newer approaches to the preparation of monoclonal antibodies, including the use of recombinant antibodies and phage display selection, and the expression of antibody genes in plants are reviewed.

Chapter 4 describes an application of antibody techniques to the prediction of cereal quality and genetic composition. Identification of several of the grain proteins correlating with end-use quality attributes makes possible the development of simple antibody methods to predict such properties as dough characteristics, baking performance or cooking quality of products such as pasta, in breeding programmes. Chromosome arm probes are useful in aiding the identification of new disease resistance loci, and in large-scale screening of seed for resistance-linked characters. The use of DNA markers (Chapter 5) has allowed the analysis of complex traits that underpin many breeding programmes, so that these traits can be more clearly understood as well as providing alternative screening strategies for early assessment of the characters.

An important application of diagnostics has been detection of diseases, borne by viral, fungal and bacterial means. Chapter 6 describes the application of a wider range of nucleic acid techniques in testing for seedborne diseases. Fungal infections account for major losses in crop yields, especially in wet seasons. Often infection in its early stages, while it is still treatable, can be difficult to detect by eye or microscope, and can be mistaken for nitrogen or water stress. Simple antibody tests are described (Chapter 7) for several major pathogens of commercially important crops. Rapid screening procedures of plant materials for a range of viral infections have been developed for assessing quarantine material. These include both antibody and nucleic acid approaches, as described in Chapters 8 and 9, respectively.

Immunoassays for plant toxins (Chapter 10) have been used for monitoring the safety of human and animal plant foods for some years. The best-known naturally occurring toxins on plant commodities are the mycotoxins, and with their human and animal toxicity at parts per billion or trillion levels, there have been a large number of specific immunoassays developed over the last decade, along with commercial test kits from almost a dozen manufacturers. Other plant-derived toxins can have significant influences on human and animal health, and more recently have provided targets for development of immunoassays. Detection and quantification of agrochemicals in several stages of plant production and processing is critical in a political and trading sense, because consumers are increasingly concerned with environmental quality and purity of foods we eat. Immunoassays for agrochemicals (Chapter 11) provide a simple means for sensitive monitoring of residues in either the field or in laboratory situations.

By use of innovative substrates, and provision of reagents in kit form, long-established enzyme assays can become more accessible to the non-specialist laboratory. Chapter 12 describes simple assays for polysaccharide-degrading enzymes in plants using chromogenic and colorimetric substrates. Analysis of isozyme variation in plants (Chapter 13) has also been a valuable technique for assessment of genetic variation in plant populations. Linkage of certain isozyme variants to disease-resistance traits provide a fast alternative to tracing the respective disease resistance genes in a crossing programme. Finally, Chapter 14 describes the use of carbon isotope discrimination analysis in plant improvement. Early screening of crop plants in a breeding programme for high yield components such as water use and carbon metabolism efficiency has a significant potential for improving selection procedures. Carbon isotope discrimination is based upon the measurement of the C-13 / C-12 ratio in young leaves, using a mass spectrometer, and is being investigated by breeding programmes worldwide. The theory underlying the measurement as well as its application is discussed.

We would like to acknowledge the assistance of our colleagues and collaborators both within and outside CSIRO in advising us on chapter content and illustrations, and providing independent reviews of the chapters.

John Skerritt and Rudi Appels
Canberra, December 1994

An Overview of the 1
Development and Application
of Diagnostic Methods in
Crop Sciences

J.H. Skerritt and R. Appels
CSIRO, Division of Plant Industry,
PO Box 1600, Canberra, ACT 2601, Australia.

1.1 Introduction

The spirit in which this book has been approached is to target broad issues rather than specific disciplines, and thereby enhance the process of multi-disciplinary inputs for solving diagnostic problems. The rapid pace of developments in molecular biology, for example, has created new opportunities in diagnostic areas that were normally tackled by isolated scientific disciplines. Advances in molecular biology highlight a more general trend in many scientific investigations of combining developments in so-called traditional areas with those in areas not usually associated with them. Diagnostic methods in modern biology, coupled with computerized data storage and manipulation, are able to generate large volumes of numerical data. A potential problem associated with this general trend is the failure to fully exploit large databases acquired using multidisciplinary methods. This may be because individual scientists do not fully understand the significance and limitations of the diversity of basic principles underlying the techniques used to acquire the information. Even when the basis of each parameter is understood completely by the user, there is the need to decide the relative importance of individual variables. For example, over a dozen different analyses are commonly performed in breeding programmes in the screening of barley lines for potential malting quality, and lines that may score favourably on one parameter may be deficient in others.

Diagnostics can be viewed as a discipline in its own right, combining a wide range of techniques in developing simple, fast and reproducible (SFR) procedures that measure a feature of the biological material in hand in a way

1

that is readily interpretable. In many cases, such as plant breeding, the numbers of assays required can usually amount to hundreds per week. If there are more than four or five separate steps required to complete the assay, one or several of the SFR criteria begin to suffer and the respective diagnostic test stops being attractive to the non-expert user.

1.2 Sample Preparation in Diagnostics

The 'ultimate' in diagnostic tests is where some form of radiation can penetrate the sample non-destructively and its wavelength and/or level is changed as a result of interacting with the sample. Near-infrared reflectance (NIR) or transmittance (NIT) technology, for example, allows information to be collected about the chemical composition of cereal seed endosperm, without grinding the sample into a powder, by comparing the properties of the reflected or transmitted radiation to the incident radiation. Other forms of radiation such as X-ray analyses can provide structural information by utilizing the different absorption or activation properties of material in the sample under study as well as information about the presence of small molecules.

Although increased specificity and resolution are obtained using nucleic acid and antibody assays, one of their main shortcomings is that they are 'wet chemical' in nature. Destruction of the sample before analysis usually involves a grinding step or squashing step. Where an assay can be accomplished on sap from a leaf, such as assaying for the presence of viral particles or specific DNA sequences, a simple set of rollers (Fig. 1.1) provide a fast procedure for working on very small samples. It is of little advantage to replace a 10 hour instrumental assay with a 1 hour immunoassay if the sample preparation is not also simplified. With regard to the application of antibody-based procedures for analysing grain samples, several approaches have been tried. These include analysis of sectioned grain (Rasmussen, 1985) and use of uncrushed quarter grains in 'Yes–No' analyses (e.g. detection of rye chromosomal translocations; Howes *et al.*, 1989). Solubilization of antigen from the cut face of the grain is usually adequate if the particular assay is of sufficient sensitivity. While these approaches have worked well in qualitative tests, they are not suitable for quantitative immunodiagnostic assays. A small scoop or syringe can be used to measure wholemeal by volume instead of needing to weigh large numbers of small samples.

Apart from time considerations, the nature of the target antigen may also give challenges. A key problem which underlies many sample preparation methods is the need to solubilize the molecules that form the basis for the assay being employed. Lipophilic molecules require the use of organic solvents; a water-soluble solvent can be used for sample preparation, followed by sample extraction with an immiscible solvent. In this case, the

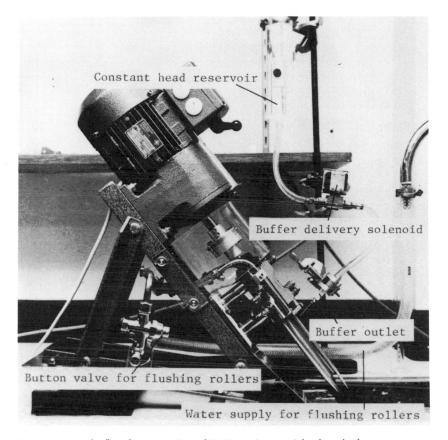

Fig. 1.1. Use of rollers for extraction of DNA or virus particles from leaf sap.

solvent can be evaporated off and the residue containing the target molecule redissolved in a miscible organic or detergent/buffer mixture. In some cases, antibodies can tolerate 10–20% of a polar miscible solvent such as methanol, without interference.

There are still some further technical limitations to the full utilization of antibodies in the ELISA analysis of complexes of macromolecules. In order to produce a solution, large aggregates may need mechanical disruption (e.g. by sonication) plus the use of detergents. Alternatively, reducing agents can be used to break disulphide bonds. While most antibodies can tolerate micromolar or millimolar concentrations of reductants, either exposure of new epitopes or loss of others can cause an undesirable change in the specificity of the assay. The 'tissue print' technique (Fig. 1.2) avoids the problems of extraction of antigens from plant tissue, and also has the advantage of being able to provide accurate information on the distribution

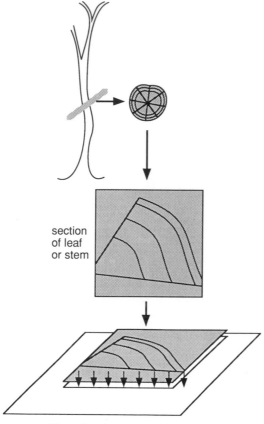

section
of leaf
or stem

Transfer of proteins and nucleic acids
in section to membrane. Molecules
immobilized on membrane can then
be assayed using antibodies or
nucleic acid probes

Fig. 1.2. Tissue print technique
for providing membrane replicas
of protein or nucleic acid
distribution in plant or stems, for
subsequent antibody or nucleic
acid probing.

of specific antigens in tissues – for example, plant virus distribution. Usually
tissue cross-sections have been used, but recently it has proven possible to
detect plant virus proteins on whole leaf blots simply by pressing the leaf on
to a dry nitrocellulose sheet, incubating in a solution of a protein such as
bovine serum albumin to block non-specific protein adsorption sites, and
then incubating with antibodies (Polston *et al.*, 1991; Taylor *et al.*, 1993). This
method should have considerable potential application to the detection of
sites of protein expression and deposition in transgenic plants (Manteuffel
and Panitz, 1993).

1.3 Instrumentation for 'Non-destructive' Measurements on Biological Samples

Discussion of some of these methods is relevant at this stage because many of them are distinguished by the relative lack of sample preparation required before analyses can be performed.

1.3.1 Broadband nuclear magnetic resonance spectroscopy (NMR)

Low resolution proton NMR has been applied to the analysis of components in plant material such as water and lipid. It provides a measure of the total hydrogen in the liquid phase of a material, and as such is especially of value in lipid analysis in plants since many plant oils are liquid.

The principle of the method (Andrew, 1955) is as follows. Molecules in liquids are in continuous motion; in water the protons are continually spinning and have some characteristics of a magnet with angular momentum. The angular momentum of the spinning proton in a gravitational field enables it to resist the magnetic torque of a magnetic field. The result of the two forces is called precession about the axis of the magnetic field. Nuclear magnetic resonance occurs when, in addition to the steady magnetic field

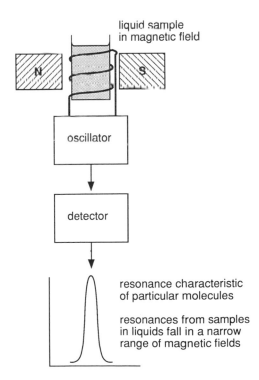

Fig. 1.3. Principle of broadband nuclear magnetic resonance (NMR) spectroscopy.

causing the precession, an oscillatory magnetic field, oscillating at the same rate as the precession, is applied at right angles to the steady field (Fig. 1.3). Under these circumstances the protons in the nucleus absorb energy from the oscillating magnetic field and this can be measured. Thus the sample tube is placed in a stable uniform magnetic field between the poles of a large magnet, and a coil around the sample tube, connected to an oscillator, is used to produce the oscillatory magnetic field. At a particular field strength and frequency of the oscillator, all of the protons in the water or oil in the sample tube resonate together, and energy is absorbed from the oscillator, according to the equation:

$$f = \tau H / 2\pi$$

where f is oscillator frequency, τ is gyromagnetic ratio for protons and H is magnetic field strength.

The energy absorbed by the oscillator is thus measured and used to measure the number of protons in the sample. The method is called low resolution NMR because it does not differentiate protons in the liquid phases from one another, even though they may be in different environments and thus have slightly different values of τ. In high resolution NMR, used for elucidation of chemical structures, the 'chemical shifts' in the resonance frequency of individual protons caused by other atoms nearby can be measured and used to provide information on the environment of each proton and hence the structure of the molecule.

NMR has the advantage that it requires no sample preparation and is non-destructive; for example whole individual oilseeds, corn kernels or cottonseeds can be first tested for oil content and then planted. Moisture can also be determined in grain, meal, flour or breadcrumb using the same instrument. Although the measurement of grain water content has been replaced by simpler moisture meters using the principle of conductivity, the NMR technology is still widely used for oil analysis (e.g. Tonnet and Snudden, 1974). Samples can either be analysed without sample preparation or following extraction with a solvent such as carbon tetrachloride (e.g. for epicuticular waxes in wheat (Johnson *et al.*, 1984)). In some situations NMR technology is simpler and faster than gravimetric or colorimetric chemical methods. Other applications include determination of rubber contents in guayule (Tonnet and Downes, 1983). The main limitation of the method is the need to thoroughly dry plant material to remove water.

1.3.2 Near-infrared technology

The near-infrared technology can, in theory, measure any molecule that absorbs radiation in the near-infrared part of the spectrum (Wetzel, 1983). The technology is very widely used in the grain, food and feed industries because of its speed, low cost per test and simplicity of routine operation;

either reflectance (NIR) or transmittance (NIT) can be measured. Its initial application was for moisture determination in seeds, although in theory NIR and NIT can specifically quantify any molecules that absorb infrared radiation, However, the application of NIR to measuring trace components in samples, such as plant hormones or chemical residues, is prevented from interferences by other molecules that have similar spectral characteristics but are present at higher concentrations. The same constraint has limited its application to quantification of specific macromolecules, e.g. particular proteins, within a mixture.

Most emphasis on the use of NIR has been on obtaining correlations between spectral characteristics and a desired parameter in the set of samples being tested, followed by using that correlation to predict the behaviour of subsequent sets. Advances in NIR hardware and computer software mean that combinations of data obtained at up to 200 different wavelengths can be analysed and calibrations developed using either multiple linear regression or principal component analysis in conjunction with partial least squares analysis. Thus it has been possible to obtain good correlations with such complex parameters as the resistance to stretching of a dough without understanding the chemical basis for the correlation.

Apart from analyses of moisture and protein content in cereal and legume grains, NIR and NIT are being increasingly evaluated as non-invasive test methods for prediction of aspects of grain milling characteristics, e.g. kernal hardness (Norris *et al.*, 1989), dough quality (Williams *et al.*, 1988), segregating US Hard Red Spring grades of wheat from other grades (Delwiche and Norris, 1993), and starch content of ground yellow corn kernels (Wehling *et al.*, 1993). Other applications include protein and fibre in silage (Hellmki and Moisio, 1983) and metabolizable energy in poultry feeds (Valdes and Leeson, 1992).

The ability to do these predictive analyses without understanding the chemical basis for such correlations is both a strength (in terms of making the method useful in the hands of routine laboratories) and a shortcoming. Relatively little research has been carried out with an aim of understanding the scientific basis of correlations. Understanding the scientific basis of specific correlations could aid in choice of 'robust' parameters, since a major shortcoming of many of the spectacular applications of NIR is that correlations involving complex parameters cannot be obtained consistently between harvest years or growth environments. With increasingly sophisticated software and enhancements in computer speed it should be possible to identify key wavelengths that consistently appear in correlation equations and relate these to general classes of molecule (e.g. glutenin polypeptides, amylopectin in starch from wheat flour). Use of genetic lines in which whole families of polypeptides are either deleted or overexpressed, together with flour blends, may assist in this endeavour. Different chemical groups (e.g. -SH, -SS, -OH) do have specific absorbances, so it may even be possible to use the technology to observe *in situ* certain interactions between molecules that are lost in the extraction step required for other biochemical diagnostic tests. Although the major application of NIR and NIT methods has thus far been for analyses of harvested crops, it may also be useful for the analysis of aspects of field performance, such as drought resistance in wheat (Flagella *et al.*, 1992).

1.3.3 Atomic absorption spectrometry

Although other spectroscopic procedures are now available, atomic absorption spectroscopy is still a widely used method for analysing plant material. The principle underlying this form of analysis is that all atoms absorb light at a wavelength that is characteristic for a particular element; the amount of light absorbed is proportional to the amount of the respective element present in the sample under examination. The electronic transitions induced by the absorption of specific wavelength of light characterizes each element because their atoms differ in electronic structure. The absorption recorded at specific wavelengths by atoms vaporized in a flame thus provides the basis for determining the concentration of specific elements in a sample, in combination with calibrators and comparison with standards. The spectral properties of a particular element are affected by the presence of other elements and corrections need to be applied in order to obtain correct estimates of the concentrations of specific elements.

1.3.4 X-ray fluorescent spectroscopy and inductively coupled plasma spectroscopy

Toxicities and deficiencies are diagnosed by measuring the concentration of an element in a tissue and comparing that value with a predetermined 'critical value' (below that value indicates deficiency) or 'threshold value' (above that value indicates toxicity). These analyses have been performed for several decades, but modern improvements include use of specific tissues instead of whole plants (e.g. youngest open leaf, grain), greater specificity of the fraction measured (e.g. inorganic phosphate rather than total phosphorus; sulphate rather than sulphur) and use of more versatile instrumentation such as inductively coupled plasma spectroscopy (ICPS) (Munter *et al.*, 1984) and X-ray fluorescent spectroscopy (XRFS; Hutton and Norrish 1977; Norrish and Hutton, 1977).

The principle of XRFS is as follows. When a sample is bombarded with X-rays the atoms absorb energy which excites electrons to move into orbitals further from the nucleus. When the atoms return to their steady states, and the electrons move back into their normal orbitals, they give up their excess energy as X-rays. The emitted energy (fluorescence) has a wavelength that is characteristic of the respective element. The X-ray fluorescence can be separated in a colimator and measured. The intensity is proportional to the amount of the respective element in the sample, as described by the equation:

$$C = K (R_p - R_b) \mu$$

where C is the concentration of the element, $(R_p - R_b)$ the measurement of the net fluorescence at a given wavelength, μ is its mass absorption coefficient and K is a constant. The value of μ is affected by sample composition and homogeneity, and for low atomic number elements can be determined from the absorption characteristics for fluorescence irradiation. Each of the elements that contribute to the absorption characteristics of a sample needs to be characterized in order to determine the matrix absorption coefficient. These matrix coefficients can then be used to correct for inter-element absorption in a plant sample by considering the sample as a pure cellulose matrix in which small proportions of the cellulose have been replaced by the various elements (Norrish and Hutton, 1977). The satisfactory grinding of plant material, to ensure a homogeneous mixture for measurement, is essential for the application of matrix absorption coefficients.

The essential features of the X-ray fluorescence technology from a diagnostic point of view include the ease of sample preparation and the fact that a large suite of elements is measured in one operation. Calibration standards are available for plant material. A drawback of the technology is that not all elements of interest in plant nutrition can be measured. Elements such as nitrogen and boron have atomic masses that are too low to produce a significant signal and the resulting lack of sensitivity prevents this technology being used for detection of changes in the amounts of these elements.

Inductively coupled plasma optical emission spectroscopy and inductively coupled plasma mass spectrometry both rely on an extremely high temperature source (inductively coupled plasma) to completely ionize elements so that singly charged positive ions predominate. The high temperature source is obtained by supplying radio frequency power to a stream of argon by means of an induction coil. The powerful magnetic and electric fields generated by the coil cause free electrons (provided by a high-voltage spark ignitor) in the argon to oscillate rapidly and result in collisions that quickly raise the temperature to where a plasma is formed. The temperatures attained are near 8000°C. The detection of the ionized elements is either by spectral analysis (Thompson and Walsh, 1989) or mass spectrometry (Date and Gray, 1989). The detection systems, particularly the mass spectrometer, are very sensitive.

1.4 Colorimetric Assays

The eventual aim of many diagnostic procedures is to produce a colour reaction. The design of the procedure usually allows the degree of colour to be directly related to the presence of a specific molecule. Many procedures employ two steps in the detection procedure, the primary reaction that is responsible for the specific detection of a molecule (substrate) and a secondary, less specific reaction that is responsible for assaying the substrate–probe complex. Dividing a diagnostic procedure into two broad steps is based on the concepts developed by Holt and O'Sullivan (1958) for detecting enzyme activities, *in situ*, in tissues prepared for microscopic examination (Fig. 1.4). The dissection of diagnostic procedures in this way is useful because it aids both the interpretation of data and the process of problem solving. Knowledge about the specificity of the primary reaction is essential for interpreting information because cross-reaction of the probe with other molecules, not directly correlated to the trait of interest, leads to spurious conclusions. The lack of specificity in the secondary reaction is significant when problems of high background readings are encountered in assays and may reflect the reaction of components used at this stage with extraneous material in the sample. Many substrate–probe complexes are detected by making use of enzymatic activities that act on modified colourless substrates and provide the basis for specificity. For example, changing the functional group on an indole residue alters the selectivity of the substrate for enzyme activities (Fig. 1.5).

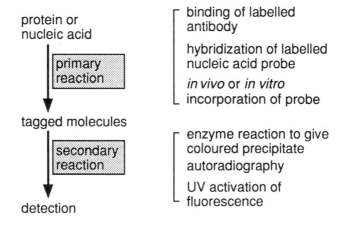

Fig. 1.4. Concept of primary and secondary reactions in diagnostic methods employing colorimetric detection.

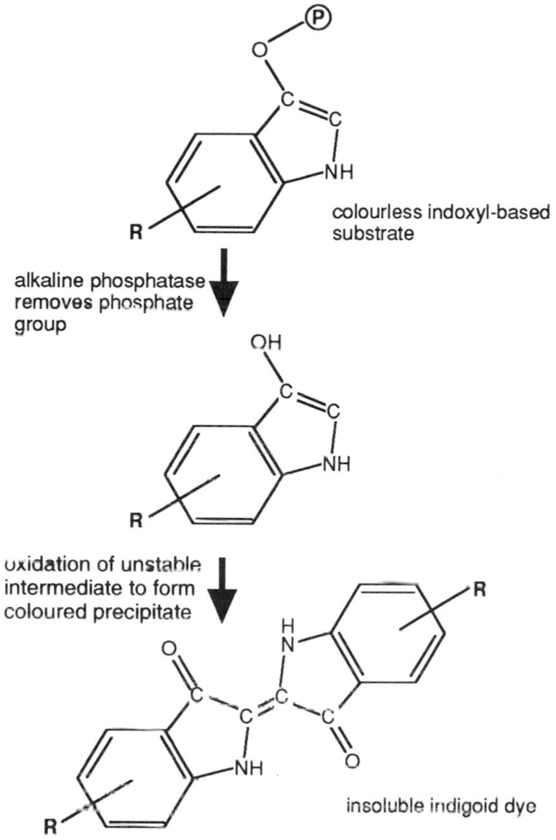

Fig. 1.5. Use of modified colourless substrates for detection of enzyme activities.

1.5 Diagnostics Using Antibodies

Although a variety of solid supports are used to perform enzyme-linked immunosorbent assays (ELISA), the assays fit into three general categories. In each assay format, there are intermediate incubation and washing steps to minimize non-specific binding of immunoreactants. A variety of other immunoassay formats has been described in the literature and these are in routine use in clinical testing. They include assays based on agglutination and chromatography, and 'homogeneous' assays, that do not require a washing step (Price and Newman, 1991). One of the main reasons why these assays have not been more extensively used in plant science and testing is potential interference from other components in the sample matrix.

1.5.1 Direct and indirect assays

These assays involve the immobilization of antigen on to the solid phase (e.g. microwell or a protein-binding membrane, such as nitrocellulose), followed in the direct ELISA by incubation with an enzyme-labelled antibody. Immobilization of protein antigens is usually by passive adsorption on to the solid phase, although covalent binding is often performed for antigens such as carbohydrates. In the indirect ELISA, an unlabelled antigen-specific antibody is first used, followed by a labelled antibody which recognizes the Fc tail (the conserved portion of the antibody molecule) of the first antibody. The extra incubation and washing steps slow the assay but remove the need to directly label the antigen-specific antibody. In some cases, amplification steps, facilitated by formation of biotin–streptavidin complexes, can be added (Fig. 1.6). The use of a colorimetric chromogen/substrate combination can reveal either the extent of antibody binding if the coloured product is soluble or the sites (and semi-quantitatively the extent) of binding, if the coloured product is insoluble. The direct and indirect assays are the simplest to perform, and are used routinely to assess antibody titre, which is a function of both concentration and affinity.

Antigens present in test samples can also be immobilized on to the solid phase and assayed using these ELISA formats. In this case they are best suited to screening for the presence or absence of specific antigens rather than

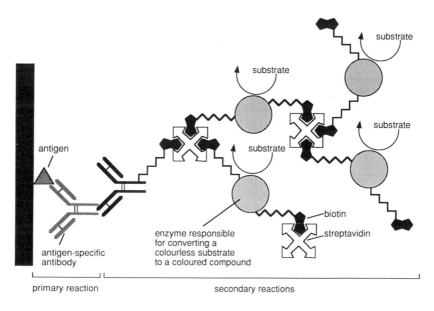

Fig. 1.6. Amplification of signal detection in immunoassay using complexes of biotin and streptavidin.

for quantitative assays. Two main factors complicate the use of direct or indirect ELISA in strict antigen quantification: firstly, the slope of the antigen concentration versus ELISA colour development curve is rather flat and, secondly, other components in the test sample may compete with the target antigen for binding to the solid phase.

1.5.2 Sandwich assays

The most commonly used format for antigen quantification in test samples is the double-antibody 'sandwich' assay (Fig. 1.7). In this assay, unlabelled antibody is immobilized on to the surface of microwells, antigen added, then after intermediate washing steps, enzyme-labelled antibody added. The latter antibody recognizes a part of the antigen molecule that is not bound to the immobilized antibody. The method is simple and robust and results easy to interpret – increasing antigen concentrations give greater colour development. However, it requires isolation of either antibodies that recognize different, spatially separate epitopes or of a single antibody that binds to an epitope which is repeated two or more times on the surface of the antigen. Antibodies must also be of moderate affinity and suitable for direct labelling. Sandwich assays are also not suitable for small molecules of less than 2000–5000 kDa in molecular mass.

1.5.3 Competition assays

These assays involve either of two general formats. Following antibody immobilization on to the solid phase, antigen in the test sample competes with enzyme-labelled antigen for binding to the immobilized antibody (Fig. 1.7). Increasing amounts of antigen lead to decreased binding of the enzyme label and thus decreasing colour in the assay. Alternatively, known amounts of a standardized or purified antigen can be immobilized on the solid phase, and antigen in the test sample incubated with labelled antibody. In the absence of antigen in the test sample, high amounts of labelled antibody can bind to the immobilized antigen. Increasing concentrations of antigen in the test sample give decreasing binding of the antibody. While competition assay formats are most commonly performed with small molecules, they function well with macromolecules such as proteins. They have the advantage in the latter case of providing a quantitative assay using a single antibody, such that variation in the amounts of (or conformation of) a single epitope can be studied.

1.5.4 Sources of antibodies

Antibodies used in routine diagnostic tests are either polyclonal or monoclonal. Polyclonal antibodies are simpler to produce, and are commonly

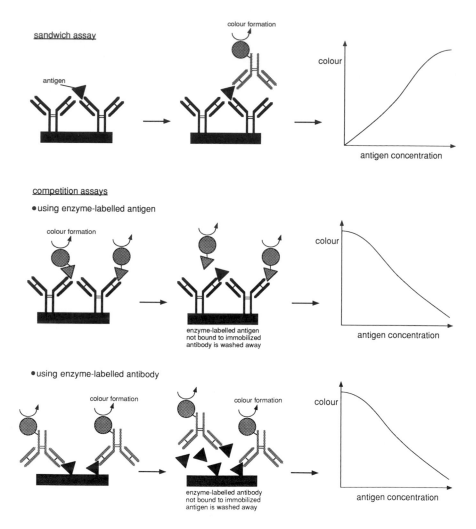

Fig. 1.7. Comparison of sandwich and competition ELISA formats for antigen detection.

obtained by repeated immunization of an experimental animal such as a rabbit, test bleeding, then fractionation of the serum. Monoclonal antibodies are usually prepared following removal of the spleen of a hyperimmunized animal and fusion of spleen cells with a myeloma cell line (see Chapter 2). Only fused cells survive, due to the use of nutritional mutant myeloma cells and the inability of spleen cells to be viable in long-term culture. The distinguishing feature of monoclonal antibodies is that they recognize single epitopes on the target antigen. This does not necessarily mean that they are more specific than polyclonal antibodies, since if the epitope that is recognized is found on several proteins or other antigens in the target sample, the monoclonal antibody may have rather broad specificity. Monoclonal antibodies are also less commonly used in immunoassay of haptens, such as agrochemicals. This is because the sensitivities of the competition assays used for hapten detection are highly dependent on antibody affinity. Typically, monoclonal antibodies are lower in affinity than polyclonal antibodies, partly because of the properties of murine antibodies (species from which most monoclonal antibodies are derived) and partly because the competition assays will select the highest affinity antibodies present in a polyclonal mixture. Monoclonal antibodies are of greatest application where detection of single antigens or a subfamily is required from a complex mixture of structurally related antigens. Another advantage of monoclonal antibodies, which will assume increasing importance, is that the antibody genes can be isolated and manipulated to express recombinant antibodies.

1.6 Whole Chromosome Analyses

In plants, the rapidly dividing tissues of the root or shoot meristems are the main source of chromosomes for somatic karyotypes because many cells are undergoing mitosis. Any tissue containing rapidly dividing cells is suitable for mitotic karyotype analysis after treatment with drugs such as colchicine or other spindle-inhibiting agents, which arrest cells in metaphase. Although mitotic metaphase chromosomes are used extensively in karyotype analysis, they are highly condensed and do not reveal structural differentiation along their length. The fluorochrome, quinacrine mustard, was the first compound to reveal a reproducible banding structure, or Q-banding, within metaphase chromosomes. Q-banding results in bright fluorescent bands against a dull fluorescent background, and although it was quickly applied to human prenatal diagnosis, it has not been widely applicable to plants. In 1971, a C-banding procedure (C = constitutive) was shown to produce reproducible bands of darkly stained regions in metaphase chromosomes. At about the same time, a number of other chemical procedures were also reported to reveal structural differentiation along the length of the chromosomes.

The technique of C-banding can be applied universally to reveal the location of condensed regions (or heterochromatin) in mitotic metaphase chromosomes. In a typical procedure, chromosomes that have been spread on glass slides are sequentially treated with acid or alkali, followed by incubation in a salt buffer and staining with Giemsa stain. Studies in diverse organisms such as humans, *Drosophila* and wheat have demonstrated that this staining technique differentiates substructure within the chromosome. The mechanism of C-banding is not clear, but it is known that the procedure preferentially stains regions of chromosomes enriched in highly repetitive sequences. In crops such as wheat, and its relatives, the chromosomes have been well studied and analyses of chromosomes using C-banding can be used on cultivars carrying distinct chromosome segments, that are to be used in setting up a crossing programme (Fig. 1.8). The technology is, however, highly specialized and is not as widely used as many of the other diagnostic tests described in this book. An extension of the C-banding technique is to

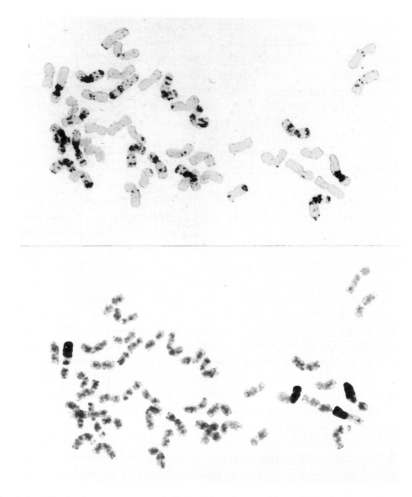

Fig. 1.8. Whole chromosome analysis by C-banding and *in situ* hybridization. C-banding (top panel) gives characteristic dark staining regions that are reproducible and diagnostic for specific chromosome segments. The C-banding can be followed by *in situ* hybridization (bottom panel) that can further identify chromosome segments. In the example shown, a wheat line carrying rye chromosome segments (dark staining in bottom panel) was analysed. (Based on Jiang and Gill (1993), photograph kindly supplied by Dr J. Jiang.)

use specific DNA probes to hybridize to chromosome preparations, a technique referred to as *in situ* hybridization. The principles involved in *in situ* hybridization are the same as in any assay utilizing the ability of single DNA strands to form a double helix with complementary sequence of bases in another single strand of DNA. A single DNA strand preparation or 'probe' that is labelled with a fluorescent dye can find its complementary sequence in the chromosomal DNA immobilized on a glass slide (Pinkel *et al.*, 1986). The position of the fluorescent dye then identifies the physical location of the sequence in the chromosome (Fig. 1.8). Several different probes, labelled with different colours, can then 'fingerprint' a chromosome for unambiguous identification. New developments in PCR technology provide potentially higher sensitivity alternatives to standard *in situ* hybridization (Godsen and Hanratty, 1993).

1.7 Nucleic-acid-based Diagnostics

Most DNA probe methods incorporate amplification steps that fall into two general classes. In the first, the nucleic acid target material is amplified in amount to increase sensitivity, as described below for the polymerase chain reaction (PCR). Alternatively, a chemical cascade can be produced, whereby a bifunctional probe, such as a biotinylated DNA probe is used in combination with avidin-peroxidase to increase the signal for detecting the hybridization of the DNA probe.

1.7.1 The PCR principle

One of the most exciting applications of knowledge about DNA replication is the development of the polymerase chain reaction (PCR), for amplifying regions of the genome without prior cloning. The PCR reaction, first described in 1985, has revolutionized the way in which molecular biology is carried out. The procedure enables small amounts of specific DNA, which may be mixed with large amounts of contaminating DNA, to be amplified approximately a millionfold.

PCR is an *in vitro* procedure for the enzymatic synthesis of DNA using two oligonucleotide primers that hybridize to opposite strands of the parental DNA template, and flank the region of interest (Fig. 1.9). A repetitive series of cycles involving template denaturation, primer annealing, and the extension of the annealed primers by DNA polymerase, results in the exponential accumulation of a specific fragment whose termini are defined by the 5'-ends of the primers. The amplification is dramatic because the extension products of one cycle serve as templates for the following reactions and thus the number of target copies, for the reaction, double at every cycle. After 20 cycles of PCR, up to a millionfold amplification of a particular

DNA sequence is achieved. The key factor in the widespread use of PCR was the introduction, in 1986, of the thermostable DNA polymerase (*Taq* polymerase) from *Thermus aquaticus*. The commercial supply of the enzyme meant that the reaction components (template, primers, *Taq* polymerase,

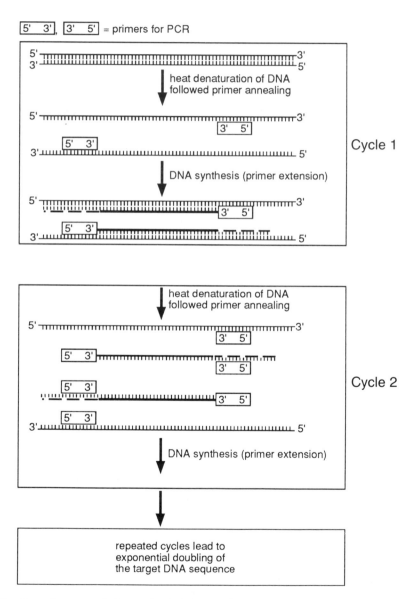

Fig. 1.9. The polymerase chain reaction (PCR) principle.

nucleoside triphosphates and buffer) could all be simply mixed and subjected to temperature cycling. The kinetics of this reaction can be monitored directly if small amounts of the DNA binding dye, ethidium bromide, are included in the reaction mix. The reaction is not exponential over the entire range of cycles, presumably because some *Taq* polymerase activity is lost in each heating cycle of the amplification reaction. However, the amplification reaction is still occurring after 50 cycles and, as expected, is dependent on the number of starting template DNA molecules.

The impact of PCR on diagnostic assays is that if a gene product is related to a character of interest then assays can be developed to assay the gene directly. In addition, techniques for DNA fingerprinting have been made much simpler with the PCR technique and thus allow variety identification to be carried out as well as quality control measures on purity of a seed sample.

Current restrictions on the use of PCR as a routine diagnostic tool include its sensitivity to inhibition by matrix components and DNA extraction solutions (Rossen *et al.*, 1992). In the non-medical area this has actually been most thoroughly studied in attempts to develop sensitive assays for pathogenic bacteria in food and environmental matrices. Problems also occur when PCR is applied to real environmental samples, for example, for detecting bacterial contamination of water or soil pathogens (Cox *et al.*, 1992). It would also be advantageous to remove the need for thermal cycling and this is an active area of research in industry (Guatelli *et al.*, 1990).

1.7.2 *Application of restriction fragment length polymorphism (RFLP) and random amplified polymorphic DNA (RAPD) techniques in breeding programmes*

The use of RFLP DNA markers has made many organisms available for genetic mapping in providing the required polymorphic markers. Plants have been extensively studied (Tanksley *et al.*, 1989) and detailed genetic maps are available for maize, tomato, lettuce and wheat. Plant breeders are using these genetic maps to underpin many breeding programmes so that agronomic traits can be more clearly understood as well as providing alternative screening strategies for early assessment of the characters. Use of RFLPs in polyploid species such as wheat has lagged behind that in diploid species because of complications in identification or co-migration produced by multiple fragments from homologous chromosomes (Sorrells, 1992). In wheat, low levels of inter-varietal polymorphism and the large genome size also complicate RFLP mapping (Gale *et al.*, 1990). Mapping in wheat is facilitated by use of linkage maps for diploid relatives and aneuploid stocks.

Assays based on amplification of random DNA segments with short primers of arbitrary nucleotide sequence (RAPD) can reproducibly amplify segments of genomic DNA (Williams *et al.*, 1990; see Chapter 5). This approach has been useful in detecting polymorphisms in diploid wheat genotypes (Vierling and Nguyen, 1992) and together with denaturing gradient gel electrophoresis, RAPD analysis has detected DNA sequence polymorphisms among wheat varieties (He *et al.*, 1992).

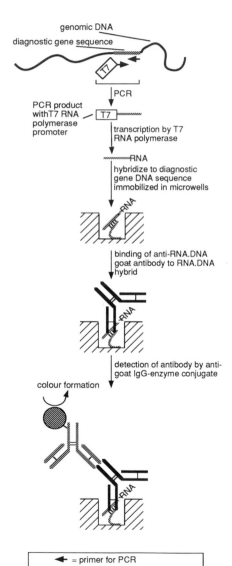

Fig. 1.10. Enhancement of the sensitivity of PCR detection systems, by *in vitro* transcription of amplicons incorporating bacteriophage T7 RNA polymerase promoter sequences attached to the 5' end of one of the PCR priming oligonucleotides.

1.7.3 Immunochemical approaches to gene probe assays

Combining immunochemical and nucleic acid methods can have advantages of increased specificity, sensitivity or assay simplicity (Stollar and Rashtchian, 1987). Blais (1994) described a method for enhancement of the sensitivity of PCR detection systems, by *in vitro* transcription of amplicons incorporating bacteriophage T7 RNA polymerase promoter sequences in one of the priming oligonucleotides (Fig. 1.10). RNA products resulting from T7 RNA polymerase transcribing amplified DNA were then detected by hybridization with a DNA probe immobilized on to a microwell. The resulting RNA–DNA hybrids were then detected by an enzyme-labelled antibody

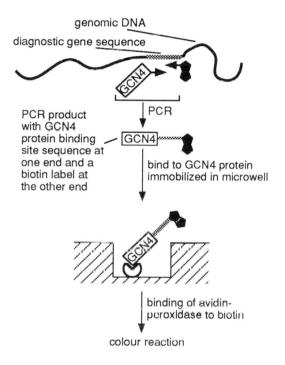

Fig. 1.11. Use of a DNA-binding protein in development of a microwell-based gene probe assay. After an initial PCR amplification, a second nested set of nucleotides is incorporated by PCR. One of these primers contains a site for the DNA-binding protein, GCN4, for immobilization to the microwell, while the other is biotinylated (for detection using avidin–peroxidase).

that binds specifically to RNA–DNA duplexes, without reference to specific bases. In addition to increased sensitivity, this approach also avoids the need for introduction of chemical labels on to either of the reacting nucleic acid strands.

A related approach has been developed by Kemp *et al.* (1989). After an initial PCR amplification, a second set of oligonucleotides, nested between the first two, is incorporated by PCR. One of these oligonucleotides is

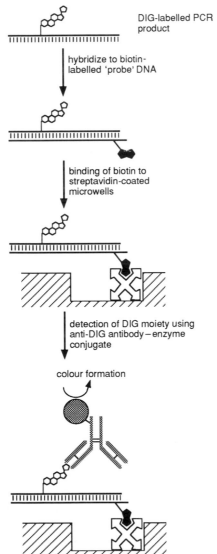

DIG-labelled PCR product

hybridize to biotin-labelled 'probe' DNA

binding of biotin to streptavidin-coated microwells

detection of DIG moiety using anti-DIG antibody – enzyme conjugate

colour formation

Fig. 1.12. Detection of PCR products using an antibody specific for digoxigenin (DIG)-labelled sequences.

biotinylated while the other contains a site for GCN4 (a double-stranded DNA-binding protein, Fig. 1.11). The amplified DNA can be immobilized on to microwell-bound GCN4, and detected using avidin–peroxidase, which binds to the biotin derivative. A kit to enable colorimetric detection of specific DNA segments is now commercially available.

Several groups have coined the term 'PCR ELISA', for procedures to detect PCR products. This method of analysis is more sensitive and quantitative than ethidium bromide staining after agarose gel electrophoresis, and is faster than DNA blotting or dot blotting. In one commercial kit the PCR product is labelled with digoxigenin (DIG), denatured and hybridized to a biotin-labelled capture probe. The hybrid so formed is bound to streptavidin-coated microwells and the DIG label detected using anti-DIG antibody–enzyme conjugate (Fig. 1.12).

1.7.4 Comparison of nucleic acid probe and antibody assays

In some cases, a decision will need to be made as to whether it is more appropriate to pursue an antibody or a nucleic acid probe in the development of a diagnostic test. Clearly, small molecules such as lipids, mycotoxins or agrochemicals are amenable only to antibody-based detection, but for macromolecules and plant breeding applications, the choice is not so clear-cut. Nucleic acid probes are generally better for genotype characterization, while antibodies could provide a better test for phenotype – especially if the target is a product of a gene whose expression is influenced by growth environment. For example, in F_1 or backcross populations where hetero-zygosity needs to be noted, DNA markers are much more powerful. However, where thousands of individuals need to be screened in more advanced selections, quarter grains can be assayed for particular character-istics, using antibody-based techniques in an assay taking only 1–2 days. The embryo ends of the grains can be stored and chosen samples planted for subsequent crossing.

These two examples emphasize that choosing a particular technique relates to the compromise that often occurs between method speed or throughput and complexity or resolution. The results of diagnostic methods differ con-siderably in their information content. For example, screening of genotypes using an ELISA (or DNA-dot hybridization) assay may reveal considerable differences in responses, but the data do not discriminate between differences in the abundance of a particular sequence (or in affinity of binding to that sequence) and binding of the probe to a sequence that may not be directly related to the genotype in question. The numerical data have a lower information content than a blot, electrophoresis gel or chromatogram.

Apart from new technologies such as PCR, one of the other main advances in nucleic acid diagnostics over the last decade has been the simplification of nucleic acid methods. Simple DNA extraction methods

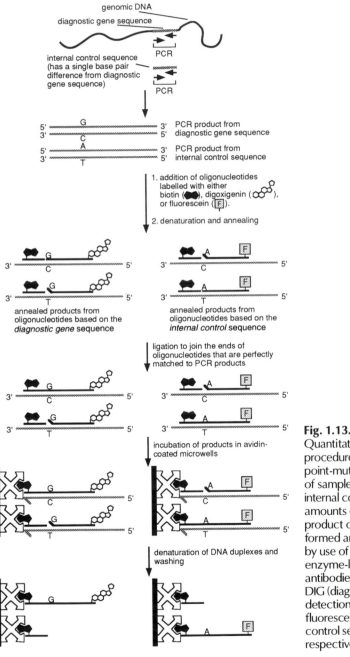

Fig. 1.13. Quantitative PCR procedure, using point-mutated version of sample DNA as an internal control. The amounts of PCR product originally formed are quantified by use of enzyme-labelled antibodies specific for DIG (diagnostic gene detection) and fluorescein (internal control sequence) respectively.

have been developed for DNA analysis in cereal endosperm or leaf pieces using PCR (Benito *et al.*, 1993). Other approaches have been developed to avoid use of radiolabels and/or electrophoretic separation steps. For example, Friedhoff *et al.* (1993) recently described a quantitative PCR procedure which used a point-mutated version of the sample DNA as an internal control. After competitive PCR, amplified sample and control DNA are distinguished using an oligodeoxynucleotide ligation assay using two oligodeoxynucleotides (Fig. 1.13). One oligodeoxynucleotide contains a 5'-biotinylation and a digoxigen at the 3' end; this sequence is specific for sample DNA. The control oligonucleotide is also 5' biotinylated, but has fluorescein incorporated near the 3' end. The biotinylated oligonucleotides are immobilized on avidin-coated microwell plates, allowing the hapten-labelled oligonucleotide ligation products to be measured using specific antibodies to the two labels. This method thus combines the specificity of nucleic acid probe with the simplicity of ELISA detection.

1.7.5 Simplification of nucleic-acid-based assays

The last few years have seen a move away from membrane hybridization assays to microwell assays for quantification of oligonucleotides. The simplest approach, conceptually, here is to immobilize target DNA directly on to microwells through glutaraldehyde and then react with an enzyme-labelled or biotinylated DNA probe (Coll, 1991). It is analogous to the direct or indirect ELISA. The disadvantages of this approach are relatively low sensitivity and competition in binding to the solid phase by other components of the plant. A more sophisticated method involves amplification of the DNA (Kawai *et al.*, 1993). Here, to immobilize the probes on to microwells, single stranded DNA containing a tandem array of probes prepared from a phagemid vector was adsorbed on to the polystyrene surface of the well (Fig. 1.14). PCR for the target sequence was performed using biotin-labelled primers. The denatured PCR product was then hybridized to the probe in the well, the wells washed, followed by the detection of hybridized PCR products using enzyme-labelled streptavidin.

Some 'field' tests based on nucleic acid hybridization have been developed commercially (Alon *et al.*, 1993). In a field assay for tomato yellow leaf curl virus, tobacco mosaic virus and potato virus Y, pieces of plant tissue (or of the insect vector) are squashed on to a membrane, followed by incubation with a haptenated virus DNA probe. Binding of this probe is detected by antibody incubation. Using this method, virus is detectable well over a week before the appearance of symptoms in the plant.

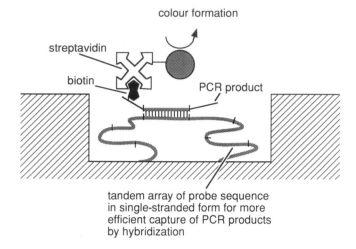

tandem array of probe sequence
in single-stranded form for more
efficient capture of PCR products
by hybridization

Fig. 1.14. Use of a tandem array of probes in single-stranded DNA to immobilize PCR amplified DNA. The biotinylated PCR product is detected using a streptavidin–enzyme conjugate.

1.8 Commercialization

There are two main situations in which the results of diagnostic research can be commercialized. These basically fall under the classification of hardware (such as systems for near-infrared reflectance) and biochemical test kits. The comments in this section refer to the latter.

From a practical standpoint, there is a tremendous difference between completing the research required to publish a research paper on a new method and having a diagnostic test available routinely, especially in a commercial framework. Often publication of PCR primer sequences will allow other plant research or testing laboratories to prepare the corresponding synthetic oligonucleotide, and to redevelop the assay for their own use. On the other hand, antibodies tend to be specialized reagents, so unless the antibody is made widely available the broader use of the diagnostic method will be necessarily restricted. Most researchers are happy to provide samples of published reagents gratis, but often the quantities available after the research and the project have been completed will be limited. The considerable turnover of scientific staff in most institutions also introduces concerns about continuity of reagent supply.

Commercial development of each type of biochemical diagnostic test has its own particular challenges, but there are several common features that distinguish development from the research that has preceded the commercialization phase. There are two main new concerns: assay precision and reagent stability. Assay precision is initially established within the developing laboratory, and takes the form of establishing within- and between-assay variation for standards and samples, as well as establishing the 'ruggedness' of the assay (i.e. how readily results are perturbed by minor alterations in incubation volumes, times and temperatures). Inter-laboratory precision can be established informally, or through formal inter-laboratory trials. Data from such trials can be reviewed by bodies such as the Association of Official Analytical Chemists International (AOAC; based in Arlington, Virginia, USA), and if the data are of sufficient quality, the method can be proposed for 'Official Methods Status'. The slowness and expense of this process has given rise to a streamlined commercial kit evaluation programme by the AOAC Research Institute. The latter procedure does not require as extensive a trial, but it also encompasses evaluation of the instructions and other manufacturing data. The two processes differ qualitatively; the latter provides some measure of 'kit approval', while the former involves 'method approval'.

Simplification of assays is also important. The commercial appeal of a new agricultural diagnostic will be much lower if expensive, specialized equipment is required or if the assay contains a large number of wash steps. In most cases, antibody incubations may be carried out at room temperature. This often improves within-assay precision (since temperature gradients between neighbouring microwells are less likely to occur), although ambient temperature variation means that between assay colour development may vary. This is usually of limited practical concern, as in most of these assays test samples are analysed with reference to a standard curve prepared at the same time.

Few researchers undertake thorough stability trialling of reagents, even if a method is used routinely within the laboratory. In the research laboratory, reagents such as enzymes, antibodies and chromogens are diluted from refrigerated concentrated stocks immediately before the assay. Solid phases such as microwell plates may be coated freshly for each assay. However, in commercial assays, components must function well for the life of the kit – typically one year at refrigerator temperatures – and as well for field assays, the possibility of several days at ambient temperature. Components are often provided in kits at or near the working concentrations to enable the assay to be simple to use and reproducible (eliminating assay to assay dilution variation), and to enable the manufacturer to conserve reagents. Immobilized components are often stabilized by air-drying or freeze-drying in the presence of additives, while other additives are added to stabilize enzymes and substrates. Recipes for such additives are rarely published in the refereed or patent literature, with commercial groups preferring to retain them as confidential 'know-how'. The stability of components is initially assessed by carrying out accelerated (elevated temperature) stability trials followed by real-time stability trials with the first batch of kits released to the market.

Commercial development of agricultural and environmental diagnostic kits has accelerated over the last five years, but given that a large range of medical diagnostic tests have been available for over three decades and veterinary tests for almost as long, the plant agricultural and environmental areas are still in comparative commercial infancy. None the less, there is significant commercial activity, particularly in the area of immunodiagnostics. This probably reflects the currently greater simplicity of assay formats for immunodiagnostics compared with DNA diagnostics, although advances in development of non-isotopic nucleic-acid-based diagnostics will mean that there will be an increasing number of commercial DNA diagnostics in plant agriculture. Obviously, RFLP and RAPD testing are already routine in genotyping and research; perhaps the ready availability of DNA synthesizers to prepare oligonucleotides limits the commercial opportunities in some areas. There are several nucleic acid diagnostics for plant disease on the market, however.

The areas of greatest commercial activity in plant and environmental immunodiagnostics include assays for plant viruses, mycotoxins in food and feeds, fungal diseases (e.g. of turfgrass) and for agricultural and industrial chemical residues. A number of barriers will remain to the wider commercialization of agricultural diagnostics. While some applications such as plant breeding may require many thousands of assays to be performed annually in individual laboratories, the cost per assay that can be borne is very low. Indeed, studies of the potential suitability of different diagnostic approaches may include a cost comparison; for example RFLP and RAPD genotyping costs in plant breeding have been systematically compared (Ragot and Hoisington, 1993). In contrast, the user of the information in medical areas is usually prepared to pay considerably more for medical diagnostic information since government subsidies or medical insurance rebates apply in many countries for medical diagnostic testing. Accordingly, the need to keep costs low has limited the use in agriculture of some of the more attractive assay formats that have recently been developed for medical applications. This applies for both automated laboratory analyses and for rapid formats that would be applicable under field situations.

A further constraint is that a number of plant agricultural diagnostics can best be described as niche markets. For example, there may be a need to discriminate between two important but morphologically similar species of a crop insect pest, but those pests may only be of major importance in one geographic region. Tests for viral and fungal diseases of minor horticultural crops may go uncommercialized unless the diseases become important in a major crop. The lack of a commercial diagnostic or even a commercialization plan does not mean that there has been no research on diagnostic development or extension to farmers. Quite often in agricultural 'niche markets', government agricultural research laboratories may develop a method and either distribute the assay to regional labs or provide an in-house testing service. This sort of extension work is critically important, and must not be lost through the pressure that many research organizations feel to work only on 'commercially-viable' projects.

There are also some examples where a decision has been made not to commercialize particular diagnostics, even though there are potentially large agricultural applications within industry. Examples include immunoassays and nucleic acid assays that may give a particular company or country a competitive advantage in breeding for quality or disease resistance. If the quality attribute or resistance factor is generic enough, it may mean that if the diagnostic were made openly available, the developer would lose a competitive advantage. Quite often the real financial return from development of a diagnostic is from its application rather than from actual sale of kits to a limited number of users. Another major group of commercial developers of diagnostics are agrochemical companies. While their research has often been very thorough, these companies have developed the assays with the major aim of using them in-house in field dissipation studies and assisting in the preparation of the large amounts of data required for product registration.

The format of the assay as well as the target analyte may influence the decision whether a particular assay is commercialized more widely. For example, it may not be commercially viable to release a microplate assay for use in barley breeding where this programme is undertaken by four or five institutions per country, in government or university stations. On the other hand, a rapid tube form of the same assay may have a lucrative market if it is to be run as a routine component of quality assurance by breweries. A major need for diagnostics exists where particular plant disease or agrochemical contamination problems are important in developing countries; the lower costs, simplicity and fieldability of such diagnostics offer considerable benefits. However, if the same problems are not also of concern in North America or Europe, there will be little commercial impetus.

Despite these constraints, the development and application of agricultural diagnostics will continue to develop as farming moves from often being a family occupation to a scientifically based business system, depending on economic and technical data of high quality for its profitability. Many of the latter data will be provided by diagnostic tests, in many cases performed on-farm, because a universal feature over the last decade has been the establishment of technology transfer companies or offices in universities and government laboratories. Often the implications of the use of a kit (e.g. in plant breeding or insect control) have much greater financial returns than those derived directly from the sales of the corresponding kits.

References

Alon, U., Levy, H. and Czosnek, H. (1993) New approaches in plant virus detection. *Agro-Food Industry Hi-Tech* 4, 27–29.

Andrew, E.R. (1955) *Nuclear Magnetic Resonance. Cambridge Monographs on Physics.* Cambridge University Press, Cambridge.

Benito, C., Figueiras, A.M., Zarogoza, C., Gallego, F.J. and de la Pena, A. (1993) Rapid identification of Triticeae genotypes from single seeds using the polymerase chain reaction. *Plant Molecular Biology* 21, 181–183.

Blais, B.W. (1994) Transcriptional enhancement of the *Listeria monocytogenes* PCR and simple immunoenzymatic assay of the product using anti-RNA : DNA assays. *Applied and Environmental Microbiology* 60, 348–352.

Coll, J.M. (1991) Hybridization of peroxidase-labeled DNA probes to microtiter solid-phase bound DNA (Hybrelisa). *Technique* 3, 29–32.

Cox, P., Fricker, C. and Hammond, L. (1994) Fathoming bacterial PCR. *Today's Life Science* 6, 30–34.

Date, A.R. and Gray, A.L. (1989) *Applications of Inductively-coupled Plasma Mass Spectrometry.* Blackie Publishers, Glasgow.

Delwiche, S.R. and Norris, K.H. (1993) Classification of hard red wheat by near-infrared diffuse spectroscopy. *Cereal Chemistry* 70, 29–35.

Flagella, Z., Pastore, D., Campanile, R.G. and Di Fonzo, N. (1992) Near infrared reflectance: a new approach for evaluating drought resistance in durum wheat. *Journal of Genetics and Breeding* 46, 21–28.

Friedhoff, P., Hahn, M., Wolfes, H. and Pingoud, A. (1993) Quantitative polymerase chain reaction with oligonucleotide ligation assay/enzyme-linked immunosorbent assay detection. *Analytical Biochemistry* 215, 9–16.

Gale, M.D., Chao, S. and Sharp, P.J. (1990) RFLP mapping in wheat – progress and problems. In: Gustafson, J.P. (ed.), *Gene Manipulation in Plant Improvement II.* Plenum Press, New York, pp. 353–364.

Godson, J. and Hanratty, D. (1993) PCR *in situ*: a rapid alternative to *in situ* hybridization for mapping short, low copy number sequences without isotopes. *BioTechniques* 15, 78–80.

Guatelli, J.C., Whitfield, K.M., Kwoh, D.Y., Barringer, K.J., Richman, D.D. and Gingeras, T.R. (1990) Isothermal, in vitro amplification of nucleic acids by a multienzyme reaction modeled after retroviral replication. *Proceedings of the National Academy of Sciences, USA* 87, 1874–1878.

He, S., Ohm, H. and Mackenzie, S. (1992) Detection of DNA sequence polymorphisms among wheat varieties. *Theoretical and Applied Genetics* 84, 573–578.

Hellmki, M. and Moisio, T. (1983) Prediction of protein and fiber contents in silage by near infrared reflectance analysis. *Milchwissenschaft* 38, 14–15.

Holstrom, K., Rossen, L. and Rasmussen, O.F. (1993) A highly sensitive and fast non-radioactive method for detection of polymerase chain reaction products. *Analytical Biochemistry* 209, 278–283.

Holt, S.J. and O'Sullivan, D.G. (1958) Studies in enzyme cytochemistry I. Principles of cytochemical staining methods. *Proceedings of the Royal Society of London, B* 148, 465–480.

Howes, N.K., Lukow, O.M., Dawood, M.R. and Bushuk, W. (1989) Rapid detection of the 1BL/1RS chromosomal translocation in hexaploid wheats using monoclonal antibodies. *Journal of Cereal Science* 10, 1–4.

Hutton, J.T. and Norrish, K. (1977) Plant analysis by X-ray spectrometry. 2. Elements of atomic number greater than 20. *X-ray Spectrometry* 6, 12–17.

Jiang, J. and Gill, B.S. (1993) Sequential chromosome banding by *in situ* hybridization. *Genome* 36, 792–795.

Johnson, D.A., Tonnet, M.L. and Richards, R.A. (1984) Estimation of epicuticular wax amounts in wheat using wide-line proton magnetic resonance. *Crop Science* 24, 679–682.

Kawai, S., Maekawajiri, S. and Yamane, A. (1993) A simple method of detecting amplified DNA with immobilized probes on microtiter wells. *Analytical Biochemistry* 209, 63–69.

Kemp, D.J., Smith, D.B., Foote, S.J., Samaras, N. and Peterson, M.G. (1989) Colorimetric detection of specific DNA segments amplified by polymerase chain reactions. *Proceedings of the National Academy of Sciences, USA* 86, 2423–2427.

Manteuffel, R. and Panitz, R. (1993) In situ localization of faba bean and oat legumin-type proteins in transgenic tobacco seeds by a highly sensitive immunological tissue print technique. *Plant Molecular Biology* 22, 1129–1134.

Munter, R.C., Halverson, T.C. and Anderson, R.D. (1984) Quality assurance for plant tissue analysis by ICP-AES. *Communications in Soil Science and Plant Analysis* 15, 1285–1322.

Norris, K.H., Hruschka, W.R., Bean, M.M. and Slaughter, D.C. (1989) A definition of wheat hardness using near-infrared spectroscopy. *Cereal Foods World* 34, 696–705.

Norrish, K. and Hutton, J.T. (1977) Plant analysis by X-ray spectrometry. 1. Low atomic number element Na-Ca. *X-ray Spectrometry* 6, 6–11.

Pinkel, D., Straume, T. and Gray, J.W. (1986) Cytogenetic analysis using quantitative, high-sensitivity, fluorescent hybridization. *Proceedings of the National Academy of Sciences, USA* 83, 2934–2938.

Polston, J.E., Burbrick, P. and Perring, T.M. (1991) Detection of plant virus coat proteins on whole leaf blots. *Analytical Biochemistry* 196, 267–270.

Price, C.P. and Newman, D.J. (1991) *Principles and Practice of Immunoassay.* Macmillan, Basingstoke, UK, 650pp.

Ragot, M. and Hoisington, D.A. (1993) Molecular markers for plant breeding: comparisons of RFLP and RAPD genotyping costs. *Theoretical and Applied Genetics* 86, 975–984.

Rasmussen, U. (1985) Immunological screening for specific protein content in barley seeds. *Carlsberg Research Communications* 50, 83–93.

Rossen, L., Norskov, P., Holstrom, K. and Rasmussen, O.F. (1992) Inhibition of PCR by components of food samples, microbial diagnostic assays and DNA extraction solutions. *International Journal of Food Microbiology* 17, 37–45.

Sorrells, M.E. (1992) Development and application of RFLPs in polyploids. *Crop Science* 32, 1086–1091.

Stollar, B.D. and Raschtchian, A. (1987) Immunochemical approaches to gene probe assays. *Analytical Biochemistry*, 161, 387–394.

Tanksley, S.D., Young, N.D., Paterson, A.H. and Bonierbale, M.W. (1989) RFLP mapping in plant breeding: new tools for an old science. *BioTechnology* 7, 257–264.

Taylor, R., Inamine, G. and Anderson, J.D. (1993) Tissue printing as a tool for observing the immunological and protein profiles in young and mature celery petioles. *Plant Physiology* 102, 1027–1031.

Thompson, M. and Walsh, J.N. (1989) *Handbook of Inductively-coupled Mass Spectrometry*, 2nd edn, Blackie, Glasgow.

Tonnet, M.L. and Downes, R.W. (1983) Estimation of the rubber content of guayule (*Parthenium argentatum*) using low-resolution proton magnetic resonance. *Journal of the Science of Food and Agriculture* 34, 169–174.

Tonnet, L. and Snudden, P.M. (1974) Oil and protein content of the seeds of some pasture legumes. *Australian Journal of Agricultural Research*, 25, 767–774.

Valdes, E.V. and Leeson, S. (1992) Near infrared reflectance analysis as a method to measure metabolizable energy in complete poultry feeds. *Poultry Science* 71, 1179–1187.

Vierling, R.A. and Nguyen, H.T. (1992) Use of RAPD markers to determine the genetic diversity of diploid wheat genotypes. *Theoretical and Applied Genetics* 84, 835–838.

Wehling, R.L., Jackson, D.S., Hooper, D.G. and Ghaedian, A.R. (1993) Prediction of wet-milling starch yield from corn by near-infrared spectroscopy. *Cereal Chemistry* 70, 720–723.

Wetzel, D.L. (1983) Near-infrared reflectance analysis. *Analytical Chemistry* 55, 1165A–1176A.

Williams, J.G.K., Kubelik, A.R., Livak, K.J., Rafalski, J.A. and Tingey, S.V. (1990) DNA polymorphisms amplified by arbitrary primers are useful as genetic markers. *Nucleic Acids Research* 18, 6531–6535.

Williams, P.C., Jaby El-Haramein, M.F., Ortiz-Ferrara, G. and Srivastaya, J.P. (1988) Preliminary observations on the determination of wheat strength by near-infrared reflectance. *Cereal Chemistry* 65, 109–114.

Varietal Identification of Crop Plants $\boxed{2}$

R.J. Cooke

National Institute of Agricultural Botany,
Huntingdon Road, Cambridge CB3 0LE, UK.

2.1 Introduction

In this chapter, 'varietal identification' will be interpreted broadly to include the following:

1. Identification in the strict sense, i.e. what variety is this sample?
2. Variety discrimination – is variety A different from variety B?
3. Varietal purity does this sample consist of more than one variety, and (the supplementary question) if so, what are the other varieties?
4. Variety characterization, that is can we examine varieties in such a way that the data obtained can be used firstly to provide a description and then to aid in the classification of collections of varieties?

These questions cover the general areas of activity in which varietal identification is important. The ability to discriminate between and identify varieties (or cultivars) of agricultural and horticultural crops can be considered central to the operation of the seed trade. This is not an especially new concept. As Carson (1957) stated:

> The basis of modern crop production is the variety. That scarcely needs emphasising, for it is well known that the difference between growing one variety and another may be the difference between a profit and a loss. In consequence, it is essential for the buyer to get the variety he asks for and not something else. And I may also add that quite apart from any question of profit and loss, the buyer is still entitled to receive what he orders.

Varietal identification as part of a philosophy of consumer protection extends through the entire seed trade and allied industries, from plant breeders to food consumers. Increasingly, countries are introducing schemes for Plant Breeders' (or Variety) Rights (PBR, PVR). These reward breeders financially for their efforts and offer protection for their varieties, but in turn require that new varieties are distinct from others and also uniform and stable in the expression of their characteristics (the so-called D, U and S criteria). Again, seed certification, which forms a link between variety registration and seed production, assures the quality of seed marketed to farmers by having rigid standards for varietal identity and purity. Finally, the ultimate consumers of the harvested produce need to know what they are buying, particularly if there is a premium paid for certain varieties or if the produce is to be used for large-scale processing (e.g. mechanized bread making). The needs for varietal identification in different industry sectors have recently been reviewed (Cooke, 1995).

The requirements of these various operations, although clearly related, also obviously differ and hence the approaches taken to address them will inevitably differ as well. Again, different types of crops and species offer varying possibilities for analysis. Thus a wide range of solutions to the problems of varietal identification can be envisaged.

Varietal identification has traditionally been carried out by what might be called a classical taxonomic approach. This requires a detailed study of seeds and/or growing plants and the observation, recording and analysis of a number of morphological characters or descriptors. In practice, such an approach is extremely successful and largely forms the basis, for instance, of current D, U and S testing procedures. However, it can be time consuming and expensive. Large areas of land are needed in addition to highly skilled and trained personnel, making what are often subjective decisions. Many of the morphological descriptors used are multigenic, quantitative or continuous characters, the expression of which is altered by environmental factors. Again, in some species the number of descriptors is limited or is no longer sufficient for discrimination between all varieties. There are thus good grounds for finding alternative procedures to augment the morphologically based approach. The observed physical appearance (phenotype) of a plant arises from an interaction between its genotype and the environment. In general, it is an increasingly desirable objective to reduce or eliminate the environmental influence, so that the genotype of a variety can be observed and utilized more directly.

The rest of this chapter is concerned with modern diagnostic techniques which can be applied to the various facets of varietal identification and which address the environmental interaction question. Such techniques are of two types:

1. The use of computerized systems to capture and process morphological information (image analysis).
2. The use of biochemical methods to analyse various components of plants (chemotaxonomy).

2.2 Image Analysis and Varietal Identification

The technique of image analysis (IA), also known as machine vision, computer vision or robot vision, offers the possibility of developing new methods of varietal identification and characterization. Put simply, IA is a measurement technique, the basic elements of which comprise the following:

1. Image capture – the samples to be measured are placed in front of an image acquisition device, such as a TV or video camera, and the electrical output from the camera is transmitted to a computer. Normal TV cameras take and transmit about 30 frames per second. The computer has to contain a device called an analog to digital (A to D) converter, which converts the signal from a single frame into a numerical form and stores this information in its memory. The image can be divided into a grid of pixels (very small squares), each of which is given a number to indicate the intensity of the image in that particular square. For instance, light areas may be given a high number, darker areas a lower number.
2. Image processing – the image in the computer memory can be modified by changing the intensity numbers, an activity known as image processing. There are many useful modifications which can be made in this way, including enhancing the contrast of the image, highlighting boundaries or other regions, removing extraneous features, simplifying the image or filling in gaps or holes.
3. Extraction of information from the processed image, which can be achieved at various levels of sophistication, including binary (or silhouette) imaging, grey-level imaging (to examine surface features) or colour imaging.
4. Pattern recognition – the development of statistical and other models to sort and compare images.
5. Decision making – the development of computer software to assess the significance of the data produced from all of the above operations (see Draper and Keefe, 1988).

Such a largely computerized technique has a number of attractions compared to conventional measuring systems, including speed, objectivity, the possibility of measuring new characters or features, electronic transmission of data and relief from operator stress or tedium. Also, IA is largely non-destructive and can be used 'remotely', i.e. photographs, negatives, photocopies or other representations can be analysed as readily as living material.

Given this, it is not surprising that a great deal of interest has been expressed in the use of IA for varietal identification. Most of the published work to date has concentrated on wheat (*Triticum* spp.) grain classification. The laboratories concerned have all used different IA equipment and computer software, differing collections of varieties and different methods for describing grain shape. In addition, various aspects of the identification question have been addressed. This makes comparison of the results rather difficult. None the less, it is reasonable to state that IA has had a measure of success in discriminating between wheat varieties and classes from North America (Zayas *et al.*, 1986, 1989; Sapirstein *et al.*, 1987), Australia (Barker *et al.*, 1992a, c) and Europe (Keefe and Draper, 1986, 1988).

Perhaps the work most directly related to wheat variety identification *per se* is that of Keefe (1992), who reported the development and use of a dedicated wheat grain analyser. This instrument utilizes a charge coupled device (CCD) camera to view individual seeds placed crease side down on a horizontal surface, with their longitudinal axes perpendicular to the camera and the embryos to the left. To take 33 separate measurements from 50 individual seeds and process the resultant data takes about 5 min. From the measurements, 69 shape parameters in total are derived (Table 2.1), from which an 'identity' can be assigned to an unknown sample. Although the approach taken by the use of this instrument was generally successful for identification purposes, problems became apparent because of the inevitable overlap between individual seeds from different varieties in the grain shape characters utilized. This meant that it was not unequivocally possible to assign any given seed to its correct variety, without compiling and maintaining a massive database of shape information relating to all of the possible varieties of interest. This clearly limits the practical application of this kind of IA technology. Recent work, especially from Australia, seems to be indicating that for a practicable system of wheat variety discrimination and identification, it will be necessary to use combinations of feature sets and differently derived descriptors and pattern recognition systems (Barker *et al.*, 1992b, d).

It may be that the most useful applications of IA in varieties and seeds work will lie not with identification in its strictly narrow sense but with taxonomy and variety characterization, particularly for the purposes of providing new characters and better descriptions for D, U and S and certification work. An indication of this comes from the work of Draper and Keefe (1989) on onion (*Allium cepa*) bulb shape description. This is an interesting example, since the IA was performed using 35 mm negatives of photographs of onion bulbs. A model for bulb shape was derived, which treated the bulb as a 16-sided polygon (see Fig. 2.1). The points on the bulb circumference were defined by reference to four cardinal points (the middle points of the top and bottom lines of the image and the furthest points left and right). Between each pair of these four points, three other points were located, which were defined in terms of the width of the bulb and selected empirically to give the best visual fit to the recorded image. This work demonstrated clearly that IA was capable not only of reproducing measurements already utilized in variety description and registration work, but also could generate novel characters which were of value for improving the discrimination between genotypes. Similar conclusions have been drawn by van de Vooren and van der Heijden (1993) using IA to measure pod size and

Table 2.1. Measurements and definitions of the 69 parameters utilized by a dedicated wheat grain image analyser.

V1	Area	$V35 = V4/V3$
V2	Length	$V36 = V5/V3$
V3	Height	$V37 = V6/V2$
V4	Brush height	$V38 = V36/V35$
V5	Germ height	$V39 = V8/V1$
V6	Germ length	$V40 = V9/V2$
V7	Germ angle	$V41 = V10/V2$
V8	Embryo end feature (undisclosed)[1]	$V42 = V11/V12$
V9	Embryo end feature (undisclosed)	$V43 = V12 + V18$
V10	Foot length	$V44 = V7 + V11$
V11	Embryo end feature (undisclosed)	$V45 = V15/V16$
V12	Brush end feature (undisclosed)	$V46 = V15/V17$
V13	High point	$V47 = V15/V18$
V14	Perimeter	$V48 = V16/V17$
V15	Brush end feature (undisclosed)	$V49 = V16/V18$
V16	Brush end feature (undisclosed)	$V50 = V17/V18$
V17	Brush end feature (undisclosed)	$V51 = V21/V3$
V18	Brush end feature (undisclosed)	$V52 = V22/V3$
V19	Overall feature (undisclosed)	$V53 = V23/V1$
V20	Overall feature (undisclosed)	$V54 = V24/V2$
V21	Embryo end feature (undisclosed)	$V55 = V55/V1$
V22	Embryo end feature (undisclosed)	$V56 = V26/V2$
V23	Embryo end feature (undisclosed)	$V57 = V27/V2$
V24	Embryo end feature (undisclosed)	$V58 = V28/V2$
V25	Brush end feature (undisclosed)	$V59 = V29/V2$
V26	Embryo end feature (undisclosed)	$V60 = V30/V2$
V27	Embryo end feature (undisclosed)	$V61 = V42/V50$
V28	Brush end feature (undisclosed)	$V62 = V39 + V53$
V29	Brush end feature (undisclosed)	$V63 = V39 + V54 + V40$
V30	Embryo end feature (undisclosed)	$V64 = V54 + V40 - V39$
V31	Embryo end feature (undisclosed)	$V65 = V44/V49$
$V32 = (4 \times \pi \times V1)/(V14 \times V14)$		$V66 = (V35 + V39 + V53)/V17$
$V33 = V3/V2$		$V67 = V22 - V5$
$V34 = V1(V3 \times V2)$		V68 Brush end feature (undisclosed)
		V69 Embryo end feature (undisclosed)

Source: Keefe (1992).
[1] Undisclosed – for commercial confidentiality reasons it is not possible to provide an exact description of these parameters.

shape in *Phaseolus vulgaris* (bean) varieties. Further applications of IA in variety description, including characterization of pod shape in other species and leaf shapes in general, can be anticipated in the near future.

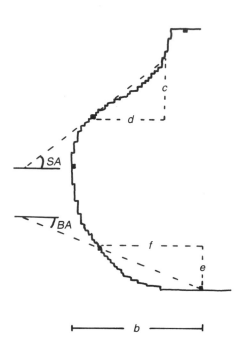

Fig. 2.1. Diagram of a digitized image of an onion, illustrating some of the parameters used for variety classification. The shoulder angle (*SA*) and bottom angle (*BA*) are first calculated by finding, for each one, the coordinates of two points on the circumference of the bulb; these points are defined in terms of the width (*a* or *b*) of the quarter of the bulb in which they lie, and are measured from the widest point on that side of the bulb; the widths *a* and *b* are not always the same because bulbs are rarely perfectly symmetrical; the tangent of *SA* = c/d, tan (*BA*) = e/f; the distances *c, d, e* and *f* are calculated from the x and y coordinates of the points on the circumferences. Adapted and re-drawn from Draper and Keefe (1989).

2.3 Biochemical Methods for Varietal Identification

There are two groups of compounds which chemotaxonomists have found useful for the classification of plants and other organisms – episemantic (secondary) compounds, such as fatty acids, anthocyanins and other pigments, and semantide (sense-carrying) molecules, i.e. proteins and nucleic acids. There exists a wide spectrum of chemical tests and techniques which can be useful for varietal identification, ranging from simple colour tests to the analysis of DNA polymorphisms.

2.3.1 Rapid identification techniques

The definition of 'rapid' techniques can never be entirely objective. However, the tests which are generally recognized under this heading utilize specific treatments to reveal chemical differences between seeds or seedlings of varieties. The tests share several important characteristics – they require very limited technical expertise or training, they can be completed in a short time, they are inexpensive to perform and do not require sophisticated equipment or facilities. In addition, the results are usually clear-cut and easy

Table 2.2. Some of the rapid chemical tests than can be used for species and varietal identification of various crops.

Test	Species
Phenol test	Wheat, rice, oats, barley, ryegrass, *Poa pratensis*, *Vicia striata*, triticale
Fluorescence tests (including examination following chemical treatment)	*Lolium* spp. (roots), oats (seed), peas (seed), soyabean (roots), fescue (roots), *Sinapis arvensis* (seeds)
Chromosome counting (root tips)	Sugar beet, *Lolium* spp., *Trifolium* spp.
NaOH/KOH tests	Wheat, sorghum, rice and red rice
$CuSO_4$–NI I_3 test	Sweet clovers
HCl tests	Oats, barley (pearled), *Sinapis arvensis*
Iodine tests	Wheat, lupins
Vanillin tests	*Vicia faba*, barley

to interpret. Since in most cases individual seeds or plants are tested, the rapid tests are particularly useful for assessing varietal purity and for identifying mixtures.

There are a number of rapid techniques which can be used for varietal identification purposes. Indeed, the International Seed Testing Association (ISTA) has recently (1992a) published a *Handbook of Variety Testing*, one section of which is devoted to rapid chemical identification techniques. Many of these techniques, such as the phenol test for wheat (and certain other graminaceous species), are well known and have been used for a number of years. Table 2.2 lists some of the rapid tests currently utilized and the crop species to which they are applicable for identification purposes. Further details and references can be found in the ISTA *Handbook of Variety Testing* (1992a) and also in van der Burg and van Zwol (1991).

Rapid tests generally rely on the presence of certain, not always known, chemicals which elicit colour changes in parts of the seed or seedling in some varieties and not in others. However, there are other rapid tests which are designed to assay specific enzyme activities. Perhaps the best known of these is the seedcoat peroxidase test in soyabeans, which is an official test recognized by the Association of Official Seed Analysts (AOSA, 1988) and can be used for purity analysis of varieties. Also in soyabeans, there is a rapid colour test which can be used to screen seeds for the presence of lip-oxygenases, which is of use not only for identification purposes but also is useful to plant breeders for the selection of lines and to commercial processors of the seed and meal (Hammond *et al.*, 1992).

Although extremely useful in certain circumstances, the limited discriminating power inevitably offered by such rapid techniques is insufficient for

many purposes. Hence there has been considerable interest in the application
of more sophisticated analytical methods for varietal identification.

2.3.2 *Chromatographic techniques*

The chromatographic separation of episemantic (secondary) compounds for
variety identification has been investigated to a considerable extent. Earlier
work utilized thin-layer chromatography, which can still be useful in certain
cases. Thus, Anderson *et al.* (1988) used thin-layer chromatography to
analyse anthocyanins and carotenoids from various flower colour phenotypes of chrysanthemums. However, most research has concentrated on gas–
liquid or high performance liquid chromatography (GLC, HPLC – see
Morgan, 1989, for review).

GLC has been particularly used for discrimination between various
Brassica varieties. Heaney and Fenwick (1980), for example, used GLC to
analyse glucosinolates from seeds of 22 varieties of Brussels sprouts and
successfully distinguished between them. Similar results have been reported
by various workers with a range of *Brassica* crops, including cabbage, turnip,
swede, calabrese, Chinese cabbage, cauliflower, broccoli, kale, mustard,
kohlrabi and radish (see Morgan, 1989, for references). Both seed and
vegetative tissues can give useful information, depending on the species of
interest. It should be noted that not all *Brassica* species showed varietal
differences in glucosinolate profile, the most notable exception being oilseed
rape (canola, *Brassica napus*). However, GLC has been used to analyse the
fatty acids present in the seed of different rape varieties grown in the UK
(White and Law, 1991). Canonical variate analysis of the fatty acid profiles
indicated that there is a varietal component which remains sufficiently stable
with respect to environment to enable the differences between profiles to be
reliably used for identification purposes.

HPLC has also been used for glucosinolate analysis, but has found an
important taxonomic role in the analysis of anthocyanins and other pigments
from the flowers of ornamental species and the use of the resultant profiles
to distinguish between varieties. Thus there are reports of the use of HPLC
for variety identification in roses, carnations, poinsettias, azaleas, *Impatiens*,
geraniums, *Gerbera* and certain irises, among others. Such analysis can be
valuable not only for identification purposes but also for the breeding of new
flower colour types. Thus Yabuya (1991) analysed the anthocyanins from the
outer perianths of 43 varieties and four wild forms of the Japanese garden iris
(*Iris ensata*). Based on the proportions of the major anthocyanins, the
varieties could be classified into six broad types. One of these types was
noteworthy because it demonstrated the presence of delphinidin-
3-(*p*-cumaroyl)-rutinosido-5-glucoside, a compound which could be used to
breed a blue-flowered variety of iris.

HPLC can also be used to analyse less-well-defined fractions of plants and to provide data which can be used for identification purposes. Thus Mailer *et al.* (1993) extracted the ethanol-soluble components from oilseed and separated these using a reversed-phase C_{18} column. Significant differences were found between the chromatograms of the 29 varieties investigated, which were stable and reproducible and which allowed unknown samples to be allocated their correct identity, given suitable statistical analysis. The precise composition of the ethanol extracts is not yet clear, but includes sinapine, phenolic compounds, some glucosinolates and breakdown products and perhaps proteins.

Hence it is evident that chromatography of secondary compounds can be useful for varietal identification. However, there is no doubt that the analysis of semantides, and in particular proteins and enzymes to date, has been generally far more widely used and successful. Proteins are ideally suited for the purposes of variety identification, since they are direct gene products and their expression is not affected by environmental factors. Thus the analysis of protein composition can be considered to be an analysis of gene expression and the comparison of compositions provides a measure of genetic differences. For this purpose, it is necessary to analyse proteins that are polymorphic and also, preferably, that are present in relatively large amounts and are easy to extract. For these reasons, seed proteins in general and seed storage proteins in particular have been widely exploited, although a range of vegetative proteins and enzymes has also found favour for identification (Cooke, 1988).

Of the various chromatographic techniques, only HPLC can be adequately used for protein separations and it is not surprising that much effort has been put into the analysis of seed and other proteins by HPLC and the subsequent identification of varieties. The first report of the successful separation of wheat seed proteins (Bietz, 1983) has been followed by the unequivocal demonstration that HPLC can be used for varietal identification in a range of crops, primarily cereals such as wheat, barley (*Hordeum vulgare*), rice (*Oryza sativa*), oats (*Avena sativa*), maize (*Zea mays*), sorghum (*Sorghum bicolor*), but also including soyabeans (*Glycine max*) and other legumes (see Morgan, 1989, Smith, 1986, for references). The proteins mostly analysed have been the alcohol-soluble prolamins of cereals – gliadins (wheat), zeins (maize), etc. – but there are published methods involving the separation of albumins, globulins and glutelins. Reversed phase HPLC on silica-based C_{18} columns with a 300Å pore size has been commonly chosen, especially for cereal proteins, although ion-exchange columns can also be useful (e.g. Wingad *et al.*, 1986). Varietal protein profiles produced by a typical HPLC separation of seed proteins are illustrated in Fig. 2.2. Varieties can be distinguished from one another by the absence or presence of particular protein peaks detectable at specific points on the profiles. In addition, the amount of protein in a particular peak (peak area or peak height) can be readily measured, by computerized integration of peak data,

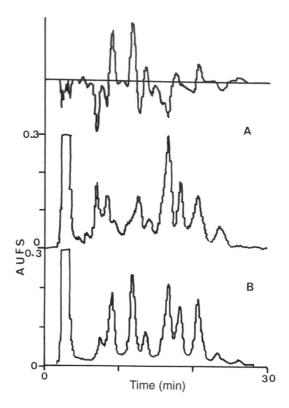

Fig. 2.2. HPLC profiles of two maize inbred lines (A and B), with the uppermost part of the figure representing a plot of their differences. Re-drawn from Smith (1986).

and hence quantitative differences between varieties can also be used as an aid to identification (see Fig. 2.2).

There can thus be no question that the HPLC-based comparison of protein profiles allows varieties to be distinguished from each other and to be identified. Indeed, the technique can be extremely discriminating and several authors have published catalogues of profiles of cereals and other crops (see, for instance, Bietz, 1983; Marchylo and Kruger, 1984; Smith, 1986; Lookhart *et al.*, 1987; Ng *et al.*, 1988; Buehler *et al.*, 1989). Analysis of plant material grown in different locations and years has confirmed that the qualitative nature of the protein chromatograms is not affected by environment, although occasionally quantitative variations have been noted (see Smith and Smith, 1992, for review and references). Thus HPLC can be used with confidence in a range of varietal identification situations, including the determination of overall germplasm relationships and pedigrees. The techniques have been used as evidence in court cases concerning alleged misappropriation of varieties (Smith *et al.*, 1988).

The attractions of HPLC, in addition to the reduction of environmental effects and greater discrimination possibilities, include speed and automation. A typical analytical separation can be generally completed within an hour, or even less, so that results are rapidly available. However, it must be remembered that this is a result for the analysis of only one sample, which could be an individual seed or plant. Any kind of varietal purity estimation clearly requires the analysis of a number of such individuals and could make the process rather lengthy. Possibilities do exist for substantial automation of HPLC separation procedures. Using programmable sample injection systems, for instance, the data produced can not only be readily quantified, which enhances discrimination possibilities, but can also be stored,

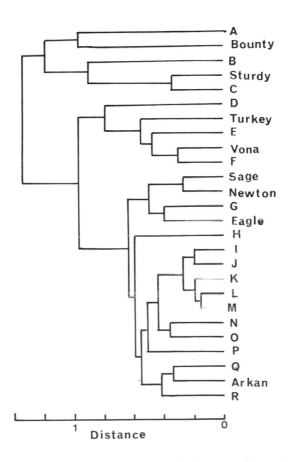

Fig. 2.3. A cluster analysis of a collection of US hard red winter wheats, using HPLC data of gliadin profiles and taking into account both qualitative and quantitative differences. A–R represent unnamed wheat genotypes. Re-drawn from Smith *et al.* (1988).

processed and statistically analysed by computer. This allows various different classification indices to be readily used, as well as relationships to quality and other characteristics to be examined (Lookhart *et al.*, 1993). Associations between varieties can be determined by cluster analysis or similar techniques (see Fig. 2.3).

Despite these advantageous features of HPLC, its use outside of the research laboratory has been somewhat limited. Probably the main reason for this is the difficulties caused by poor column stability, leading to lack of reproducibility of the protein profiles. Such reproducibility problems have serious ramifications for varietal identification work, especially where computerized pattern-matching procedures are employed to assist identification, and could lead to incorrect identification decisions being made. The conventional bonded-phase columns used for reversed phase HPLC are known to be unstable over relatively short periods of time when used at low pH in the presence of trifluoroacetic acid (precisely the conditions utilized in most protein separations). A new class of silane-modified silica has been developed recently and appears, at least for the analysis of wheat seed storage proteins, to provide columns of greater stability and to reduce column-to-column variability (Marchylo *et al.*, 1992). It remains to be seen whether such columns are of more general utility for plant protein separations. It is thus evident that HPLC is extremely useful for the varietal identification of crop plants. However, by far the most widely used and successful techniques which have made the largest impact in this area to date are the various types of gel electrophoresis.

2.3.3 Electrophoresis and varietal identification

The analysis of seed and seedling proteins or isozymes by electrophoresis and the subsequent use of the information provided for varietal identification purposes is now a well established topic and has been reviewed comprehensively in recent years (e.g. see Cooke, 1988; Smith and Smith, 1992). The success of electrophoresis depends on the extensive polymorphism of many plant proteins and the fact that, as with HPLC, the proximity of proteins to the primary genetic information limits the extent of environmental interaction on their expression. Electrophoresis methods are generally fairly easy to carry out and do not require extravagant or expensive laboratory facilities, making them potentially attractive for 'routine', as well as research, purposes.

The ways in which electrophoresis can be applied for varietal identification depend largely on the crop species in question. Two principal approaches can be recognized:

1. A direct comparison of protein compositions between varieties, which usually (but not necessarily) requires the analysis of polymorphic proteins which are encoded at multiple loci. The prolamins of cereal seeds (gliadins, hordeins, zeins, etc.) provide good examples of such proteins. These seed storage proteins are generally encoded at several multigenic loci, and the products of one locus can comprise a group of electrophoretically separable protein bands. With this approach, the criterion for a difference between varieties is usually taken as the presence or absence of a particular band or group of bands at a defined position on the gel.

2. An indirect comparison of the frequency of occurrence of protein band phenotypes within different varieties. This generally (but again not exclusively) necessitates the analysis of proteins (usually enzymes) derived from a single locus, i.e. isozymes.

The choice between these two approaches is largely determined by the genetic structure of the varieties in question, which in turn varies primarily according to the mode of reproduction of the particular crop species. It is conventional to consider two broad categories of crops: those which are self-pollinating (autogamous), which is taken to include vegetative and other asexual means of propagation; and those which are cross-pollinating, or allogamous.

Electrophoresis of self-pollinating crops

Some of the most important crops in world agricultural terms, such as rice, wheat, barley, potatoes (*Solanum tuberosum*), are largely self-pollinating within the definition above, and hence it is perhaps not surprising that so much attention has been paid to this group in terms of varietal identification. Generally, the direct comparative approach has been taken, using storage and other multi-locus proteins. Many different electrophoretic techniques have been successfully applied – Cooke (1988) listed 27 published methods for wheat variety identification and 16 for barley, for instance, and several more have been reported since (e.g. see Lookhart, 1991). Such a wide range of available methodology has caused problems, particularly when trying to compare results from different laboratories, and various organizations have sought to rationalize the situation by adopting 'standard' methods, at least for the more important crops. Chief among these internationally have been the International Association of Cereal Science and Technology (ICC) and the International Seed Testing Association (ISTA) and both have standard reference methods for the identification of wheat varieties by electrophoresis. In principle, the methods are similar, involving the use of polyacrylamide gel electrophoresis (PAGE) at acid pH to separate gliadins. However, they do differ somewhat and can give different separations in certain circumstances (although it should be noted that when the methods are used comparatively, i.e. is the unknown sample the same as the reference sample of a given variety, the actual *results* given by the various methods should not disagree). Some national

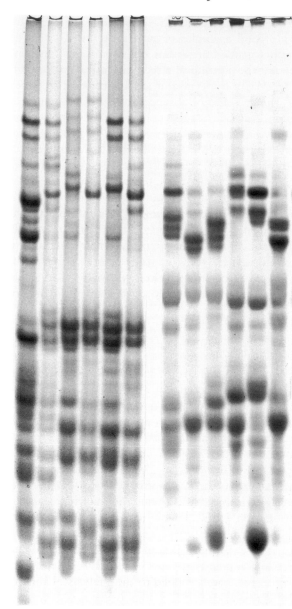

Fig. 2.4. The use of the ISTA standard reference acid PAGE method to analyse the storage proteins of varieties of wheat (left) and barley (right). The varieties are (left to right) Apollo, Avocet, Alexandria, Broom, Brimstone, Copain, Igri, Marko, Digger, Goldspear, Cameo and Capri. Note the polymorphism of the gliadins and hordeins and the differences between varieties, which allow for variety discrimination and identification. The origin and anode are at the top.

bodies, such as the Royal Australian Chemical Institute (RACI) also have officially adopted methods for wheat identification, which differ from both the ICC and ISTA methods. The ISTA has also been active in promoting the adoption of standard reference methods for other crops, such as barley and peas (*Pisum sativum*), and has published a *Handbook of Electrophoresis*

Testing (1992b). Typical results for various varieties of wheat and barley analysed using the ISTA acid PAGE procedure are shown in Fig. 2.4.

Methods such as these are capable of revealing extensive polymorphism in cereal seed storage proteins and have been used, with modifications, in oats, rice, maize, triticale (*Triticosecale*) and sorghum. This polymorphism allows impressive levels of discrimination between varieties to be achieved. For example, White and Cooke (1992) found that in a collection of 353 mostly European barley varieties, 70 different groups could be recognized on the basis of the hordein composition. Various systems of nomenclature have been used to identify the individual bands detected on gels (see Wrigley *et al.*, 1982; Cooke, 1988, 1989; White and Cooke, 1992) and authors from laboratories in many countries have published catalogues of electrophoretic patterns, particularly for wheat and barley (see Cooke, 1988, for references).

Other kinds of electrophoresis have also proved to be extremely useful for varietal identification in self-pollinating crops. PAGE in the presence of sodium dodecyl sulphate (SDS–PAGE) is particularly suitable for legume species such as peas and *Phaseolus* beans, but has been widely applied to the analysis of barley hordeins and wheat glutenin polypeptides. Again, catalogues of protein patterns have been produced by various authors to facilitate identification. Isoelectric focusing (IEF) of seed and other proteins and enzymes has also been used (see Wrigley *et al.*, 1982; Cooke, 1984, 1988; Smith and Smith, 1992, for references).

The use of different kinds of electrophoresis in tandem can improve the degree of discrimination within a given collection of varieties. For example, at NIAB, we have shown (unpublished) that IEF of seed esterases can distinguish between some barley varieties with identical hordein compositions. Electrophoresis methods can also be combined in a two-dimensional (2-D) approach. This involves the analysis of a sample first by one technique (e.g. IEF), with the resultant gel then forming the sample for a second separation by a different technique (e.g. SDS–PAGE), carried out at right angles to the first. The 'maps' produced by 2-D analysis can contain many hundreds of protein spots, providing considerable potential for discrimination between even closely related genotypes. This is particularly so if immobilized pH-gradient gels are used in the first dimension and the 2-D gel is subjected to highly sensitive silver staining (see Gorg, A., *et al.*, 1992, for instance). Methods such as this have been applied to the analysis of various self-pollinating crops (see Cooke, 1984, 1988, for references), but the technical complexity of 2-D analysis and difficulties in data interpretation have limited the widespread use of 2-D electrophoresis for most routine variety identification purposes. However, there may be a specialized role for 2-D mapping procedures in areas such as D, U and S testing and varietal registration.

Due to breeding methods and selection, varieties of self-pollinating crops usually consist of a single homozygous line and are uniform, both phenotypically and genotypically. However, in cases where this has been investigated, it is always found that a proportion of varieties consist of two or more electrophoretically distinguishable lines, sometimes called biotypes (Cooke, 1984, 1988, 1989). This is a consequence of the lack of conscious selection for protein or isozyme homogeneity and does not unduly detract from the utility of electrophoresis for varietal identification. It is important, though, for investigators to be aware of the potential lack of uniformity and to analyse individual seeds or plants, especially from standard varieties used for reference purpose, in order to detect and record biotypes.

The direct comparative approach using multiple encoded proteins is also very successful when applied to vegetatively propagated or asexually reproducing crops. Varieties of these species can be considered to be essentially clones of phenotypically and genotypically identical individuals, each containing the same fixed complement of genes. Several crops of this type have been investigated electrophoretically, including bluegrass (*Poa pratensis L.*), bananas (*Musa* spp.), roses (*Rosa* spp.), strawberries (*Fragaria* spp.), pears (*Pyrus* spp.) and other fruits (Cooke, 1988; Gilliland, 1989). Undoubtedly, though, the most thoroughly researched is the potato. Due primarily to the pioneering work of the group in Braunschweig, over 15,000 potato varieties, species, subspecies and wild types have been examined by electrophoresis (Huaman and Stegemann, 1989). The most generally applied method is PAGE at pH 7.9 or 8.9 of the soluble tuber proteins and/or

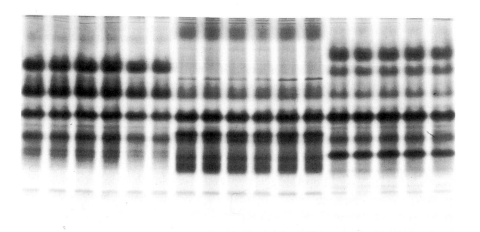

Fig. 2.5. The use of PAGE of soluble tuber proteins to distinguish between and identify varieties of potato. Each track represents the proteins from a single tuber of the varieties King Edward (tracks 1–6), Cara (7–12), and Maris Bard (13–17). Note the uniformity of the protein profiles within a variety and the differences between varieties.

esterases expressed from potato sap (see Fig. 2.5). IEF and various 2-D approaches have also been widely used and with the appropriate combination of methods, potato varieties can invariably be distinguished from one another (Stegemann, 1984). Similar techniques can successfully be used for other root crops such as sweet potato (*Ipomoea* spp.) (Stegemann *et al.*, 1992).

Electrophoresis of cross-pollinating crops

Cross-pollinated species present special problems from the electrophoretic varietal identification point of view. Because of their genetic structure, varieties of such species are populations of individuals, expressing a range of phenotypic characters. The individuals can be genetically distinct, containing different combinations of homozygous and heterozygous genes, including those encoding storage or other proteins and enzymes. This variation is maintained in equilibrium over generations. There are two ways of tackling the problem of identification in these species:

1. Analyse a bulked (pooled) extract of seeds or plants, to obtain an overall varietal protein profile. This means that varieties can then be treated as self-pollinated crops and the direct comparative approach, examining seed storage proteins for instance, can be used.

2. Analyse individual seeds or plants from a variety, examining single locus proteins or isozymes and determining the variability within and between varieties statistically.

Both of these approaches have been successfully used in cross-pollinating species, with a good example being provided by the work which has been carried out with ryegrass (*Lolium* spp., see Cooke, 1988; Gilliland, 1989). The seed storage protein composition of bulked (approximately 2g) seed samples of ryegrass varieties has been analysed by SDS–PAGE, which provides a high level of discrimination between varieties. The ISTA has recently adopted SDS–PAGE as a standard reference method for identifying commercial seedlots of ryegrass varieties and species (ISTA, 1992b).

For the analysis of single locus isozymes in ryegrass, most studies have used starch gel electrophoresis to examine young leaf enzymes, with PGI (phosphoglucoisomerase) being particularly well researched (see, for instance, the review by Nielsen, 1985, and Greneche *et al.*, 1991). There are reported to be four common alleles of PGI in ryegrass (called *a*, *b*, *c* and *d*), which give rise to ten electrophoretic phenotypes. Plants which are homozygous for PGI (e.g. *aa*, *bb*) display a single band following starch gel electrophoresis, whereas heterozygotes (*ab*, *bc*, etc.) have three bands, one corresponding to each allele and a heterodimer of intermediate mobility, arising from the dimeric structure of PGI. For variety identification, extracts of leaves from 50 to 100 individual plants are taken, electrophoresed and the

R.J. Cooke

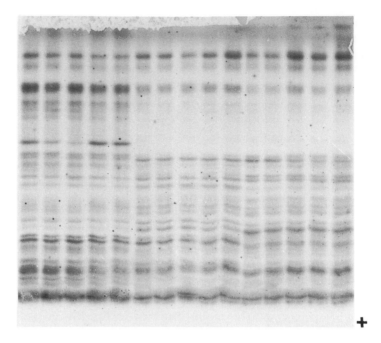

Fig. 2.6. The use of IEF of seed esterases to distinguish between lines and hybrids of pearl millet. Each track represents the profile from a different bulked sample of 25 individual seeds of a male-sterile line (tracks 1–5), a hybrid (6–10) and a restorer line (11–15). Note the reproducibility of the profile within a genotype and the differences between genotypes. From Varier and Cooke (1992).

gels stained for PGI activity. The frequency of occurrence of each phenotype (which can be directly related to genotype) in a variety is determined and these frequencies in different varieties are then compared statistically, using a chi-squared or other suitable analysis. Thus varieties are discriminated from one another on the basis of differences in the frequency of occurrence of alleles. This method, using a single enzyme, can be very powerful. For example, it has been shown that about 70% of a collection of 149 perennial and Italian ryegrass varieties could be distinguished ($P<0.05$) on the basis of PGI allele frequencies (Gilliland, 1989). By utilizing more isozymes, even greater discrimination is achievable (Greneche *et al.*, 1991). Also, it has been reported that by varying the pH conditions of the starch gel separations, further allelic variants of PGI can be detected, which add to the discriminatory possibilities of the technique (Loos and Degenaars, 1993). As with the cereal species mentioned above, authors have published catalogues of the allelic compositions of collections of varieties (e.g. Lallemand *et al.*, 1991).

Many other cross-pollinating species have been investigated using either the bulked approach or the analysis of individuals. In addition to starch gel electrophoresis and SDS–PAGE, IEF and native PAGE separations of a wide range of protein and enzyme systems have been carried out. An example is shown in Fig. 2.6. Among the many hundreds of cases that could be quoted are reports of the successful use of electrophoresis for varietal identification in crops such as rye (*Secale cereale*), Faba beans, oilseed rape, sugar beet (*Beta vulgaris*), alfalfa (*Medicago sativa*), pasture and forage species including fescue (*Festuca* spp.), timothy (*Phleum* spp.) and lovegrass (*Eragrostis curvula*), sunflowers (*Helianthus annuus*) and various horticultural species such as onions and chicory (*Cichorium* spp.), vegetables and fruits (see for examples, Cooke, 1985, 1988; Goodrich *et al.*, 1985; Arulsekar and Parfitt, 1986; Gardiner and Forde, 1988; Mundges *et al.*, 1990; Anisimova *et al.*, 1991; Baes and van Cutsem, 1992; di Renzo *et al.*, 1992).

Practical applications of electrophoresis

Electrophoretic techniques have been used extensively in the seed trade and associated industries. All of the types of question regarding varietal identity posed in the introduction to this chapter can be addressed in many species by the application of an appropriate electrophoresis method or methods. The questions of identity and purity are important commercially, especially in those industries requiring particular quality varieties for mechanized processing, e.g. flour milling, bread baking and malting, and many such companies operate quality control procedures based on electrophoresis. Varietal identity is also important in industries where varieties are required for particular end-uses, potato trading being a good example, and again electrophoresis is widely used to check the material being traded. Varietal purity is difficult to investigate in cross-pollinating species and electrophoresis in these crops is normally limited to the issues of identification and/or discrimination. A particularly important use of electrophoresis in the seed industry is in the measurement of F_1 hybrid purity, particularly in maize but increasingly in other crops as well (see Cooke, 1988). Other practical applications include plant breeding, varietal registration and the documentation of genetic resources. There can thus be no doubt that gel electrophoresis is at present the most widely used and important of the diagnostic methods available for plant variety identification.

2.3.4 DNA profiling and varietal identification

HPLC and gel electrophoresis are methods which reveal variability at a number of loci which encode for storage and other proteins. If such variability exists, then it follows that there must also be variation in the underlying genetic material (DNA) itself. Indeed, it is probable that there will be more DNA variability, since not all differences will occur in expressed regions of DNA. Thus an extremely powerful diagnostic tool for plant variety identification would be available if DNA polymorphisms could be revealed and utilized. Advances in molecular biology methodology, especially over the past five years or so, now enable variations at the DNA level to be examined and such methods are increasingly being applied to the questions of varietal identification. A detailed consideration of the various types of method available is beyond the scope of this chapter. However, in terms of identification work, there are basically two types of methods which have been used – probe-based technologies and amplification technologies.

Probe-based technologies

Restriction fragment length polymorphisms (RFLPs) are the most widely reported means of revealing DNA sequence variations in a diverse range of organisms, including plant varieties (see Fig. 2.7). The potential of RFLPs for identification purposes has been reviewed by Ainsworth and Sharp (1989) and it is clear that the large number of available restriction enzyme/probe combinations make this a powerful approach.

Conventional RFLP analysis utilizes randomly selected single-copy or low-copy genomic and/or cDNA clones as probes (as in Fig. 2.7). This is an effective way of revealing differences between varieties and there are reports of successful RFLP analysis in many species, including wheat, barley, rice, maize, oats, brassicas, peppers (*Capsicum* spp.), roses, grapes and apples (Wang and Tanksley; 1989; Graner *et al.*, 1990; Harcourt and Gale, 1991; Smart *et al.*, 1991; Smith and Smith, 1991; Watillon *et al.*, 1991; Prince *et al.*, 1992; Rajapakse *et al.*, 1992; Bowers *et al.*, 1993; Nienhuis *et al.*, 1993; Pecchioni *et al.*, 1993; Vaccino *et al.*, 1993). These methods can be very discriminating. For instance, Graner *et al.* (1990) analysed 48 barley varieties using 23 single copy probes, distributed over the seven barley chromosomes. Polymorphisms were detected at a frequency of 43%, when three different restriction enzymes were used. Again, Smith and co-workers have clearly demonstrated that an array of probes and probe/enzyme combinations can readily distinguish between and identify a large number of US maize hybrids and inbred lines (e.g. Smith and Smith, 1991).

In cases where single copy probes have been found to provide only low levels of polymorphism, the use of multiple copy probes, either random or of known derivation, can be advantageous. Thus, Gorg, R., *et al.* (1992) reported that they were able to identify uniquely 122 out of 134 potato varieties using *Taq-1* restricted DNA probed with a marker known as GP-35. Similarly both the 5S ribosomal gene clusters and the intergenic spacer sequences in wheat can reveal more polymorphisms and discriminate between more genotypes when used as probes than single copy clones (May and Appels, 1992; Roder *et al.*, 1992).

There are alternative sources of probes in RFLP analysis. For instance, some of the probes used for DNA 'fingerprinting' in humans and other

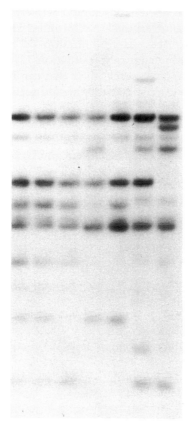

Fig. 2.7. An example of RFLP analysis in oilseed rape. DNA from seven different varieties was restricted with the enzyme *Hind*III, electrophoresed on an agarose gel, blotted and the blot subsequently probed using a *Brassica napus*-derived probe. Unpublished data of Dr David Lee (NIAB).

animals can reveal useful polymorphisms in plants and be used for variety identification purposes. The M13 repeat probe has been particularly used by Nybom and her colleagues to distinguish between and identify varieties of several horticultural species such as apples, blackberries and raspberries (e.g. Nybom *et al.*, 1990). Other workers have used synthetic repetitive oligonucleotides as probes. The (GATA)$_n$-type probes have proven to be particularly effective and appear to be applicable to a range of plant species (e.g. Weising *et al.*, 1991; Beyermann *et al.*, 1992). There are few detailed reports of the use of such oligonucleotides for variety identification, although they have been demonstrated to be very useful in tomatoes, a crop where genetic diversity as revealed by conventional RFLP analysis is very limited (Vosman *et al.*, 1992).

Most probe-based work has used radioactively labelled probes followed by autoradiography to reveal the DNA profiles. This widespread use of ^{32}P would undoubtedly place a large limitation on the more routine use of RFLPs in many variety identification situations. However, there are various fluorescent and chemiluminescent labels that can be used as alternatives to ^{32}P and such non-radioactive methods would seem to hold much promise. For instance, Parent and Page (1992) used a commercially available non-radioactively labelled probe to distinguish between 15 varieties of raspberries. The potential advantages of non-radioactive labelling of plant DNA probes have been recently discussed (Neuhaus-Url and Neuhaus, 1993).

Amplification technologies

Amplification technologies based on the polymerase chain reaction (PCR) offer some apparent advantages over RFLPs for variety identification purposes. Generally, PCR-based methods are quicker, need very little target

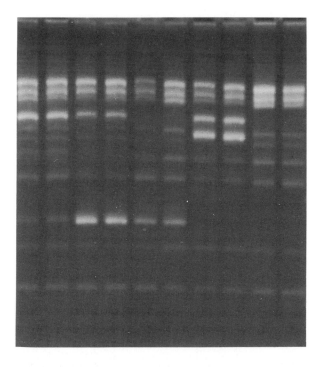

Fig. 2.8. An example of RAPDs analysis in oilseed rape. DNA from five different varieties was extracted and amplified in a PCR reaction, using an arbitrary 10-mer oligonucleotide as primer. The amplification products were separated by electrophoresis on a polyacrylamide gel (in pairs) and stained with ethidium bromide. Unpublished data of Dr David Lee (NIAB).

DNA, avoid the need for radioactive labels and are more amenable to automation. The discovery that arbitrarily chosen primers, made with no knowledge of any specific target DNA, can act as templates for the amplification of several fragments of genomic DNA (Welsh and McClelland, 1990; Williams *et al.*, 1990) has led to a new generation of diagnostic techniques, known as RAPDs (random amplified polymorphic DNA) and AP-PCR (arbitrarily primed-PCR). The nature of the amplified fragments depends upon the primer sequence and on the DNA being analysed. Different primers give rise to different amplified bands and polymorphisms at the priming sites result in the disappearance of an amplified band (Fig. 2.8). Thus PCR with arbitrary primers is a method for detecting polymorphisms distributed throughout the genome, with a primer usually amplifying several bands, each of which will probably originate from a different locus.

While the preliminary work showed not much more than the existence of differences between a few rice and soyabean genotypes, there has recently been an explosion of interest in the use of RAPDs for variety discrimination and identification. Thus there are reports of the successful application of RAPDs/AP-PCR to distinguish between varieties, genotypes, lines and species of cereal crops (Dweikat *et al.*, 1993; Ko *et al.*, 1994), barley (Tinker *et al.*, 1993), maize (Welsh *et al.*, 1991), various brassica species (Demeke *et al.*, 1992; Boury *et al.*, 1992; Mailer *et al.*, 1994), apples (Koller *et al.*, 1993), papaya (*Carica* spp.) (Stiles *et al.*, 1993), cocoa (*Theobroma* spp.) (Russell *et al.*, 1993), sugarbeet (Jung *et al.*, 1993), onions (Wilkie *et al.*, 1993), chrysanthemums (Wolff and Peters van Rijn, 1993) and rhododendrons (Iqbal *et al.*, 1994), among others. Often, in addition to demonstrating the existence of observable differences between varieties, authors have used the RAPDs-generated data to establish taxonomic relationships and to estimate genetic distances between varieties. Such estimates of associations and genetic distance should be more reliable than those produced from protein data (e.g. as in Fig. 2.3), since there will usually be a larger number of loci sampled by RAPDs (or by the use of multiple probes in RFLP analysis) than by HPLC or electrophoresis.

Other techniques involving PCR are also being investigated. For instance, PCR-based analysis of simple sequence repeats (microsatellites) has been reported in *Brassica* (Lagercrantz *et al.*, 1993), which may prove to be particularly useful for variety identification work. The use of PCR methods in cereals has recently been summarized (Henry, 1994).

Clearly these are early days for a proper evaluation of the use of DNA profiling for plant variety identification. However, enough is known of the emerging technologies to indicate the great diagnostic potential of this approach. By analysing the genetic material itself, environmental interaction should be absent and examination of the genotype of the material in question is ever closer. What is now required is a systematic and comparative study of the various technologies available, to compare the degrees of discrimination achievable within wider collections of material, to obtain data on the question of intra-varietal polymorphisms (cf. 'biotypes' in protein electrophoresis) and to produce simpler yet robust analytical procedures.

2.4 Future Trends

It should be evident from the foregoing that there has been considerable interest in the topic of varietal identification of crop plants from a range of scientific disciplines. What does the future hold in this area?

It is likely that in the medium term, electrophoresis will remain the diagnostic method of choice, at least for many species. There will be an increasing tendency towards the use of smaller and thinner, commercially available, pre-cast gels and also to a 'systems' approach, integrating electrophoretic separation and computer-assisted profile matching (e.g. see Wrigley *et al.*, 1992; Hsam *et al.*, 1993; Moller and Spoor, 1993). The first applications of capillary zone electrophoresis (CZE) for variety identification have been reported (Dinelli and Bonetti, 1992) and it is reasonable to assume more activity within this field, offering as it does possible advantages of time and cost over both HPLC and gel electrophoresis.

Methods of protein detection based on immunological reactions and especially the use of monoclonal antibodies have been suggested as being potentially useful for variety identification (e.g. Skerritt *et al.*, 1988). However, the relative lack of specificity of most storage protein antibodies probably limits such a use to distinguishing quality types or classes of grain, rather than for the identification of varieties. None the less, this could still be valuable commercially and further developments here can be expected. The wider availability of image-analysis-based variety identification systems draws nearer, as does the application of this technology to a greater range of species, although the major use may eventually be found to be in description and taxonomy rather than identification. In the rather more long term, DNA profiling appears to offer the greatest potential. Already there have been reports of the use of both RFLPs and RAPDs for the evaluation of F_1 hybrid purity (Livneh *et al.*, 1990; Boury *et al.*, 1992) and further practical applications now currently utilizing protein electrophoresis will surely follow. Extraction of DNA from dry seeds, along with more rapid methods for DNA extraction in general and the ability to analyse individual seeds or plants, will improve the prospects for the use of such technologies in routine testing systems (Brunel, 1992; Benito *et al.*, 1993, McDonald *et al.*, 1994). New approaches to the identification of gel patterns, applicable to both protein and DNA profiles, offer considerable promise as a means of addressing the thorny problems of gel evaluation and data collection. Digital gel image processing and artificial neural networks have been used to examine gels and the application of such an approach to varietal identification is certainly feasible (Sondergaard *et al.*, 1994). Again, if the need to run gels could be avoided, for instance by the use of dot-blot or fluorescence assays using amplified single-copy probes, a new era in identification technology could be envisaged.

Thus there are many exciting prospects for new diagnostic methods for crop plant variety identification. The existing methods have already found widespread acceptance within the seed trade and allied industries. It is important that resources are made available to continue research work into all diagnostic methodologies for identification of varieties. The use of modern techniques of genetic manipulation in plant breeding, as well as the imminent availability of transgenic crop varieties and the need for overall improved seed quality control, together with continuing pressure to utilize cost-effective analytical procedures, all ensure that more use will need to be made of the appropriate technologies in the future.

Acknowledgements

The contributions of various friends and colleagues at NIAB and elsewhere to the development of the work and ideas contained in this chapter are gratefully acknowledged. Thanks also to Andy Tiley for the photography and to the NIAB typists for their efforts.

References

Ainsworth, C.C. and Sharp, P.J. (1989) The potential role of DNA probes in plant variety identification. *Plant Varieties and Seeds* 2, 27 34.

Anderson, N.O., Ascher, P.D. and Widmer, R.E. (1988) Thin-layer chromatographic analysis of flower color phenotypes in *Dendranthema grandiflorum* Ramatuelle inbreds and clonal cultivars. *Euphytica* 37, 229–239.

Anisimova, I.N., Gavrilyuk, I.P. and Konarev, V.G. (1991) Identification of sunflower lines and varieties by helianthinin electrophoresis. *Plant Varieties and Seeds* 4, 133–141.

Arulsekar, S. and Parfitt, D.E. (1986) Isozyme analysis procedures for stone fruits, almond, grape, walnut, pistachio and fig. *Hortscience* 21, 928–933.

Association of Official Seed Analysts (1988) Rules for testing seeds (1989 revision). *Journal of Seed Technology* 12, 1–126.

Baes, P. and van Cutsem, P. (1992) Chicory seed lot identification by leucine aminopeptidase and esterase zymogram analysis. *Electrophoresis* 13, 885–886.

Barker, D.A., Vuori, T.A. and Myers, D.G. (1992a) The use of slice and aspect ratio parameters for the discrimination of Australian wheat varieties. *Plant Varieties and Seeds* 5, 47–52.

Barker, D.A., Vuori, T.A. and Myers, D.G. (1992b) The use of Fourier descriptors for the discrimination of Australian wheat varieties. *Plant Varieties and Seeds* 5, 93–102.

Barker, D.A., Vuori, T.A., Hegedus, M.R. and Myers, D.G. (1992c) The use of ray parameters for the discrimination of Australian wheat varieties. *Plant Varieties and Seeds* 5, 35–45.

Barker, D.A., Vuori, T.A., Hegedus, M.R. and Myers, D.G. (1992d) The use of Chebychev coefficients for the discrimination of Australian wheat varieties. *Plant Varieties and Seeds* 5, 103–111.

Benito, C., Figueiras, A.M., Zarogoza, C., Gallego, F.J. and de la Pena A. (1993) Rapid identification of Triticeae genotypes from single seeds using the polymerase chain reaction. *Plant Molecular Biology* 21, 181–183.

Beyermann B., Nurnberg, P., Weike, A., Meiner, M., Epplen, T.J. and Borner, T. (1992) Fingerprinting plant genomes with oligonucleotide probes specific for single repetitive DNA sequences. *Theoretical and Applied Genetics* 83, 691–694.

Bietz, J.A. (1983) Separation of cereal proteins by reversed-phase high performance liquid chromatography. *Journal of Chromatography* 255, 219–238.

Boury, S., Lutz, I., Gavalda, M-C., Guidet, F. and Schlesser, A. (1992) Empreintes génétiques du chou-fleur par RAPD et verfication de la pureté hybride F1 d'un lot de semences. *Agronomie* 12, 669–681.

Bowers, J.E., Bandman, E.B. and Meredith, C.P. (1993) DNA fingerprint characterization of some wine grape cultivars. *American Journal of Enology and Viticulture* 44, 266–274.

Brunel, D. (1992) An alternative, rapid method of plant DNA extraction for PCR analyses. *Nucleic Acids Research* 20, 4676.

Buehler, R.E., McDonald, M.B., van Toai, T.T. and St Martin, S.K. (1989) Soybean cultivar identification using high performance liquid chromotography of seed proteins. Crop Science 29, 32–37.

Carson, G.P. (1957) Variety and its identification. *Journal of the National Institute of Agricultural Botany* 8, 194–197.

Cooke, R.J. (1984) The characterisation and identification of crop cultivars by electrophoresis. *Electrophoresis* 5, 59–72.

Cooke, R.J. (1985) Characterisation of cultivars of *Allium cepa L.* (onions) by ultrathin-layer isoelectric focussing of seed esterases. *Electrophoresis* 6, 572–573.

Cooke, R.J. (1988) Electrophoresis in plant testing and breeding. *Advances in Electrophoresis* 2, 171–261.

Cooke, R.J. (1989) The use of electrophoresis for the distinctness testing of varieties of autogamous species. *Plant Varieties and Seeds* 2, 3–13.

Cooke, R.J. (1995) Introduction: the reasons for variety identification. In: Wrigley, C.W. (ed.), *Identification of Food Grain Varieties.* American Association of Cereal Chemists, St Paul, MN, USA, pp. 1–17.

Demeke, T., Adams, R.P. and Chibber, R. (1992) Potential taxonomic use of random amplified polymorphic DNA (RAPD): a case study in *Brassica. Theoretical and Applied Genetics* 84, 990–994.

Dinelli, G. and Bonetti, A. (1992) Capillary electrophoresis in species and cultivar determination. *Seed Science and Technology* 20, 561–569.

di Renzo, M.A., Poverene, M.M. and Medina, M.I. (1992) Identification of lovegrass (*Eragrostis curvula*) cultivars by electrophoresis of seed isozymes. *Seed Science and Technology* 20, 101–110.

Draper, S.R. and Keefe, P.D. (1988) Electrophoresis of seed storage proteins and whole-seed morphometry using machine vision: alternative or complementary methods for cultivar identification? *Proceedings ISTA Symposium*, Leningrad, pp. 27–35.

Draper, S.R. and Keefe, P.D. (1989) Machine vision for the characterisation and identification of cultivars. *Plant Varieties and Seeds* 2, 53–62.

Dweikat, I., Mackenzie, S., Levy, M. and Ohm, H. (1993) Pedigree assessment using RAPD-DGGE in cereal crop species. *Theoretical and Applied Genetics* 85, 497–505.

Gardiner, S.E. and Forde, M.B. (1988) Identification of cultivars and species of pasture legumes by sodium dodecyl sulphate polyacrylamide gel electrophoresis of seed proteins. *Plant Varieties and Seeds* 1, 13–26.

Gilliland, T.J. (1989) Electrophoresis of sexually and vegetatively propagated cultivars of allogamous species. *Plant Varieties and Seeds* 2, 15–26.

Goodrich, W.J., Cooke, R.J. and Morgan, A.G. (1985) The application of electrophoresis to the characterisation of *Vicia faba* L. *FABIS Newsletter* 13, 8–11.

Gorg, A., Postel, W., Baumer, M. and Weiss, W. (1992) Two-dimensional gel electrophoresis, with immobilised pH gradients in the first dimension, of barley seed proteins: discrimination of cultivars with different malting grades. *Electrophoresis* 13, 192–203.

Gorg, R., Schachtschabel, U., Ritter, E., Salamini, F. and Gebhardt, C. (1992) Discrimination among 136 tetraploid potato varieties by fingerprints using highly polymorphic DNA markers. *Crop Science* 32, 815–819.

Graner, A., Siedler, H., Jahoor, A., Herrman, R.G. and Wenzel, G. (1990) Assessment of the degree and type of restriction fragment length polymorphism in barley (*Hordeum vulgare*). *Theoretical and Applied Genetics* 80, 826–832.

Greneche, M., Lallemand, J. and Migaud, O. (1991) Comparison of different enzyme loci as a means of distinguishing ryegrass varieties by electrophoresis. *Seed Science and Technology* 19, 147–158.

Hammond, E.G., Duvick, D.N., Fehr, W.R., Hildebrand, D.F., Lacefield, E.C. and Pfeiffer, T.W. (1992) Rapid screening techniques for lipoxygenases in soybean seeds. *Crop Science* 32, 820–821.

Harcourt, R.L. and Gale, M.D. (1991) A chromosome-specific DNA sequence which reveals a high level of RFLP in wheat. *Theoretical and Applied Genetics* 81, 397–400.

Heaney, R.K. and Fenwick, G.R. (1980) Glucosinolates in brassica vegetables. Analysis of 22 varieties of Brussels sprout (*Brassica oleracea* var. *gemmifera*). *Journal of the Science of Food and Agriculture* 31, 785–793.

Henry, R.J. (1994) Molecular methods for cereal identification. *Australasian Biotechnology* 4, 150–152.

Hsam, S.L.K., Schickle, H., Westermeier, R. and Zeller, F.J. (1993) Identification of cultivars of crop species by polyacrylamide electrophoresis. 1. Commercial barley (*Hordeum vulgare* L.) cultivars grown in Germany. *Monatsschrift für Brauwissenschaft* 46, 86–94.

Huaman, Z. and Stegemann, H. (1989) Use of electrophoretic analyses to verify morphologically identical clones in a potato collection. *Plant Varieties and Seeds* 2, 155–161.

International Seed Testing Association (1992a) *Handbook of Variety Testing – Rapid Chemical Identification Techniques* (Payne, R.C. ed.), Zurich, pp. 1–52.

International Seed Testing Association (1992b) *Handbook of Variety Testing –*

Electrophoresis Testing (Cooke, R.J., ed.), Zurich, pp. 1–42.

Henry, R.J. (1994) Molecular methods for cereal identification. *Australasian Biotechnology* 4, 150–152.

Iqbal, M.J., Gray, L.E., Paden, D.W. and Rayburn, A.L. (1994) Feasibility of rhododendron DNA profiling by RAPD analysis. *Plant Varieties and Seeds* 7, 59–63.

Jung, C., Pillen, K., Frese, L., Fahr, S. and Melchinger, A.E. (1993) Phylogenetic relationships between cultivated and wild species of the genus *Beta* revealed by DNA 'fingerprinting'. *Theoretical and Applied Genetics* 86, 449–457.

Keefe, P.D. (1992) A dedicated wheat grain image analyser. *Plant Varieties and Seeds* 5, 27–33.

Keefe, P.D. and Draper, S.R. (1986) The measurement of new characters for cultivar identification in wheat using machine vision. *Seed Science and Technology* 14, 715–724.

Keefe, P.D. and Draper, S.R. (1988) An automated machine vision system for the morphometry of new cultivars and plant genebank accessions. *Plant Varieties and Seeds* 1, 1–11.

Ko, H.L., Henry, R.J., Graham, G.C., Fox, G.P., Chadbone, D.A. and Haak, I.C. (1994) Identification of cereals using the polymerase chain reaction. *Journal of Cereal Science* 19, 101–106.

Koller, B., Lehmann, A., McDermott, J.M. and Gessler, C. (1993) Identification of apple cultivars using RAPD markers. *Theoretical and Applied Genetics* 85, 901–904.

Lagercrantz, U., Ellegren, H. and Andersson, L. (1993) The abundance of various microsatellite motifs differs between plants and vertebrates. *Nucleic Acids Research* 21, 1111–1115.

Lallemand, J., Michaud, O. and Greneche, M. (1991) Electrophoretical description of ryegrass varieties : a catalogue. *Plant Varieties and Seeds* 4, 11–16.

Livneh, O., Nagler, Y., Tal, Y., Harush, S.B., Gafni, Y., Beckmann, J.S. and Sela, I. (1990) RFLP analysis of a hybrid cultivar of pepper (*Capsicum annuum*) and its use in distinguishing between parental lines and in hybrid identification. *Seed Science and Technology* 18, 209–214.

Lookhart, G.L. (1991) Cereal proteins: composition of their major fractions and methods for identification. In: Lorenz, K.L. and Kulp, K. (eds), *Handbook of Cereal Science and Technology*. Marcel Dekker, New York, pp. 441–468.

Lookhart, G.L., Albers, L.D., Pomeranz, Y. and Webb, B.D. (1987) Identification of US rice cultivars by high-performance liquid chromatography. *Cereal Chemistry* 64, 199–206.

Lookhart, G.L., Cox, T.S. and Chung, O.K. (1993) Statistical analyses of reversed-phase high performance liquid chromatography (RP-HPLC) patterns of hard red spring and hard red winter wheat cultivars grown in a common environment: classification indices. *Cereal Chemistry* 70, 430–434.

Loos, B.P. and Degenaars, G.H. (1993) pH-dependent electrophoretic variants for phosphoglucoisomerase in ryegrass (*Lolium* spp.): a research note. *Plant Varieties and Seeds* 6, 55–60.

Mailer, R.J., Daun, J. and Scarth, R. (1993) Cultivar identification in *Brassica napus* L. by RP-HPLC of ethanol extracts. *Journal of the American Oil Chemists' Society* 70, 863–866.

Mailer, R.J., Scarth, R. and Fristensky, B. (1994) Discrimination among cultivars of rapeseed (*Brassica napus* L.) using DNA polymorphisms amplified from arbitrary primers. *Theoretical and Applied Genetics* 87, 697–704.

Marchylo, B.A. and Kruger, J.E. (1984) Identification of Canadian barley cultivars by reversed-phase high-performance liquid chromatography. *Cereal Chemistry* 61, 295–301.

Marchylo, B.A., Hatcher, D.W., Kruger, J.E. and Kirkland, J.J. (1992) Reversed-phase high-performance liquid chromatographic analysis of wheat proteins using a new, highly stable column. *Cereal Chemistry* 69, 371–378.

May, C.E. and Appels, R. (1992) The nucleolus organiser regions (*Nor* loci) of hexaploid wheat cultivars. *Australian Journal of Agricultural Research* 43, 889–906.

McDonald, M.B., Elliot, L.J. and Sweeney, P.M. (1994) DNA extraction from dry seeds for RAPD analyses in varietal identification studies. *Plant Varieties and Seeds* 22, 171–176.

Moller, M. and Spoor, W. (1993) Discrimination and identification of *Lolium* species and cultivars by rapid SDS-PAG electrophoresis of seed storage proteins. *Seed Science and Technology* 21, 213–223.

Morgan, A.G. (1989) Chromatographic applications in cultivar identification. *Plant Varieties and Seeds* 2, 35–44.

Mundges, H., Kohler, W. and Friedt, W. (1990) Identification of rape seed cultivars (*Brassica napus*) by starch gel electrophoresis of enzymes. *Euphytica* 45, 179–187.

Neuhaus-Url, G. and Neuhaus, G. (1993) The use of the nonradioactive digoxigenin chemiluminescent technology for plant genomic southern blot hybridization: a comparison with radioactivity. *Transgenic Research* 2, 115–120.

Ng, P.K.W., Scanlon, M.G. and Bushuk, W. (1988) A catalog of biochemical fingerprints of registered Canadian wheat cultivars by electrophoresis and high-performance liquid chromatography. Publication No. 139 of the Food Science Department, University of Manitoba, pp. 1–83.

Nielsen, G. (1985) The use of isozymes as probes to identify and label plant varieties and cultivars. *Current Topics in Biological and Medical Research* 12, 1–32.

Nienhuis, J., Slocum, M.K., DeVos, D.A. and Muren, R. (1993) Genetic similarity among *Brassica oleracea* L. genotypes as measured by restriction fragment length polymorphisms. *Journal of the American Society of Horticultural Science* 118, 298–303.

Nybom, H., Rogstad, S.H. and Schaal, B.A. (1990) Genetic variation detected by use of the M13 'DNA fingerprint' probe in *Malus, Prunus* and *Rubus* (Roseaceae). *Theoretical and Applied Genetics* 79, 153–156.

Parent, J-G. and Page, D. (1992) Identification of raspberry cultivars by non-radioactive DNA fingerprinting. *HortScience* 27, 1108–1110.

Pecchioni, N., Stanca, A.M., Terzi, V. and Cattivelli, L. (1993) RFLP analysis of highly polymorphic loci in barley. *Theoretical and Applied Genetics* 85, 926–930.

Prince, J.P., Loaiza-Figueroa, F. and Tanksley, S.D. (1992) Restriction fragment length polymorphism and genetic distance among Mexican accessions of *Capsicum*. *Genome* 35, 726–732.

Rajapakse, S., Hubbard, M., Kelly, J.W., Abbott, A.G. and Ballard, R.E. (1992) Identification of rose cultivars by restriction fragment length polymorphism *Scientia Horticulturae* 52, 237–245.

Roder, M.S., Sorrells, M.E. and Tanksley, S.D. (1992) 5S ribosomal gene clusters in wheat: pulsed field gel electrophoresis reveals a high degree of polymorphism. *Molecular and General Genetics* 232, 215–220.

Russell, J.R., Hosein, F., Johnson, E., Waugh, R. and Powell, W. (1993) Genetic differentiation of cocoa (*Theobroma cacao* L.) populations revealed by RAPD analysis. *Molecular Ecology* 2, 89–97.

Sapirstein, H.D., Neuman, M., Wright, E.H., Shwedyk, E. and Bushuk, W. (1987) An instrumental system for cereal grain classification using digital image analysis. *Journal of Cereal Science* 6, 3–14.

Skerritt, J.H., Wrigley, C.W. and Hill, A.S. (1988) Prospects for the use of monoclonal antibodies in the identification of cereal species, varieties and quality types. *Proceedings ISTA Symposium*, Leningrad, pp. 110–123.

Smart, G.F., Rawles, H. and Chojecki, J. (1991) Application of RFLP technology to oat cultivar identification. *Cereal Research Communications* 19, 459–464.

Smith, J.S.C. (1986) Biochemical fingerprints of cultivars using reversed-phase high-performance liquid chromatography and isozyme electrophoresis: a review. *Seed Science and Technology* 14, 753–768.

Smith, J.S.C. and Smith, O.S. (1991) Restriction fragment length polymorphisms can differentiate among US maize hybrids. *Crop Science* 31, 893–899.

Smith, J.S.C. and Smith, O.S. (1992) Fingerprinting crop varieties. *Advances in Agronomy* 47, 85–140.

Smith, J.S.C., Smith, O.S. and Martin, B.A. (1988) The identification of wheat (*Triticum* spp.) by reversed-phase high-performance liquid chromatography: a review of potentials and prospects. *Proceedings ISTA Symposium*, Leningrad, pp. 35–46.

Sondergaard, I., Jensen, K. and Krath, B.N. (1994) Classification of wheat varieties by isoelectric focusing patterns of gliadins and neural network. *Electrophoresis* 15, 584–588.

Stegemann, H. (1984) Retrospect on 25 years of cultivar identification by protein patterns and prospects for the future. In: Draper, S.R. and Cooke, R.J. (eds), *Biochemical Tests for Cultivar Identification*. International Seed Testing Association, Zurich, pp. 20–31.

Stegemann, H., Shah, A.A., Krogerrecklenfort, E. and Hamza, M.M. (1992) Sweet potato (*Ipomoea batatas* L.): genotype identification by electrophoretic methods and properties of their proteins. *Plant Varieties and Seeds* 5, 83–91.

Stiles, J.L., Lemme, C., Sondur, S., Morshidi, M.B. and Manshardt, R. (1993) Using randomly amplified polymorphic DNA for evaluating genetic relationships among papaya cultivars. *Theoretical and Applied Genetics* 85, 697–701.

Tinker, N.A., Fortin, M.G. and Mather, D.E. (1993) Random amplified polymorphic DNA and pedigree relationships in spring barley. *Theoretical and Applied Genetics* 85, 976–984.

Vaccino, P., Accerbi, M. and Corbellini, M. (1993) Cultivar identification in *T. aestivum* using highly polymorphic RFLP probes. *Theoretical and Applied Genetics* 86, 833–836.

Varier, A. and Cooke, R.J. (1992) Discrimination between cultivars and lines of pearl millet by isoelectric focussing. *Seed Science and Technology* 20, 711–713.

van der Burg, W.J. and van Zwol, R.A. (1991) Rapid identification techniques used in laboratories of the International Seed Testing Association : a survey. *Seed Science and Technology* 19, 687–700.

van de Vooren, J.G. and van der Heijden, G.W.A.M. (1993) Measuring the size of French beans with image analysis. *Plant Varieties and Seeds* 6, 47–53.

Vosman, B., Arens, P., Rus-Kortekaas, W. and Smulders, M.J.M. (1992) Identification of highly polymorphic DNA regions in tomato. *Theoretical and Applied Genetics* 85, 239–244.

Wang, Z.Y. and Tanksley, S.D. (1989) Restriction fragment length polymorphism in *Oryza sativa* L. *Genome* 32, 1113–1118.

Watillon, B., Druart, P., Du Jardin, P., Kettmann, R., Boxus, P. and Burny, A. (1991) Use of random cDNA probes to detect restriction fragment length polymorphisms among apple clones. *Scientia Horticulturae* 46, 235–243.

Weising, K., Beyermann, B., Ramser, J. and Kahl, G. (1991) Plant DNA fingerprinting with radioactive and digoxigenated oligonucleotide probes complementary to simple repetitive DNA sequences. *Electrophoresis* 12, 159–169.

Welsh, J. and McClelland, M. (1990) Fingerprinting genomes using PCR with arbitrary primers. *Nucleic Acids Research* 18, 7213–7218.

Welsh, J., Honeycutt, R.J., McClelland, M. and Sobral, B.W.S. (1991) Parentage determination in maize hybrids using the arbitrarily primed polymerase chain reaction (AP-PCR). *Theoretical and Applied Genetics* 82, 473–476.

White, J. and Cooke, R.J. (1992) A standard classification system for the identification of barley varieties by electrophoresis. *Seed Science and Technology* 20, 663–676.

White, J. and Law, J.R. (1991) Differentiation between varieties of oilseed rape (*Brassica napus*) on the basis of the fatty acid composition of the oil. *Plant Varieties and Seeds* 4, 125–132.

Wilkie, S.E., Isaac, P.G. and Slater, R.J. (1993) Random amplified polymorphic DNA (RAPD) markers for genetic analysis in *Allium*. *Theoretical and Applied Genetics* 86, 497–504.

Williams, J.G.K., Kubelik, A.R., Livak, K.J., Rafalski, J.A. and Tingey, S.V. (1990) DNA polymorphisms amplified by arbitrary primers are useful as genetic markers. *Nucleic Acids Research* 18, 6531–6535.

Wingad, C.E., Iqbal, M., Griffin, M. and Smith, F.J. (1986) Separation of hordein proteins from European barley by high-performance liquid chromatography: its application to the identification of barley cultivars. *Chromatographia* 21 49–54.

Wolff, K. and Peters-van Rijn, J. (1993) Rapid detection of genetic variability in chrysanthemum (*Dendranthema grandiflora* Tzvelev) using random primers. *Heredity* 71, 335–341.

Wrigley, C.W., Autran, J.C. and Bushuk, W. (1982) Identification of cereal varieties by gel electrophoresis of the grain proteins. *Advances in Cereal Science and Technology* 5, 211–259.

Wrigley, C.W., Batey, I.L., Bekes, F., Gore, P.J. and Margolis, J. (1992) Rapid and automated characterisation of seed genotype using Micrograd electrophoresis and pattern matching software. *Applied and Theoretical Electrophoresis* 3, 69–72.

Yabuya, T. (1991) High performance liquid chromatographic analysis of anthocyanins in Japanese garden iris and its wild forms. *Euphytica* 52, 215–219.

Zayas, I., Lai, F.S. and Pomeranz, Y. (1986) Discrimination between wheat classes and varieties by image analysis. *Cereal Chemistry* 63, 52–56.

Zayas, I., Pomeranz, Y. and Lai, F.S. (1989) Discrimination of wheat and non-wheat components in grain samples by image analysis. *Cereal Chemistry* 66, 233–237.

Monoclonal Antibody Technology $\boxed{3}$

A. Schots,[1] R. Griep[1] and J. Bakker[2]

[1]*Laboratory for Monoclonal Antibodies, PO Box 9060, 6700 GW Wageningen, The Netherlands;* [2]*Wageningen Agricultural University, Department of Nematology, PO Box 8123, 6700 ES Wageningen, The Netherlands.*

3.1 Introduction

Over the past decade, monoclonal antibody-based immunoassays have been widely applied in agriculture to detect and quantify crop diseases, pesticides and naturally occurring compounds. The information obtained from such immunoassays can be used for advisory systems with regard to crop rotation, cultivar selection, pesticide application, harvest dates, postharvest handling, certification, etc. Initially, the development of immunoassays was hindered by difficulties with standardization and raising of specific antibodies. Standardization was hampered by the variety and complexity of the samples met in agriculture, which range from relatively simple ground water to more complex matrices of soils and homogenized tissues or organisms (Lankow *et al.*, 1987). Obtaining specific antibodies was impeded by the complex structure of plant pathogens like fungi, bacteria and nematodes.

The hybridoma technology developed by Köhler and Milstein (1975) offers possibilities to overcome the problems associated with the development of immunoassays. With this technique, antibody-producing B cells are isolated from an immunized animal and fused *in vitro* with a lymphoid tumour cell (myeloma). In the resulting hybrid cell (hybridoma), the characteristics of both parental cell types are combined, i.e. the production of specific (monoclonal) antibodies and the ability to multiply *in vitro*. In this way, the humoral immune response against complex molecules containing many antigenic sites, each of which give rise to an individual antibody, can be dissected into separate component antibodies. Moreover, rare antibody specificities, whose potential reactivities never become manifest *in vivo*, might be generated *in vitro* by the hybridoma technique (Metzger *et al.*,

1984). This has resulted in monoclonal antibodies discriminating between isoenzymes and other closely related proteins (Berzofsky *et al.*, 1980; Hollander and Katchalski-Katzir, 1986).

Monoclonal antibodies used in agriculture are mainly directed against plant viruses. However, the number of monoclonal antibodies directed against bacteria, fungi and nematodes is rapidly increasing (Schots *et al.*, 1990; Dewey *et al.*, 1991).

The major drawback of the hybridoma technology is that it is best suited for the generation of murine antibodies, while the main interest has always been human-like antibodies for medical applications, e.g. tumour killing. In this application murine monoclonal antibodies offer no alternative, because they give rise to an undesirable immune response in humans (Winter and Milstein, 1991). Therefore other alternatives were developed by using DNA technology. Initially, variable (V) domains of murine hybridomas were grafted on human constant domains (see Fig. 3.4; Orlandi *et al.*, 1989). Later, mouse complementarity determining regions, the parts of the V domains responsible for binding the antigen, were put in a human framework (Fig. 3.4; Jones *et al.*, 1986). More recently, human antibodies were developed by using antibody phage display libraries (Clackson *et al.*, 1991).

Although these techniques are developed for medical use, the crop sciences can benefit from these developments. Antibodies developed with these new techniques will undoubtedly be of great value for diagnostic purposes. Another application is the expression of antibodies in plants, which offers possibilities to endow plants with new properties like resistance against pathogens (Schots *et al.*, 1992).

3.2 Immune Selection

The ability to respond to an apparently limitless array of foreign antigens is one of the most remarkable features of the vertebrate immune system. Almost all antibody molecules contain a unique stretch of amino acids in their variable region. It is estimated that the immune system is capable of generating more than 10^8 different antibody molecules. This diversity is possible because during B-cell differentiation in the bone marrow, immuno-globulin gene segments are randomly shuffled by a dynamic genetic system. This process is carefully regulated, such that B-cell differentiation from an immature pre-B cell to a mature cell involves an ordered progression of immunoglobulin gene rearrangements. The result is a mature immuno-competent B cell which contains a *single* functional variable-region DNA sequence for its heavy and its light chain. After antigenic stimulation, further rearrangement of constant-region gene segments can generate changes in the isotype expressed and consequently change the associated biological effector functions without changing the specificity of the immunogloblin molecule. Antigenic stimulation also results in the formation of plasma and memory cells. Upon repeated contact with antigen, memory cells produce antibodies with increased affinity, which is a consequence of the affinity maturation process leading to somatic mutations in the variable regions of the heavy and light chains.

3.3 Antigens

The starting point for the development of monoclonal antibodies is a well defined antigen. Pure or partially purified antigens can be obtained in three ways (Dewey *et al.*, 1991):

1. They can be obtained from the target organism or from an organism which produces an antigen of sufficient homology.
2. They can be synthesized if the antigen is (bio)chemically well characterized, e.g. plant hormones or peptides.
3. They can be produced in bacteria if a cDNA encoding the antigen of interest is available.

In the screening phase, regardless of whether antibodies are produced with the hybridoma technology or using antibody phage display libraries, it is necessary to have the availability of a highly purified antigen preparation unless it is possible to screen for binding against the impurities present (e.g. a partially purified plant virus versus homogenized plant tissue from uninfected plants).

3.4 Monoclonal Antibodies from Hybridomas

Until the early 1980s antibodies were, in agriculture, mainly used for the detection of plant viruses using conventional polyclonal antisera. This was a consequence of the ability to purify plant viruses to a high grade, thus avoiding an immune response from impurities like plant proteins. Another advantage is their relative simple morphology – plant viruses have in most cases only one specific coat protein. Cross-reactions are, therefore, hardly observed. The structure of plant viruses, containing an assembly of a large number of identical protein molecules, makes them highly immunogenic.

The hybridoma technology increased possibilities to use antibodies with a defined high specificity as it enabled the detection and identification of other pathogens, plant compounds (e.g. plant hormones) and chemical substances like pesticides (Dewey *et al.*, 1991). Hybridomas result from the fusion of a single antibody-producing B cell with a lymphoid tumour cell. Because the parental B cell produced identical antibodies directed against one epitope, the hybridoma produces truly monoclonal antibodies. Therefore, one can select for antibodies which bind specifically to the target molecules. Antibodies which show cross-reactions or bind to impurities can be discarded.

In Fig. 3.1 a schematic representation is given of the production of monoclonal antibodies with the hybridoma technology: 3–4 days after the final immunization the spleen of an immunized mouse or rat is taken out and splenocytes (B lymphocytes) are fused with tumour cells of lymphoid origin (myeloma cells). Polyethylene glycol is usually used as fusing agent. The

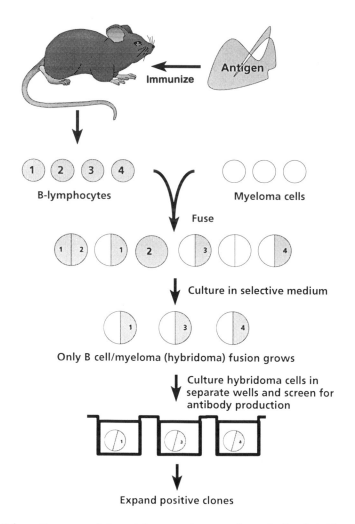

Fig. 3.1. Schematic representation of the procedure to make and select hybridomas. A more detailed description of the procedure is given in the text.

fused cells are plated out in microwell plates using a selective medium which kills the parental tumour cells. Selection takes place by blocking the main route for purine synthesis using aminopterin or azaserin. Only fused cells survive as they, in contrast to the parental tumour cells, can use the escape pathway for purine synthesis. The non-fused splenocytes die anyway because they cannot multiply. Two weeks after fusion the cell cultures can be screened for specific antibody production. Numerous screening methods are used, including ELISA, RIA, dot-blot, immunoblot and a variety of immunofluorescence techniques. The positive hybridomas are subcloned, this implies that they have to start afresh from the one cell per culture stage to check whether or not they have a stable genome. Because hybridomas are hybrid cells containing initially two genomes, instability is frequently observed leading to chromosome loss. Consequently this can lead to a loss in the ability to produce antibodies. After another 2 weeks, a second screening takes place to check whether the hybridoma cells still produce antibodies and whether or not the cell line is stable. If not, a second round of subcloning may be necessary. Stable clones are expanded to larger cell culture volumes.

For large-scale production of monoclonal antibodies, nowadays, *in vitro* systems are often used. The cells are cultured in spinner flasks, hollow fibre units or other systems. These *in vitro* systems have replaced the production of large quantities of monoclonal antibodies via ascites cultures. An ascites is a peritoneally growing tumour which can be induced by injecting a mouse with hybridoma cells. Because of ethical concerns this technique is in many countries replaced by *in vitro* systems. Purification of antibodies from the large-scale cultures takes place using different chromatography techniques. Purified antibodies can, if desired, be conjugated to enzymes, biotin or radioactive labels in order to visualize binding of the antibody.

3.5 Screening for Positive Hybridomas

The most critical step in the production of monoclonal antibodies is the initial screening about 2 weeks after cell fusion. The general rule for the type of assay one has to choose is that it should be as similar as possible to the assay wherein the monoclonal antibody finally will be used. The different types of assays used are: ELISAs, RIAs, dot-blotting, immunoblotting, immunofluorescence on cells and immunofluorescence or immunoenzymatic staining of tissue sections or tissue parts.

ELISAs and RIAs can be used in different formats (Fig. 3.2A). The two most widely used general formats are the antigen-coated plate method (ACP method) and the double-antibody-sandwich method (DAS method). With the ACP method, antigen is coated on the microtitre plate whereafter the antibody is added. For the DAS method, antibodies are coated on the

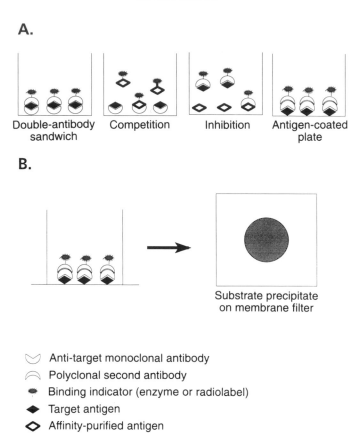

Fig. 3.2. A. Four general methods (ELISA or RIA) for the detection of target antigens using monoclonal antibodies. **B.** Principle of the dot-blot method.

microtitre plate to capture the antigen, this is followed by the addition of a second antibody, usually labelled, which can be the same or another one. Binding can in both cases be visualized by using a directly labelled antibody or by adding a secondary labelled antibody. One can also use these assays in a competitive format. DAS-ELISAs have been used for the selection of monoclonal antibodies against plant viruses and plant pathogenic fungi (Halk and De Boer, 1985; Van De Koppel and Schots, 1994). Thereto, a polyclonal antibody was used as capture antibody to trap antigen from infected plant material. The hybridoma supernatants were then added followed by an alkaline phosphatase conjugated anti-mouse antibody. ACP-ELISAs have been used for the selection of monoclonal antibodies against bacteria and the potato cyst nematode species *Globodera pallida* and *G. rostochiensis* (Schots *et al.*, 1989; Franken *et al.*, 1992). Bacteria or nematode antigens were coated, culture supernatants were then added and a secondary antibody to visualize binding.

Dot-blotting (Fig. 3.2B) is essentially the same as an ACP-ELISA. The only difference is that instead of a microtitre plate a membrane filter is used for immobilization of the antigen. Visualization takes place using a substrate which precipitates. With immunoblotting the antigens are first separated using an electrophoretic method followed by transfer to a membrane filter. Visualization takes place in a similar fashion as with dot-blotting. Dot-blotting has been used for the selection of hybridomas producing monoclonal antibodies against plant pathogenic bacteria (Franken *et al.*, 1992).

In some cases, screening of hybridoma supernatants for the presence of specific antibodies takes place by immunofluorescent or immunoenzymatic visualization of the binding to cells or tissues. Although these methods are laborious because microscopical examination of each preparation is required, the information obtained on where and how the antibody binds is very detailed. We have used immunofluorescence with success for the selection of specific antibodies directed against conidia of the fungus *Botrytis cinerea* (Salinas and Schots, 1994) and a variety of structures of *G. rostochiensis* (Fig. 3.3).

3.6 Antibody Engineering

Since 1985, gene technology has revolutionized hybridoma technology. Until then only few functional properties of monoclonal antibodies could be changed, for instance by switching heavy chain constant regions (Kipps, 1985) or by making bispecific antibodies (Milstein and Cuello, 1984). Antibody genes from hybridoma cell lines or lymphocytes can be cloned into plasmid vectors and expressed in bacteria, yeast, mammalian cells or plants (Fig. 3.5). An IgG antibody is a Y shaped molecule (Fig. 3.4), in which the domains (V_L, V_H) forming the tips of the arms bind to antigen. The domains forming the stem (Fc fragment) are responsible for triggering effector functions to eliminate the antigen from the animal. An IgG molecule consists of four polypeptide chains: two heavy (H) and two light (L). Each domain consists of two beta-sheets which pack together to form a sandwich, with exposed loops at the end of the strands.

The domain structure in combination with the polymerase chain reaction facilitates protein engineering of antibodies (Orlandi *et al.*, 1989; Winter and Milstein, 1991; Fig. 3.4):

1. Fv and Fab fragments can be used separately from the Fc fragments without loss of affinity. By using PCR for cloning, Fv and Fab fragments and single chain Fvs (scFvs) can be expressed in heterologous systems. In a scFv the variable parts of the light (V_L) and heavy chain (V_H) are linked by a

peptide. Fv fragments, Fab fragments and scFvs have been successfully expressed in bacteria, yeast and murine myeloma cell lines.

2. Toxins or enzymes can be fused to the antigen-binding domains to endow them with new properties, for example immunotoxins have been constructed

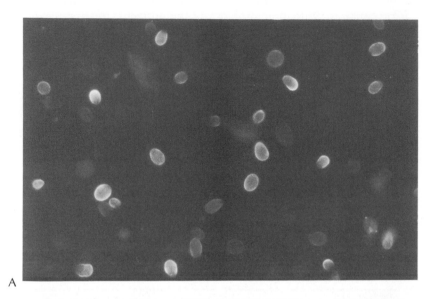

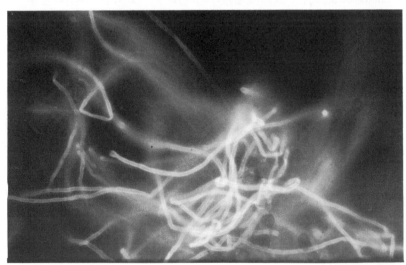

Fig. 3.3. (and opposite) Immunofluorescence pictures **A.** Conidia of *Botrytis cinerea*. **B.** Mycelium of *Botrytis cinerea*. **C.** Muscles of *G. rostochiensis*. **D.** Salivary glands of the potato cyst nematode *Globodera rostochiensis*.

as Fv fusion proteins (Batra *et al.*, 1990; Chaudhary *et al.*, 1990) and expressed in *Escherichia coli*.

3. Murine antibodies can be humanized by grafting the variable domains on the constant domains of human IgG isotypes; chimeric antibodies can be made which are expressed in myeloma cells (Orlandi *et al.*, 1989).

The humanization of rodent antibodies shows that engineering of

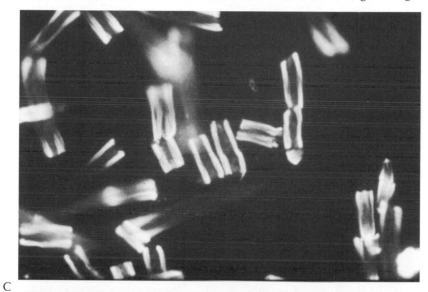

C

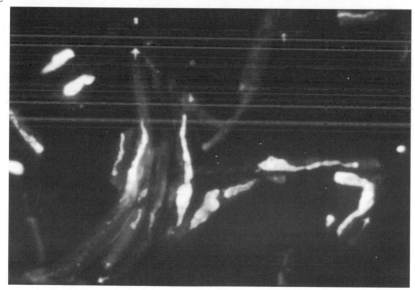

D

A. Schots et al.

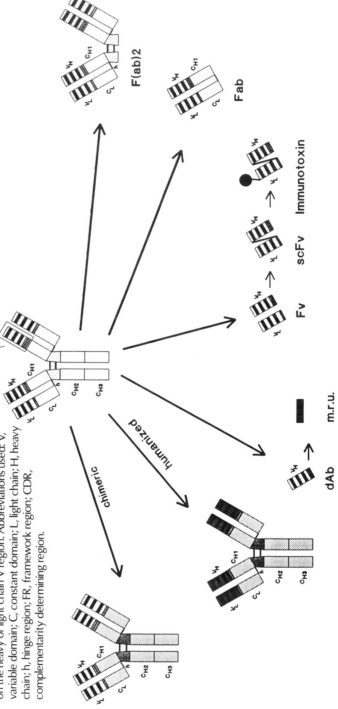

Fig. 3.4. The domain structure of the antibody molecule facilitates antibody engineering. The different possibilities are indicated. F(ab)₂ and Fab fragments can be obtained through proteolytic cleavage of antibody molecules or through gene technology. Gene technology must be used to obtain Fv fragments, single-chain Fv fragments (both polypeptide chains linked by a 15-mer peptide), single domain antibodies (dAbs), chimeric (humanized) antibodies or reshaped antibodies. A minimal recognition unit (m.r.u.) is a synthetic peptide resembling the amino acid sequence of a hypervariable region on the heavy or light chain V region. Abbreviations used: V, variable domain; C, constant domain; L, light chain; H, heavy chain; h, hinge region; FR, framework region; CDR, complementarity determining region.

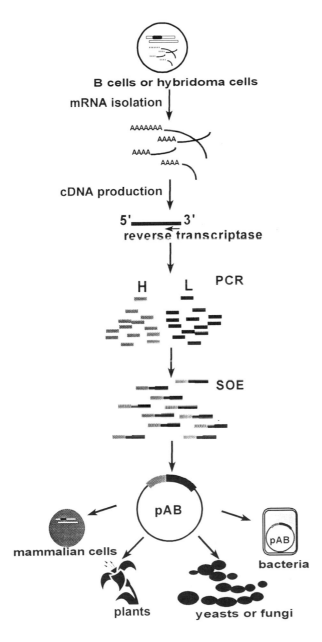

B cells or hybridoma cells

mRNA isolation

cDNA production

5' _____ 3'
reverse transcriptase

H   L   **PCR**

SOE

pAB

mammalian cells

bacteria

plants

yeasts or fungi

Fig. 3.5.
Representation of the general strategy to clone antibody genes. mRNA is isolated from B cells or hybridoma cells followed by cDNA production. The antibody genes are specifically amplified using the polymerase chain reaction (PCR). The parts encoding the heavy and light chain variable domains and the linker peptide are 'ligated' using the method 'Splicing by Overlap Extension' (SOE; Horton et al., 1989). After cutting with restriction enzymes, the chimeric gene is ligated in an appropriate vector and the target organism is transformed.

antibodies is not restricted to the use of whole domains. The antigen-binding loops in a V-region framework of a human antibody can be replaced by those of a rodent antibody, thus transferring the antigen binding site (Jones *et al.*, 1986).

3.7 Expression of Antibodies in Bacteria

Bacterial production of antibody fragments (Better *et al.*, 1988; Skerra and Plückthun, 1988) can be achieved in three ways. The first involves the expression of proteins intracellularly, the second as proteins secreted into the periplasmic space or culture medium and the third as fusion proteins expressed on the surface of phages.

Intracellular production of antibodies leads to non-functional proteins sequestered in inclusion bodies. As a result, they need to be refolded *in vitro*, procedures which often have to be individually tailored to a particular antibody and the folded products always have to be separated from misfolded forms for use in accurate quantifications of antigens. Secretion, on the other hand, offers several advantages compared with intracellular production: efficient purification, theoretically higher yields, no aggregation of the product, correctly folded proteins, the possibility for disulphide bond formation and the possibility for continuous cultivation and production. The third possibility is the display on the surface of phages as fusion protein with a coat protein (McCafferty *et al.*, 1990; Chang *et al.*, 1991). Mainly the filamentous phage fd has been used. Fd is a non-lytic bacteriophage which has been used to display peptides, enzymes and antibodies. This is usually done as fusions with gene III, encoding the minor coat protein. Fusions with gene VIII, encoding the major coat protein, have also been described (Chang *et al.*, 1991).

Although many antibody fragments have been successfully produced in *E. coli* by means of secretion, the production levels are not optimal. The cause of the low production levels can mainly be found in the use of suboptimal promoters, inefficient translation, folding and assembly and the inability to use a suitable host. Attempts to improve folding and assembly included, among others, the co-expression of disulphide isomerase, *cis–trans* isomerase and chaperones (Knappik *et al.*, 1993; Söderlind *et al.*, 1993). In these experiments, the helper proteins were successfully produced but the effect on antibody production was minimal. Another host bacterium has also been tried. An scFv was successfully produced in *Bacillus subtilis* which possibly is a more suitable host than *E. coli* (Wu *et al.*, 1993). The production levels obtained were similar when compared to *E. coli*; however, the functional yield was much better, about 98% of the protein yield. Although in these experiments no real improvement of antibody production levels was observed, this type of research should continue as it will sooner or later reveal what factor hinders the production of heterologous proteins in bacteria.

3.8 Antibody Phage Display Libraries

3.8.1 Library development

The natural antibody selection system can be mimicked by displaying the immune repertoire on phage and selection with antigen (Clackson *et al.*, 1991). To achieve this, a library of rearranged V genes, representing the immune repertoire, can be constructed with genomic DNA or mRNA or B lymphocytes or by making a synthetic library (Fig. 3.6). Such a library is constructed by amplifying a range of heavy and light chain V genes using 'universal' PCR primers, followed by direct cloning into expression vectors. Universal 3' PCR primers can be based on the sequences of the J segments to copy the sense strand, while they can also be used for first strand cDNA synthesis. The nucleotide sequence at the 5' end of the V exon emerged to be sufficiently conserved to design a 'degenerate' primer to copy the antisense strand. Thus, 'natural' mouse and human V gene libraries have been constructed. In addition, synthetic V gene repertoires have been made by randomizing CDR3 of a V_H segment of a bacterially expressed antibody fragment (Barbas *et al.*, 1992; Hoogenboom and Winter, 1992). This was achieved by designing a highly degenerate PCR primer bridging CDR3 which randomly changed the sequence of the amino acid residues. The randomization could only result in one (amber) stop codon, TAG, suppressible in *sup E E. coli* strains.

Starting from 'natural' V gene libraries, it is necessary to link the V_H and V_L genes together in order to obtain functional antibody fragments. This will result either in a chimeric gene encoding a scFv fragment or Fab fragment. The genes can be linked together by recombination, for instance, by cloning the light and heavy chain gene repertoire separately into the left and right arms of a modified lambda zapII vector and to create a random combinatorial library by recombining the arms (Huse *et al.*, 1989). An alternative and more general way of linking the heavy and light chains is 'splicing by overlap extension' (Horton *et al.*, 1989; Fig. 3.5). After amplification of the V genes by PCR, the V_H and V_L repertoires are combined with 'linker' DNA which has regions of sequence homology with the 3' end of the amplified V_H gene and the 5' end of the amplified V_L gene or vice versa. The 'linker' can be either the linker peptide in the case of scFv fragments or a DNA fragment comprising the ribosomal binding site and leader sequence in the case of Fab fragments. A further PCR amplification with outer flanking PCR primers splices together the two repertoires. Including restriction sites appended to the 5' ends of the flanking primers enables direct cloning into an expression vector, usually a phagemid vector with the antibody fragments fused to the gene III protein.

A. Schots et al.

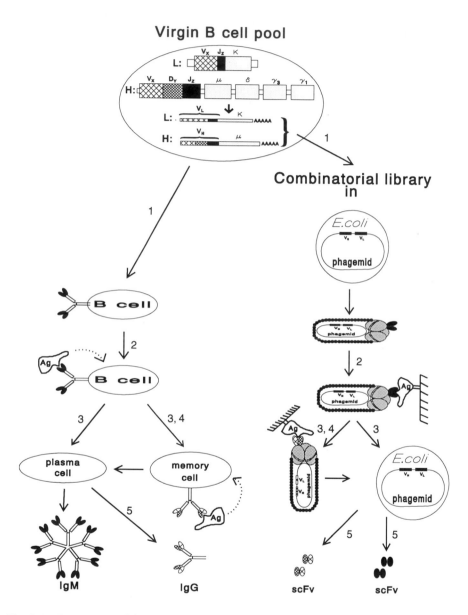

Fig. 3.6. Comparison of the strategy of the immune system *in vivo* and using phage. V genes are rearranged or assembled (step 1) and displayed on the surface (step 2). Antigen driven or affinity selection then takes place (step 3). The selected cells or phages then undergo affinity maturation (step 4) or produce soluble antibody (step 5).

3.8.2 Selection of binders

Phages displaying antibody fragments showing affinity for antigen can be selected by direct binding of the phage to immobilized antigen (McCafferty *et al.*, 1990; Clackson *et al.*, 1991; Marks *et al.*, 1991). Antigens can be immobilized on columns; on dishes or tubes; or on the surface of cells. Alternatively, phages can be allowed to bind to biotinylated antigen in solution followed by capture on paramagnetic beads (Hawkins *et al.*, 1992). After washing to remove non-binding phage, binders are eluted with antigen, acid or alkali buffers or with a reducing agent when the antigen is biotinylated with a cleavable disulphide biotin. Eluted phages are used to reinfect *E. coli* cells, whereafter the whole procedure is repeated once or a few times more to enrich the fraction of binders.

3.8.3 V gene source

Rearranged V gene repertoires for phage display libraries can be obtained from either immunized or unimmunized donors. If an immunized donor is used, a relatively large number of binders is found. Consequently, the size of the library can be relatively small (10^6 10^7). In contrast, the number of binders obtained from a naive library is low. Therefore, the library has to be as large as possible, preferably larger than 10^8. It is difficult but nevertheless worthwhile to construct such a library. Large naive libraries contain the whole immune repertoire and can, in principle, be used to select antibodies against almost any molecular structure. Recently such a library was successfully used to select antibody fragments against 12 different antigens, haptens, proteins and cellular antigens (Hoogenboom *et al.*, 1992).

To obtain antibodies with higher affinities, it will be necessary to construct larger libraries. At present, the transformation efficiency of bacteria limits the construction of larger libraries ($> 10^8$ combinations). An alternative is possible using *in vivo* recombination which is enabled by the *lox*-Cre site-specific recombination system of bacteriophage P1 (Waterhouse *et al.*, 1993). Within an infected bacterium, this system locks together the heavy and light chain genes from two different replicons.

3.8.4 Increasing the affinity

The affinity of an antibody selected from an antibody phage display library is usually low and can be improved by introducing mutations. Three systems can be used to generate point mutations. The first possibility is to use reaction conditions which decrease the reliability of *Taq* polymerase and thus produce a product with randomly introduced errors. A reamplified, mutant scFv has been obtained in this way, and expressed on phage, showing a fourfold improved affinity (Hawkins *et al.*, 1992). A second alternative is growth of the phage in an *E. coli mutD* strain in which the spontaneous mutation frequency is 10^3–10^5 times higher than in wild-type strains (Gram *et al.*, 1992). The third system enables the introduction of multiple point mutations, and is a system based on the use of oligonucleotides (Hermes *et al.*, 1989).

An alternative to increase the affinity is to use the somatic mutations introduced by the immune system. Therefore, certain loops or entire heavy or light chains are shuffled with a repertoire of *in vivo* somatically mutated V genes (Marks *et al.*, 1992). For instance, the light chain of a scFv is replaced by light chains from the library followed by selection. In a similar way parts of the heavy chain can be shuffled. However, care should be taken to avoid disrupting key features of antigen binding; in general V_H-CDR3 should be left undisturbed.

3.9 Plantibodies

Functional IgG molecules, as well as functional Fab and scFv fragments, can also be produced in plants. These findings offer possibilities to endow plants with new properties like resistance against diseases, altering metabolic routes or introducing new metabolic routes using catalytic antibodies.

Düring (1988) and Hiatt *et al.* (1989) were the first to describe the production of antibodies in plants. Constructs of coding-length cDNAs or the γ- and κ-chain, with and without leader sequences, were ligated into a plant expression vector. To transform tobacco plants, Hiatt *et al.* (1989) constructed plasmids containing the gene for either the heavy or light chain, both with and without an immunoglobulin leader sequence, under control of the constitutive 35S promoter. The transformants expressing individual immunoglobulin chains were then sexually crossed to produce progenies expressing both chains. Only plants expressing immunoglobulin chains with the original leader peptide contained assembled gamma–kappa complexes. Immunoglobulins, assessed using ELISA, were reported to accumulate up to 1.3% of the total leaf protein. However, nothing was reported on correct assembly inside the plant cell and the *in situ* localization of assembled antibody molecules.

Alternatively, Düring *et al.* (1990) constructed a vector containing both heavy and light chain genes and the barley aleurone α-amylase leader peptide coding sequence under control of the pT_R and pNOS promoters. The barley aleurone α-amylase leader peptide is thought to direct proteins to the intercellular spaces. The *in situ* localization and correct assembly of functional antibodies was studied using anti-idiotype antibodies. Assembled antibodies were detected in the endoplasmic reticulum (ER) and, surprisingly, also inside chloroplasts. Nothing was reported on the presence of antibodies in the intercellular spaces.

Recently, scFv (Owen *et al.*, 1992) and Fab fragments (De Neve *et al.*, 1993) have been successfully expressed. The scFv described by Owen *et al.* (1992) was directed against phytochrome which was inhibited intracellularly. Seeds from transgenic progeny displayed aberrant phytochrome-dependent germination. The scFv is likely to be active in the cytoplasm as the chimeric gene contained no leader peptide. This might also explain the low expression levels ($\leq$ 0.1% of total protein) obtained. Fab fragments were expressed at up to 1.3% of the total protein in *Arabidopsis* leaves (De Neve *et al.*, 1993). However, the same fragment accumulated only to around 0.05% in *Nicotiana* leaves. No satisfactory explanation can be given, although this feature may be related to the plant species.

To endow plants with new properties it is essential that several aspects concerning antibody expression in plants are optimized. Striking differences are observed in the amount of accumulated antibody. Hiatt *et al.* (1989), using *Nicotiana*, and De Neve *et al.* (1993), using *Arabidopsis*, report expression levels up to 1.3% of total leaf protein. Much lower amounts (< 0.1%) were found by Düring *et al.* (1990) and again De Neve *et al.* (1993), both using tobacco. This discrepancy can be the result of different promoters (35S versus pT_R and pNOS) used possibly in combination with different leader peptides. Both aspects as well as the differences in the amount of protein expressed in different plant species have to be studied in more detail. Furthermore, De Neve *et al.* (1993) report abberant assembly or degradation patterns of expressed whole IgG molecules. In the range between 100 kDa and 150 kDa, a series of bands is observed on an SDS-PAGE gel under non-reducing conditions in both *Nicotiana* and *Arabidopsis*. In addition some bands with a molecular mass between 40 kDa and 65 kDa are observed in *Nicotiana*. Immuno-blotting experiments revealed that all bands contain parts of the immunoglobulin molecule.

3.10 Prospects

The hybridoma technology has solved many of the problems associated with polyclonal antibodies, like standardization and specificity. However, certain questions remain, especially in those cases where the purification of the antigen is difficult and the impurities have an immunodominant character. These problems can potentially be solved using antibody phage display libraries. A proper antibody phage display library consists of up to 10^8 different antibody molecules. The immune repertoire displayed with the hybridoma technology is limited to a maximum of a few thousand V gene combinations, i.e. the number of different hybridomas obtained. Rare specificities are only occasionally observed. This is in contrast to what is observed using phage display libraries. Moreover, the random association of V_H and V_L genes leads possibly to an even larger number of combinations than observed in nature, especially when the *in vivo* recombination techniques (Waterhouse *et al.*, 1993) come into practice for the construction of libraries.

The question arises 'Will traditionally prepared monoclonal and polyclonal antibodies continue to suffice for diagnostic challenges in the plant sciences?' The answer depends on the prerequisites demanded. For the diagnosis of many plant viruses polyclonal antibodies work satisfactorily. However, standardization, especially in the long run, is difficult if not impossible. International trade and international bodies like the European Union request standardization. A change towards the sole use of monoclonal antibodies can likely not be avoided in the future. It has to be remarked that it can be questioned as to whether this is an advantage because in the case of viruses a polyclonal antiserum safeguards against missing new strains where the one target epitope detected by the monoclonal antibodies may have changed. For some other plant pathogens, plant compounds and agrochemicals only monoclonal antibodies are of sufficient specificity and the answers can be simply that polyclonal antibodies will not continue to suffice.

Although many limitations concerning the development of antibodies for application in agriculture can now be solved, most limitations for the use in appropriate screening assays do still exist. This is a consequence of the type of matrix (soil or homogenized tissue or organism) met in agriculture. Innovative ideas as with the detection of *Rhizoctonia solani* in soil samples (Thornton *et al.*, 1994) do sometimes bring solutions.

One of the most promising spin-offs of antibody technology is the expression of antibodies in plants. Plantibodies are suitable to obtain resistance against pathogens. For instance, resistance against viruses can be obtained as described by Tavladoraki *et al.* (1993). They found that intracellular expression, in plants, of an scFv derived from a monoclonal antibody recognizing the Ca^{2+} binding region of the artichoke mottled crinkle virus (AMCV) led to reduced viral infectivity. Furthermore, plants can be endowed with new properties in analogy with the recently described *in vivo* catalysis of a metabolic reaction by an antibody (Tang *et al.*, 1991). A cytoplasmically expressed Fab fragment derived from a catalytic monoclonal antibody that displays chorismate mutase activity was shown to confer a growth advantage, under auxotrophic conditions, on a chorismate-mutase-deficient mutant of yeast.

Antibody technology has an enormous potential and rapid progress is being made. Practical applications in the medical field and biological sciences are likely to emerge soon. These include developments which will lead to: (i) *in vivo* applications, *e.g.* in cancer therapy; (ii) the construction of new, improved, immunoassays; and (iii) better heterologous expression systems. These findings will be beneficial for many aspects of agricultural research and practice.

References

Barbas III, C.F., Bain, J.D., Hoekstra, D.M. and Lerner, R.A. (1992) Semisynthetic combinatorial antibody libraries: a chemical solution to the diversity problem. *Proceedings of the National Academy of Sciences, USA* 89, 4457–4461.

Batra, J.K., Chaudhary, V.K., Fitzgerald, D. and Pastan, I. (1990) TGF-alpha-anti-Tac(Fv)-PE40: a bifunctional toxin cytotoxic for cells with EGF or IL2 receptors. *Biochemical and Biophysical Research Communications* 171, 1–6.

Berzofsky, J.A., Hicks, G., Fedorko, J. and Minna, J. (1980) Properties of monoclonal antibodies specific for determinants on a protein antigen, myoglobin. *Journal of Biological Chemistry* 255, 11,188–11,191.

Better, M., Chang, C.P., Robinson, R.R. and Horwitz, A.II. (1988) *Escherichia coli* secretion of an active chimeric antibody fragment. *Science* 240, 1041–1043.

Chang, C.N., Landolfi, N.F. and Queen, C. (1991) Expression of antibody Fab domains on bacteriophage surfaces. Potential use of antibody selection. *Journal of Immunology* 147, 3610–3614.

Chaudhary, V.K., Gallo, M.G., Fitzgerald, D.J. and Pastan, I. (1990) A recombinant single-chain immunotoxin composed of anti-Tac variable regions and a truncated diphtheria toxin. *Proceedings of the National Academy of Sciences, USA* 87, 9491–9494.

Clackson, T., Hoogenboom, H.R., Griffiths, A.D. and Winter, G. (1991) Making antibody fragments using phage display libraries. *Nature* 352, 624–628.

De Neve, M., De Loose, M., Jacobs, A., Van Houdt, H., Kaluza, B., Weidle, U., Van Montagu, M. and Depicker, A. (1993) Assembly of an antibody and its derived antibody fragment in *Nicotiana* and *Arabidopsis*. *Transgenic Research* 2, 227–237.

Dewey, M., Evans, D., Coleman, J., Priestley, R., Hull, R., Horsley, D. and Hawes, C. (1991) Antibodies in plant science. *Acta Botanica Neerlandica* 40, 1–27.

Düring, K. (1988) Wundinduzierbare Expression und Sekretion von T4 Lysozym und monoklonalen Antikörpern in *Nicotiana tabacum*. PhD thesis, University of Cologne, Germany.

Düring, K., Hippe, S., Kreuzaler, F. and Schell, J. (1990) Synthesis and self-assembly of a functional monoclonal antibody in transgenic *Nicotiana tabacum*. *Plant Molecular Biology* 15, 281–293.

Franken, A.A.J.M., Zilverentant, J.F., Boonekamp, P.M. and Schots, A. (1992) Specificity of polyclonal and monoclonal antibodies for the identification of *Xanthomonas campestris* pv. *campestris*. *Netherlands Journal of Plant Pathology* 98, 81–94.

Gram, H., Marconi, L.-A., Barbas III, C.F., Collet, T.A., Lerner, R.A. and Kang, A.S. (1992) *In vitro* selection and affinity maturation of antibodies from a naive combinatorial immunoglobulin library. *Proceedings of the National Academy of Sciences, USA* 89, 3576–3580.

Halk, E.L. and De Boer, S.H. (1985) Monoclonal antibodies in plant disease research. *Annual Reviews in Phytopathology* 23, 321–350.

Hawkins, R.E., Russell, S.J. and Winter, G. (1992) Selection of phage antibodies by binding affinity. Mimicking affinity maturation. *Journal of Molecular Biology* 226, 889–896.

Hermes, J.D., Parekh, S.M., Blacklow, S.C., Pullen, J.K. and Pease, L.R. (1989) A reliable method for random mutagenesis: the generation of mutant libraries using spiked oligodeoxyribonucleotide primers. *Gene* 84, 143–151.

Hiatt, A., Cafferkey, R. and Bowdish, K. (1989) Production of antibodies in transgenic plants. *Nature* 342, 76–78.

Hollander, Z. and Katchalski-Katzir, E. (1986) Use of monoclonal antibodies to detect conformational alterations in lactate dehydrogenase isoenzyme 5 on heat denaturation and on adsorption to polystyrene plates. *Molecular Immunology* 23, 927–933.

Hoogenboom, H.R. and Winter, G. (1992) Bypassing hybridomas: human antibodies from synthetic repertoires of germ-line VH-gene segments rearranged *in vitro*. *Journal of Molecular Biology* 227, 381–388.

Hoogenboom, H.R., Marks, J.D., Griffiths, A.D. and Winter, G. (1992) Building antibodies from their genes. *Immunological Reviews* 130, 41–68.

Horton, R.M., Hunt, H.D., Ho, S.N., Pullen, J.K. and Pease, L.R. (1989) Engineering hybrid genes without the use of restriction enzymes: gene splicing by overlap extension. *Gene* 77, 61–68.

Huse, W.D., Sastry, L., Iverson, S.A., Kang, A.S., Alting, M.M., Burton, D.R., Benkovic, S.J. and Lerner, R.A. (1989) Generation of a large combinatorial library of the immunoglobulin in phage lambda. *Science* 246, 1275–1281.

Jones, P.T., Dear, P.H., Foote, J., Neuberger, M.S. and Winter, G. (1986) Replacing the complementarity-determining regions in a human antibody with those from a mouse. *Nature* 321, 522–525.

Kipps, T.J. (1985) Switching the isotype of monoclonal antibodies. In: Springer, T.A. (ed.), *Hybridoma Technology in the Biosciences and Medicine*. Plenum Press, New York, pp. 89–101.

Knappik, A., Krebber, C. and Plückthun, A. (1993) The effect of folding catalysts on the *in vivo* folding process of different antibody fragments expressed in *Escherichia coli*. *Bio/Technology* 11, 77–83.

Köhler, G. and Milstein, C. (1975) Continuous cultures of fused cells secreting antibody predefined specificity. *Nature* 256, 495–497.

Lankow, R.K., Grothaus, G.D. and Miller, S.A. (1987) Immunoassays for crop management systems and agricultural chemistry. In: Lebaron, M., Mumma, R.O., Honeycut, R.C., Duesing, J.H., Philips, J.F. and Haas, M.J. (eds), *Biotechnology in Agricultural Chemistry*. American Chemical Society, Washington DC, pp. 228–260.

Marks, J.D., Hoogenboom, H.R., Bonnert, T.P., McCafferty, J., Griffiths, A.D. and Winter, G. (1991) By-passing immunization: human antibodies from V-gene libraries displayed on phage. *Journal of Molecular Biology* 222, 581–597.

Marks, J.D., Griffiths, A.D., Malmqvist, M., Clackson, T., Bye, J.M. and Winter, G. (1992) By-passing immunization: building high affinity human antibodies by chain shuffling. *Bio/Technology* 10, 779–783.

McCafferty, J., Griffiths, A.D., Winter, G. and Chiswell, D.J. (1990) Phage antibodies: filamentous phage displaying antibody variable domains. *Nature* 348, 552–554.

Metzger, D.W., Ch'ng, L.-K., Miller, A. and Sercarz, E.E. (1984) The expressed lysozyme specific B-cell repertoire. I. Heterogeneity in the monoclonal anti-hen egg white lysozyme specificity repertoire, and its difference from the *in situ* repertoire. *European Journal of Immunology* 14, 87–93.

Milstein, C. and Cuello, A.C. (1984) Hybrid hybridomas and the production of bi-specific monoclonal antibodies. *Immunology Today* 5, 299–304.

Orlandi, R., Gussow, D.H., Jones, P.T. and Winter, G. (1989) Cloning immunoglobulin variable domains for expression by the polymerase chain reaction. *Proceedings of the National Academy of Sciences, USA* 86, 3833–3837.

Owen, M., Gandecha, A., Cockburn, B. and Whitelam, G. (1992) Synthesis of a functional anti-phytochrome single-chain F_V protein in transgenic tobacco. *Bio/Technology* 10, 790–794.

Salinas, J. and Schots, A. (1994) Monoclonal antibodies based immunofluorescence test for the detection of conidia of *Botrytis cinerea* on cut flowers. *Phytopathology* 84, 351–356.

Schots, A., Hermsen, T., Schouten, S., Gommers, F.J. and Egberts, E. (1989) Serological differentiation of the potato-cyst nematodes *Globodera pallida* and *G. rostochiensis*: II. Preparation and characterization of species specific monoclonal antibodies. *Hybridoma* 8, 401–413.

Schots, A., Gommers, F.J., Bakker, J. and Egberts, E. (1990) Serological differentiation of plant-parasitic nematode species with polyclonal and monoclonal antibodies. *Journal of Nematology* 22, 16–25.

Schots, A., De Boer, J., Schouten, A., Roosien, J. Zilverentant, J.F., Pomp, H., Bouwman-Smits, L., Overmars, H., Gommers, F.J., Visser, B., Stiekema, W.J. and Bakker, J. (1992) 'Plantibodies': a flexible approach to endow plants with new properties. *Netherlands Journal of Plant Pathology* 98, supplement 2, 183–191.

Skerra, A. and Plückthun, A. (1988) Assembly of a functional immunoglobulin Fv fragment in *Escherichia coli*. *Science* 240, 1038–1040.

Söderlind, E., Simonsson Lagerkvist, A.C., Dueñas, M., Malmborg, A.-C., Ayala, M., Danielsson, L. and Borrebaeck, C.A.K. (1993) Chaperonin assisted phage display of antibody fragments on filamentous bacteriophages. *Bio/Technology* 11, 503–507.

Tang, Y., Hicks, J.B. and Hilvert, D. (1991) *In vivo* catalysis of a metabolically essential reaction by an antibody. *Proceedings of the National Academy of Sciences USA*, 82, 8784–8786.

Tavladoraki, P., Benvenuto, E., Trinca, S., De Martinis, D., Cattaneo, A. and Galeffi, P. (1993) Transgenic plants expressing a functional single-chain Fr antibody are specifically protected from virus attack. *Nature* 366, 469–472.

Thornton, C.R., Dewey, F.M. and Gilligan, C.A. (1994) Development of monoclonal antibody-based assays for the detection of live propagules of *Rhizoctonia solani* in the soil. In: Schots, A., Dewey, F.M. and Oliver, R. (eds), *Modern Assays for Plant Pathogenic Fungi: Identification, Detection and Quantification*. CAB International, Wallingford, UK, pp. 29–36.

Van De Koppel, M.M. and Schots, A. (1994) A double (monoclonal) antibody sandwich ELISA for the detection of *Verticillium* species in roses. In: Schots, A., Dewey, F.M. and Oliver, R. (eds), *Modern Assays for Plant Pathogenic Fungi: Identification, Detection and Quantification*. CAB International, Wallingford, UK, pp. 99–104.

Waterhouse, P., Griffiths, A.D., Johnson, K.S. and Winter, G. (1993) Combinatorial infection and *in vivo* recombination: a strategy for making large phage antibody repertoires. *Nucleic Acids Research* 21, 2265–2266.

Winter, G. and Milstein, C. (1991) Man-made antibodies. *Nature* 349, 293–299.
Wu, X.-C., Ng, S.-C., Near, R.I. and Wong, S.-L. (1993) Efficient production of a functional single-chain antibody via an engineered *Bacillus subtilis* expression–secretion system. *Bio/Technology* 11, 71–76.

Antibody Probes in Cereal Breeding for Quality and Disease Resistance

4

N.K. Howes

Agriculture and Agri-Food Canada Research Station, 195 Dafoe Road, Winnipeg, Manitoba, Canada R3T 2M9.

Cereal breeding consists of selecting complementary parents for crosses, achieving homozygosity in progeny and identifying those lines having desirable gene combinations. While selecting the correct parents for a cross is probably the most important factor, identifying gene combinations is the most time consuming and difficult. Traditionally, where cereal breeders have performed quality evaluation, it has been performed on the more advanced lines which have already passed selection for agronomic type, disease resistance and yield. Large amounts of seed are usually available because several field replicates have typically been pooled and quality tests have been performed on the composite. Early generation selection for quality is usually not performed because some quality traits are difficult to identify in segregating populations and there is generally insufficient seed available for many milling and processing tests. Furthermore, many tests are expensive to perform.

4.1 Early Generation Tests for Quality

Tests including near-infrared reflectance spectroscopy (NIR) for protein content and grain hardness evaluation (Williams, 1979), SDS sedimentation for dough strength and loaf volume prediction (Axford *et al.*, 1978) and small-scale (2g) mixographs (Gras and O'Brien, 1992) for dough mixing evaluation have been perfected to enable less than 10g samples to be tested. While these tests give high correlations with more elaborate large-scale quality tests, the environmental variability of results with non-replicated early generations (F_2 and F_3) limits the usefulness of these tests as very early generation selection tools.

The development of monoclonal antibody (MAb) technology (Kohler and Milstein, 1975) and the application of MAbs to enzyme-linked immuno-sorbent assays (ELISA) has made possible the direct detection of specific seed proteins in cereals that have a role in end-use quality. This technology was first applied by Skerritt *et al.* (1984) to wheat gliadins and was soon applied by others to proteins from barley (Ullrich *et al.*, 1986), maize (Esen *et al.*, 1989) and oats (Zawistowski and Howes, 1990). The advantage of MAbs is the sensitivity of ELISA assays, requiring much less than one kernel, thus enabling many tests to be performed on the endosperm end of one kernel and planting out the embryo end for further multiplication (Howes *et al.*, 1989b). Once produced, MAbs are inexpensive to use, the methods are simple to learn and require little specialized laboratory equipment. Furthermore, it is possible to develop a range of specificities by choosing MAbs specific to an epitope that is either unique to one protein or is present in a group of proteins having similar effects upon quality.

Other methods including polyacrylamide gel electrophoresis (PAGE) (Bushuk and Zillman, 1978) and high performance liquid chromatography (HPLC) (Burnouf and Bietz, 1984; Huebner and Bietz, 1985) share the sensitivity and specificity of MAb-based tests. Together with MAbs, these methods are qualitatively different from small-scale predictive tests, because they can measure products of specific genotypes (i.e. the genes controlling the synthesis of proteins) rather than the genotype/environmental interaction (i.e. hardness, grain protein content, test mass). Generally the tests using MAbs, PAGE or HPLC are not influenced by environmental variation, and so will be more reliable for early generation screening. The value of these tests will depend upon the underlying contribution that the specific protein subunit makes to quality characteristics.

In wheat, both high molecular weight glutenin subunits (HMWGS) and low molecular weight glutenin subunits (LMWGS) or the gliadin alleles linked to LMWGS have been shown to be good predictors of quality in some breeding programmes (Wrigley *et al.*, 1981; Payne *et al.*, 1984; Gupta *et al.*, 1989) but not in others (Hamer *et al.*, 1992). In barley, malting quality has been shown to be associated with groups of storage proteins as well as with carbohydrates and hydrolytic enzymes (Stuart *et al.*, 1988).

So far, few MAb-based tests have been used in early generation screening of breeders' lines, but those that have been used have been successful in identifying specific protein subunits associated with quality (Howes *et al.*, 1989a). The MAb test for gluten strength in durums is based upon differentiating lines with γ-gliadin band 45 from γ-gliadin band 42. Earlier studies showed the association of these gliadins with durum quality (Damidaux *et al.*, 1980; Kosmolak *et al.*, 1980), even though high gluten strength is probably determined by the LMWGS associated with γ-gliadin band 45 (Pogna *et al.*, 1990). Thus, the MAb test is replacing an electrophoresis-based test. Similarly the 1AL/1RS (Graybosch *et al.*, 1992) and 1BL/1RS (Howes *et al.*, 1989b; Graybosch *et al.*, 1993) tests for wheat–

rye (AL or BL) chromosomal translocations are based on replacing electro-phoretic methods of detecting either rye secalins (Graybosch *et al.*, 1993) or the absence of wheat gliadins (Howes *et al.*, 1989b). These tests can be applied to flour, wholemeal or half kernels. ELISA-based selection can be as reliable as gel electrophoresis, but enables 3–5 times as many samples to be analysed with equivalent labour. Other rapid tests such as SDS sedimentation volume (Axford *et al.*, 1978) are also more labour intensive and require a larger sample size while giving similar prediction in F_2 or F_4 performance (Clarke *et al.*, 1993), (see Fig. 4.1).

Tests applied to early generations do not have to be as highly correlated to quality characteristics as later generation tests, since large numbers of F_2 and F_3 families are available. A rank (Spearman) correlation between the test and quality and thus being able to classify progeny into either a higher or lower quality group generally enables useful selection to be performed when the correlation coefficient (r) is above 0.6. Where single kernels are being screened for the segregation of one protein type, merely identifying either

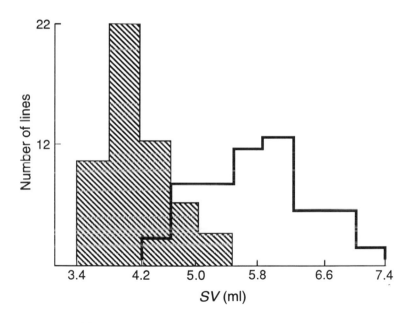

Fig. 4.1. Heritability of immunoassay results obtained using an antibody probe for a gliadin polypeptide, when used to screen flours to predict sodium-dodecyl-sulphate-containing buffer (SDS) sedimentation volume, a measure of durum wheat quality: SDS sedimentation volume (*SV*) of F_2-derived F_4 lines classified in the F_2 generation using a monoclonal antibody (P24B) test for γ-gliadin 45 (which is associated with high SDS sedimentation volume). Lines homozygous for γ-gliadin 42 (allele associated with low SDS sedimentation volume) indicated by striped bars and lines homozygous for γ-gliadin 45 indicated by empty bars.

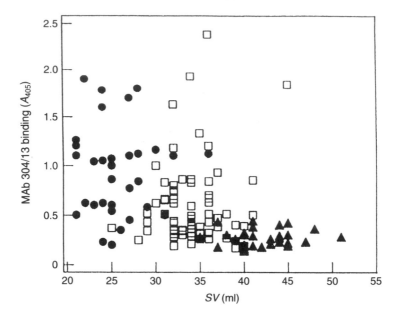

Fig. 4.2. Relationship between monoclonal antibody (304/13) affinity to chromosome 1B-encoded high Mr glutenin subunits (HMWGS) of durum wheat and SDS sedimentation volume performed on wholemeal of F_2-derived F_4 lines having HMWGS subunits 6+8 (▲), HMWGS 20 (●) or heterozygous for both alleles (□). The lines homozygous for HMWGS 6 and 8 provided low antibody binding and high SV while the others displayed a range of antibody binding. (Reproduced with permission from Kovacs et al., 1993.)

homozygous class should enable either very poor quality to be discarded, or very good quality to be retained. The heterozygous kernels can be discarded or retested in the next generation. Kovacs et al. (1993) showed that an ELISA test using a HMWGS-binding MAb was able to identify wholemeal samples derived from lines homozygous for HMWGS 20 or 6 + 8. Most lines homozygous for HMWGS 20 had poor SDS sedimentation volume and high MAb binding while lines homozygous for HMWGS 6 + 8 had higher sedimentation volumes and lower MAb binding (Fig. 4.2). There was little correlation among subunit 6 + 8 lines for MAb binding and SDS sedimentation volume, showing that the MAb in the assay used was not useful for quantifying a specific protein, but rather useful for identifying a specific genotype. Lines heterozygous for HMWGS 20 and 6 + 8 could not be clearly identified from lines homozygous for HMWGS 20 using a wholemeal sample, but could be identified by examining five single kernels.

Where single kernels are being analysed, the main error is due to sampling. For the γ42/γ45 gliadin test, examining two F_3 kernels for each F_2-derived line will identify homozygous γ42 gliadin lines correctly in 96% of cases, but in only 50% of cases will the γ45 or segregating lines be correctly identified because it is difficult to distinguish between kernels having two doses or three doses of the γ45 protein. Sampling an additional two kernels, however, in the γ45 lines will correctly classify 88–94% of them (Clarke *et al.*, 1993).

Early generation MAb tests can be useful if they measure a different component of quality that cannot be measured by other early generation tests. For example, Brett *et al.* (1993) showed that for one MAb specific to chromosome 1D coded LMWGS, its binding to 21 cultivars correlated with higher loaf volume ($r = 0.49$). The alveograph test (P) also was moderately correlated with loaf volume ($r = 0.49$), but reanalysis of their data showed that both tests were independent ($r = 0.07$) so that when combined these tests gave a good prediction of loaf volume ($r^2 = 0.45$). Similar improved correlations between MAb binding to 50 wheat cultivars and extensograph maximal resistance were obtained by combining glutenin score (Payne *et al.*, 1987) based on electrophoretic analyses of HMWGS together with binding of two MAbs, one specific for HMWGS and the other specific for LMWGS (Howes *et al.*, 1991). Individually, each measurement gave a moderate correlation ($r = 0.4$–0.5) but combined gave an improved correlation ($r^2 = 0.43$).

In barley, hordeins appear to be important in malting quality (Shewry *et al.*, 1980; Smith and Lister, 1983; Smith and Simpson, 1983) including influencing malt yield and chill haze (Asano and Hashimoto, 1980). Skerritt and Henry (1988) showed that binding of MAbs recognizing B and C hordeins correlated well with measurements of malt quality such as percentage modification at full malting. These MAbs, however, did not differentiate good and poor malting cultivars on the basis of binding to non-malted barley samples. Skerritt and Janes (1992), while showing that malting quality was related to the amount of SDS-extractable proteins, suggested that malting quality was also influenced by the orientation of disulphide bonds between B and D hordein subunits. In maize and oats, traditional breeding emphasis was placed upon protein, oil and test weight. More recently, emphasis has been placed upon modified starches and other carbohydrates which have unique industrial applications. In the future MAbs may be used to screen for enzymes responsible for synthesis of carbohydrates in these crops.

4.2 Advanced Generation Tests for Quality

Advanced breeders' lines are usually grown at a number of replicated sites, so that a large sample is available for quality tests. Tests, however, must be more accurate in predicting end-use quality since a large investment has been made in each line. In wheat breeding, antibody-based tests have a role in replacing more expensive tests such as loaf volume and extensograph measurements or replacing the more subjective tests such as crumb appearance, noodle surface texture and chewiness. Unlike early generation tests, where selection aims to identify major genes that influence quality, in later generations breeders are more interested in tests that identify genes that are influenced by environment and may involve much smaller changes in quality. Such tests would also be useful in the grain industry as an aid to monitoring deliveries of grain for quality defects caused by adverse growing or harvesting environments.

With a later generation test in mind, Skerritt (1991a) evaluated a group of MAbs that mainly recognized HMWGS for predicting dough mixing parameters. He used flour from a group of 15 cultivars, representing diverse genotypes having a wide range of dough strengths, to optimize the assay conditions. The test was designed to be easily performed by laboratory technicians with little training in cereal chemistry tests or immunology and using 500 50+0 mg of flour. The test gave high predictions of flour extensograph maximal resistance (R_{max}) and farinograph dough breakdown in these 15 cultivars grown under a range of nitrogen fertilizer levels (Skerritt, 1991b) (Fig. 4.3). More importantly, the MAb binding was giving predictive value for these quality tests in addition to the relationship of these tests to protein content.

Andrews et al. (1993a) made further improvements in the protocol and tested a number of different cultivar sets and breeders' lines. They showed that one of the MAbs gave the best correlation ($r = 0.885$ after normalizing for protein) with extensograph maximal resistance. Furthermore, they showed that the assay could be performed on 50 mg samples of flour or whole meal, using either fixed weight or fixed volume samples. Thus this

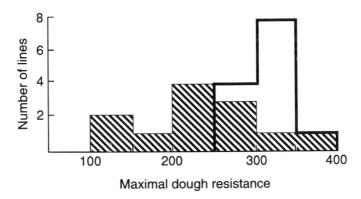

Fig. 4.3. Screening of lines for dough strength using double-antibody-sandwich ELISA; results of extensograph maximal resistance measurement of cultivars classified as strong (+ ELISA absorbance greater than 0.95) or weak (absorbance less than 0.95) using the immunoassay with monoclonal antibody 412/01.

modified test would also be suitable for early generation screening where flour samples are not available. Other dough tests such as mixograph development time and extensibility were not always correlated with the MAb test, depending upon the genotypes and the growing location of samples.

Since this dough strength test is the most intensively studied and well documented MAb test used in cereal quality screening it deserves further consideration as to the possible underlying chemical basis of the test. Andrews and Skerritt (1994) showed that the MAbs evaluated in the dough strength test bound most strongly to synthetic peptides having amino acid sequences overlapping the consensus repeat sequences TGQGQQ and Gy/HYPTSP/LQQ present in the central repeat domains of HMWGS. Gupta *et al.* (1992), using the same set of genotypes that were used to optimize the MAb dough strength test, found that R_{max} was best correlated with the percent of polymeric gluten within total protein (PG %) while extensibility and farinograph development time were best correlated with total polymeric glutenin percent of flour (FG %). This suggests that the MAb dough strength test is discriminating between strong and weak cultivars by 'measuring' the proportion of protein subunits that can combine into polymeric glutenin. Since a sandwich ELISA format was used, protein subunits or aggregates having multiple binding epitopes such as HMWGS would be favoured over protein having single or small numbers of sites binding such as LMWGS and gliadins. This would explain the improvements in discrimination by: (i) using the 0.5% SDS pre-wash, which would remove gliadins and enrich glutenins; and (ii) the addition of unlabelled free MAb to the flour extracts which would preferentially remove from the assay proteins having low numbers of MAb binding sites.

Hay (1993) evaluated this MAb dough strength test for a different application: the prediction of puff pastry height and volume. In this case the test was not correlated with R_{max} or pastry height, but was moderately correlated with pastry volume and highly correlated with protein, HMWGS and extensibility. Since many of the cultivars were not bread wheat genotypes, gluten strength in these lines may depend more upon gliadin or LMWGS composition.

4.3 Enzymes and Enzyme Inhibitors

Amylases, glucanases, proteases and inhibitors of these enzymes contribute both to barley malting quality and to wheat pasta noodle and baking quality. Measurements of enzyme activity are subject to interference from substrates present in flour, from enzyme inhibitors or from other enzymes that give products that interfere with assay of the enzyme product. The immunochemistry of cereal enzymes has been reviewed by Daussant, J and Bureau (1988). Direct measurement of enzyme protein avoids these complications as has been demonstrated with a MAb specific to (1-3, 1-4) β-glucanase (Hoj *et al.*, 1990). Stuart *et al.* (1988) showed that β-glucanase potential was positively correlated with malt extract percentage ($r = 0.85$) and negatively correlated with malt β-glucan, but not correlated with grain β-glucan. Quantification of β-glucanase activity by a MAb-based ELISA test enabled this enzyme to be assayed more conveniently. The environmental influence on this enzyme is however high, so malting is still required to measure its expression (Hoj *et al.*, 1990). Inhibitors of α-amylase can be even more difficult to assay than hydrolytic enzymes. Masojc *et al.* (1993b) used a MAb specific to the barley and wheat α-amylase inhibitor coded by chromosomes 2H in barley and 2A, 2B and 2D in wheat (Masojc *et al.*, 1993a) to quantify the levels of these inhibitors in different cultivars. Although there was a tenfold range in levels in wheat cultivars they were not able to correlate these values with preharvest dormancy. Proteins having α-amylase inhibitor activity have been associated with the starch granule protein in wheat (P. Greenwell, unpublished) and may be involved in controlling wheat kernel hardness.

4.4 Disease Resistance

Using MAb or PAbs to screen for disease resistance in cereals is similar to the use of early generation tests to detect specific quality-associated protein alleles. Likely candidates are proteins that have been identified to be genetically associated with disease resistance loci (Howes, 1986; Whelan and Lukow, 1990). While disease resistance is usually simply inherited, resistance is not expressed in the absence of the pathogen and its expression is variable in heterozygous plants. Furthermore, inoculating early generations with the pathogen is usually avoided because of the adverse effect of the pathogen even in resistant lines, and only a small number of pathogen races can be tested on each plant. Thus for testing backcross F_1 (BCF_1) and F_2 plants, antibody-based tests would be most useful. In later generations MAb could have a role in screening for multiple resistance genes and for more difficult to evaluate genes such as root rot, virus or smut resistance.

   Antibody-based ELISA tests have also been developed to detect the concentration of the pathogens in plant tissues and thus detect resistance to the pathogen. An example is the ELISA test for barley yellow dwarf virus (Torrance *et al.*, 1986), which is used to detect the expression of the Yd2 gene in barley (Schaller *et al.*, 1964). These tests, however, are still measuring the host response to infection and thus are similar to and have most of the drawbacks of traditional plant pathology methods.

   Dramatic improvements in cereal disease resistance have been made by the introgression of resistance (R) genes from related alien grass species. Usually this involves the transfer of whole corresponding chromosomes, and later a chromosome arm or segment (translocation) which displaces the chromosome region in the cultivated cereal. These R genes can be identified by DNA or proteins associated with the alien segment or by the absence of proteins coded by the chromosome region that has been displaced in the cultivated plant. Protein differences associated with the 6Ags 6AL and 6AS/6DL Aagropyron/wheat translocations (Whelan, 1988) which confers resistance to the wheat curl mite (Whelan and Lukow, 1990) have been the basis of a MAb-based test for the 6Ag (6A) substitution and the 6AgS/6AL translocation using a MAb shown to be specific to gliadins coded by chromosome 6A (Howes *et al.*, 1994). When used in an ELISA format, this MAb was specific to chromosome 6A coded proteins (Howes *et al.*, 1994). Other MAb-based ELISA assays have been shown to be specific for chromosome 6D (Howes, Thomas and Skerritt, unpublished).

In barley, a protein marker associated with the presence of the Yd2 gene was shown to be present in barley coleoptite tissue, using two-dimensional electrophoresis (Halloway and Heath, 1992). For early generation screening, the electrophoresis technique is not practical but Halloway and Heath (1992) showed that both R and susceptible (S) forms of this protein could be visualized by isoelectric focusing and probing of the immunoblot with PAb raised to the Yd2 associated protein. Considering the importance of the Yd2 gene with barley breeding programmes (Burnett, 1990), an ELISA-based test using MAbs would be warranted to be able to rapidly screen for this resistance.

In oats, a M_r 56,000 high molecular weight avenin has been shown to be associated with oat stem rust resistance gene Pg13 (Howes *et al.*, 1992) derived from *Avenasterilis* (McKenzie *et al.*, 1970) and oat crown rust resistance has been shown to be associated with alcohol-soluble avenins (Chong *et al.*, 1994). The M_r 56,000 polypeptide was used to prepare MAbs specific for this protein (Zawistowski and Howes, 1990). All MAbs, however, cross-reacted with a limited number of other high molecular weight avenins, so that single kernels could not be screened in an ELISA test, although the MAbs helped to identify the protein subunits resolved by gel electrophoresis and transferred to immunoblots (Zawistowski and Howes, 1990).

In wheat, the 1BL/1RS wheat–rye translocations have been extensively used to introduce mildew (Mettin *et al.*, 1973), stem and leaf rust resistance (Dyck *et al.*, 1987), while the 1AL/1RS translocation has been used for greenbug and mite resistance. MAb-based tests using probes specific to the 1BS gliadins (Howes *et al.*, 1989b; Andrews *et al.*, 1993b) have been used to detect the presence of these translocations in flour whole meal or half kernels. The advantage of the half kernel assay for rapid recovery and recovery of F_2 plants has been discussed earlier.

In hexaploid cereals such as common wheat and cultivated oats, MAbs have the potential to be used as probes to detect translocations as well as nullisomic and monosomic plants instead of the more time-consuming process of cytological identification of chromosomes (Skerritt *et al.*, 1991; Knox *et al.*, 1992). Since few disease-resistance genes with the exception of rust resistance have been mapped to chromosomes using monosomic analysis, this method has great promise in identifying the chromosomal location of many disease-resistance loci. At least one resistance gene for loose smut (*Urticum tritici*) has been located to chromosome 6A using this approach (Knox and Howes, 1994). Other applications of chromosome-specific MAbs would be to identify alien addition lines. To be fully effective, however, these methods require a complete MAb set – one for each addition line. Since difference storage proteins or isoenzymes are coded by genes located on each chromosome this would appear to be an achievable goal.

References

Andrews, J.L. and Skerritt, J.H. (1994) Quality related epitopes of high Mr subunits of wheat glutenin. *Journal of Cereal Science* 19, 219–230.

Andrews, J.L., Blundell, M.J. and Skerritt, J.H. (1993a) A simple antibody-based test for dough strength. III. Further simplification and collaborative evaluation for wheat quality screening. *Cereal Chemistry* 70, 241–246.

Andrews, J.L., Blundell, M.J. and Skerritt, J.H. (1993b) Wheat–rye translocation detection using antibodies to rye prolamins. In: Wrigley, C.W. (ed.), *Proceedings of the 43rd Australian Cereal Chemistry Conference*, RACJ, Melbourne, pp. 134–137.

Asano, K. and Hashimoto, N. (1980) Isolation and characterization of foaming proteins in beer. *Journal of the American Society for Brewing Chemistry* 38, 129–137.

Axford, D.W.E., McDermott, E.E. and Redman, D.G. (1978) Small-scale tests for breadmaking quality. *Milling Feed and Fertilizer* 66, 18–21.

Brett, G.M., Mills, E.N.C., Tatham, A.S., Fido, R.J., Shewry, P.R. and Morgan, M.R.A. (1993) Immunochemical identification of LMW subunits of glutenins associated with bread-making quality of wheat flours. *Theoretical and Applied Genetics* 86, 442–448.

Burnett, P.A. (ed.) (1990) World perspectives in barley yellow dwarf. *Proceedings of the International Workshop*, Udine, Italy, 6–11 July 1987, CIMMYT, Mexico City.

Burnouf, T. and Bietz, J.A. (1984) Reverse-phase high-performance liquid chromatography of durum wheat gliadins: relationships to durum wheat quality. *Journal of Cereal Science* 2, 3–14.

Bushuk, W. and Zillman, R.R. (1978) Wheat cultivar identification by gliadin electrophoregrams. I. Apparatus, method, and nomenclature. *Canadian Journal of Plant Science* 58, 505–515.

Chong, J., Howes, N.K., Brown, P.D. and Harder, D.E. (1994) Association of endosperm proteins with crown rust resistance gene *PcX* and stem rust resistance genes Pg9 and Pg13 in Dumont oat. *Genome* 3, 440–447.

Clarke, J.M., Howes, N.K., McLeod, J.G. and DePauw, R.M. (1993) Selection for gluten strength in three durum wheat crosses. *Crop Science* 33, 956–958.

Damidaux, R., Autran, J.-C. and Feillet, P. (1980) Gliadin electrophoregrams and measurements of gluten viscoelasticity in durum wheats. *Cereal Foods World* 25, 754–756.

Daussant, J. and Bureau, D. (1988) Immunochemistry of cereal enzymes. *Advances in Cereal Science and Technology* 10, 47–90.

Dyck, P.L., Kerber, E.R. and Lukow, O. (1987) Chromosome location and linkages of a new gene (*Lr33*) for reaction to *Puccinia recondita* in common wheat. *Genome* 29, 463–466.

Esen, A., Mohammad, K., Schurig, G.G. and Aycock, H.S. (1989) Monoclonal antibodies to zein discriminate certain maize inbreds and genotypes. *Journal of Heredity* 80, 17–23.

Gras, P.W. and O'Brien, L. (1992) Application of a 2 gram mixograph to early generation selection for dough strength. *Cereal Chemistry* 69, 254–257.

Graybosch, R., Seo, Y.W. and Peterson, C.J. (1993) Detection of wheat–rye chromosomal translocation using an antisecalin monoclonal antibody. *Cereal Chemistry* 70, 458–463.

Gupta, R.B., Singh, N.K. and Shepherd, K.W. (1989) The cumulative effect of allelic variation in LMW and HMW glutenin subunits on physical dough properties in progeny of two bread wheats. *Theoretical and Applied Genetics* 77, 57–64.

Gupta, R.B., Batey, I.L. and MacRitchie, F. (1992) Relationship between protein composition and functional properties of wheat flours. *Cereal Chemistry* 69, 125–131.

Halloway, R.J. and Heath, R. (1992) Identification of polypeptide markers of barley yellow dwarf virus resistance and susceptibility genes in non-infected barley (*Hordeum vulgare*) plants. *Theoretical and Applied Genetics* 85, 346–352.

Hamer, R.J., Weegles, P.L. and Marselle, J.P. (1992) Prediction of the bread quality of wheat: the use of HMW glutenin-subunit based quality scoring system. *Journal of Cereal Science* 15, 91–102.

Hay, R.L. (1993) Effect of flour quality characteristics on puff pastry baking performance. *Cereal Chemistry* 70, 392–396.

Hoj, P.B., Hoogenrad, N.J., Hartman, P.J., Vannakena, H. and Fincher, G.B. (1990) Identification of individual (1-3, 1-4)-beta-D-glucanase isozymes in extracts of germinated barley using specific monoclonal antibodies. *Journal of Cereal Science* 11, 261–268.

Howes, N.K. (1986) Linkage between the *Lr10* gene conditioning resistance to leaf rust, two endosperm proteins and hairy glumes in hexaploid wheat. *Canadian Journal of Genetics and Cytology* 28, 595–600.

Howes, N.K., Kovacs, M.I.P., Leisle, D., Dawood, M.R. and Bushuk, W. (1989a) Screening of durum wheats for pasta-making quality with monoclonal antibodies for gliadin 45. *Genome* 32, 1096–1099.

Howes, N.K., Lukow, O.M., Dawood, M.R. and Bushuk, W. (1989b) Rapid detection of the 1BL/1RS chromosome translocation in hexaploid wheats by monoclonal antibodies. *Journal of Cereal Science* 10, 1–4.

Howes, N.K., Skerritt, J.H. and Zawistowski, J. (1991) Selection of hexaploid wheats for gluten strength using monoclonal antibodies which discriminate Glu-B1 and Glu-B3 alleles. *Cereal Foods World* 36, 699 (Abstract 130).

Howes, N.K., Chong, J. and Brown, P.D. (1992) Oat endosperm proteins associated with resistance to stem rust of oats. *Genome* 35, 120–125.

Howes, N.K., Whelan, E.D.P., Lukow, O.M. and Skerritt, J.H. (1994) Identification of alien chromosome substitutions that confer resistance to colonization by the wheat curl mite using monoclonal antibodies specific to the short arm of chromosome 6A of common wheat. *Genome* (in press).

Huebner, F.R. and Bietz, J.A. (1985) Detection of quality differences among wheats by high-performance liquid chromatography *Journal of Chromatography* 327, 333.

Knox, R.E. and Howes, N.K. (1994) A monoclonal antibody chromosome marker analysis used to locate a loose smut resistance gene in wheat chromosome 6A *Theoretical and Applied Genetics* 89, 787–793.

Knox, R.E., Howes, N.K. and Aung, T. (1992) Application of chromosome-specific monoclonal antibodies in wheat genetics. *Genome* 35, 831–837.

Kohler, G. and Milstein, C. (1975) Continuous cultures of fused cells secreting antibodies of predefined specificity. *Nature* 256, 495–497.

Kosmolak, F.G., Dexter, J.E., Matsuo, R.R., Leisle, D. and Marchylo, B.A. (1980) A relationship between durum wheat quality and gliadin electrophoregrams. *Canadian Journal of Plant Science* 60, 427–432.

Kovacs, M.I.P., Howes, N.K. Leisle, D. and Skerritt, J.H. (1993) The effect of high molecular weight glutenin subunit composition on tests used to predict durum wheat quality. *Journal of Cereal Science* 18, 43–51.

Masojc, P., Zawistowski, J., Howes, N.K., Aung, T. and Gale, M.D. (1933a) Polymorphism and chromosome location of an endogenous alpha-amylase inhibitor in common wheat. *Theoretical and Applied Genetics* 85, 1043–1048.

Masojc, P., Zawistowski, J., Zawistowska, U. and Howes, N.K. (1993b) A combined monoclonal and polyclonal antibody sandwich ELISA for quantitation of the endogenous alpha-amylase inhibitor in barley and wheat. *Journal of Cereal Science* 17, 115–124.

McKenzie, R.I.H., Martens, J.W. and Rajhathy, T. (1970) Inheritance of oat stem rust resistance in a Tunisian strain of *Avena sterilis*. *Canadian Journal of Genetics* 12, 501–505.

Mettin, D., Bluthner, W.D. and Schlegel, G. (1973) *Proceedings of the 4th International Wheat Genetics Symposium*, p. 179.

Payne, P.I., Jackson, E.A. and Holt, L.M. (1984) The association between gliadin 45 and gluten strength in durum wheat varieties: a direct causal effect of the result of genetic linkage? *Journal of Cereal Science* 2, 73–81.

Payne, P.L., Nightingale, M.A., Krattiger, A.F. and Holt, L. (1987) The relationship between HMW glutenin subunit composition and the bread-making quality of British-grown wheat varieties. *Journal of the Science of Food and Agriculture* 40, 51–60.

Pogna, N.E., Autran, J.C., Mellini, F., Lafiandra, D. and Feillet, P. (1990) Chromosome 1B-encoded gliadins and glutenin subunits in durum wheat: genetics and relationship to gluten strength. *Journal of Cereal Science* 11, 15–34.

Schaller, C.W., Qualset, C.O. and Rutger, J.M. (1964) Inheritance and linkage at the Yd2 gene conditioning resistance to the yellow dwarf disease in barley. *Crop Science* 4, 544–578.

Shewry, P.R., Gaulks, A.J., Parmar, S. and Miflin, B.J. (1980) Hordein polypeptide pattern in relation to malting quality and the varietal identification of malted barley grain. *Journal of the Institute of Brewing* 86, 138–141.

Skerritt, J.H. (1991a) A simple antibody-based test for dough strength. I. Development of the method and choice of antibodies. *Cereal Chemistry* 68, 467–474.

Skerritt, J.H. (1991b) A simple antibody-based test for dough strength. II. Genotype and environmental effects. *Cereal Chemistry* 68, 475–481.

Skerritt, J.H. and Henry, R.J. (1988) Hydrolysis of barley endosperm storage proteins during malting. II. Quantification by enzyme- and radio-immunoassay. *Journal of Cereal Science* 7, 265–281.

Skerritt, J.H. and Janes, P.W. (1992) Disulphide-bonded gel protein aggregates in barley: quality-related differences in composition and reductive dissociation. *Journal of Cereal Science* 16, 219–235.

Skerritt, J.H., Smith, R.A., Wrigley, C.W. and Underwood, P.A. (1984) Monoclonal antibodies to gliadin proteins used to examine cereal grain protein homologies. *Journal of Cereal Science* 2, 215–224.

Skerritt, J.H., Martinuzzi, O. and Metakovsky, E. (1991) Chromosomal control of wheat gliadin protein epitopes: analysis with specific monoclonal antibodies. *Theoretical and Applied Genetics* 82, 44–53.

Smith, D.B. and Lister, P.R. (1983) Gel-forming proteins in barley grain and their relationships to malting quality. *Journal of Cereal Science* 1, 229–239.

Smith, D.B. and Simpson, P.A. (1983) Relationships of barley proteins soluble in sodium dodecyl sulphate to malting quality and varietal identification. *Journal of Cereal Science* 1, 185–197.

Stuart, I.M., Loi, L. and Fincher, G.B. (1988) Varietal and environmental variations in (1-3, 1- 4)-β-glucan levels and (1-3, 1-4)-β-glucanase potential in barley: relationships to malting quality. *Journal of Cereal Science* 7, 61–71.

Torrance, L., Pead, M.T., Larkins, A.P. and Butcher, G.W. (1986) Characterization of monoclonal antibodies to a UK isolate of barley yellow dwarf virus. *Journal of General Virology* 67, 549–556.

Ullrich, S.E., Rusmussen, U., Hoyer-Hansen, G. and Brandt, A. (1986) Monoclonal antibodies to hordein polypeptides. *Carlsberg Research Communications* 51, 381–399.

Whelan, E.D.P. (1988) Transmission of a chromosome from the decaploid *Agropyron elongatum* that confers resistance to the wheat curl mite in common wheat. *Genome* 30, 293–298.

Whelan, E.D.P. and Luckow, O.M. (1990) The genetics and gliadin protein characteristics of a wheat-alien translocation that confers resistance to colonizations by the wheat curl mite. *Genome* 33, 400–404.

William, P.C. (1979) Screening wheat for protein and hardness by near infra reflectance spectroscopy. *Cereal Chemistry* 56, 169–172.

Wrigley, C.W., Robinson, P.J., and Williams, W.T. (1981) Association between electrophoretic patterns of gliadin proteins and quality characteristics of wheat cultivars. *Journal of the Science of Food and Agriculture* 32, 433–442.

Zawistowski, J. and Howes, N.K. (1990) Production of monoclonal antibodies against high Mr avenin polypeptides. *Journal of Cereal Science* 12, 235–244.

The Interface between RFLP Techniques, DNA Amplification and Plant Breeding

5

5

P.M. Gresshoff

Plant Molecular Genetics, Institute of Agriculture and Center for Legume Research, The University of Tennessee, Knoxville, Tennessee 37901-1071, USA.

5.1 Introduction

Genome analysis generates information about the coding and non-coding regions of organisms. It provides an information framework for a variety of applications. For example, the evolutionary history of a set of organisms can be traced using molecular markers not confined by functional constraints common to coding sequences. Molecular markers can be profiled and used as identification tags in plant variety rights legislation, forensics, or paternity analysis (Gresshoff et al., 1990; Gresshoff, 1992). Pathogen identification, either prior to the onset of disease symptoms in a host or of microbes that are not cultured by normal laboratory techniques, can be carried out using molecular markers. The increased focus on the genome as a whole has also generated techniques which permit the comparison of near-isogenic organisms. This application finds utility in the detection of molecular markers closely linked to or part of hitherto undefined genes, which control medically or agronomically important traits.

The detection of closely linked molecular markers is part of a general scheme for the isolation of genes, for which no more than the inheritance and the phenotype are known. This strategy is known as positional cloning or map-based cloning (Wicking and Williamson, 1991). This chapter will give the background to the search for the supernodulation (*nts*) locus in order to illustrate the molecular technology that is used in developing diagnostic tests with nucleic-acid-based probes. The technical and genetic details of the newly developed DNA amplification fingerprinting (DAF) method (Caetano-Anollés et al., 1991a,b, 1993) will then be presented.

5.2 Legume Nodulation

Soyabean, a legume able to nodulate and fix nitrogen, is one of the major food crops. It supplies not only plant protein for human and animal consumption but also vegetable oils. Its utilization in the health food market is increasing as there is a move away from animal fats to low cholesterol–high fibre diets. Additionally, soyabean seeds may contain substances of pharmaceutical value.

Soyabean possesses the ability to form root nodules in symbiosis with the soil bacterium *Bradyrhizobium japonicum* as well as with some other bacterial types. Infection of the nascent nodule by bacteria results in nitrogen fixation, i.e. the conversion of atmospheric nitrogen gas to ammonia, which in turn is assimilated by the soyabean to form nitrogenous compounds used in growth and protein biosynthesis. Nodulation and nitrogen fixation contribute substantially to the nitrogen balance of soyabean crops, and distinguish soyabean from cereals (like rice and wheat) which require fertilizer application. The genetic analysis of bacterial genes involved in the symbiosis and the resultant nitrogen fixation is well advanced (see Stacey *et al.*, 1992, for review). The plant analysis of nodulation has been restricted by the absence of sufficient genetic variation in the nodulation phenotype, and inaccurate methods for the quantitative measurement of nitrogen fixation in the whole plant. The task of analysing a legume genome is also difficult, as the organism possesses a large genome, usually is diploid or tetraploid, and harbours repeated gene sequences and duplicated biochemical pathways.

The study of plant genes involved in the symbiosis advanced rapidly with the introduction of molecular biological techniques. This resulted in the isolation of several cDNA and genomic clones, which code for nodulins (see Sanchez *et al.*, 1991, for review). Nodulins are defined as proteins that are preferentially found in nodule tissue. One of the most studied nodulins is leghaemoglobin, an oxygen transport protein of infected nodule cells (Marcker *et al.*, 1984; Verma, 1992). Other nodulins are found in special nodule compartments, such as the symbiosome membrane, the uninfected cells, the nodule parenchyma, and the plant cell wall. Many of these nodulins are only defined by their structure, and function is only presumed because of the localization of the gene product (usually by *in situ* hybridization of a mRNA) or by sequence comparison.

The genetic analysis of nodulation started with work by Nutman in the 1950s (Nutman, 1952). His work demonstrated that plant variability is available in several legumes (predominantly the clovers as investigated by him and associates). Natural variants and induced mutants for symbiotic characteristics are further described (see Caetano-Anollés and Gresshoff, 1991, for review). In 1985, plant mutations were discovered in soyabean that either increased (supernodulation; Carroll *et al.*, 1985a,b) or decreased (non-nodulation; Carroll *et al.*, 1986; Mathews *et al.*, 1987, 1989a,b) the nodule

Table 5.1. Steps in positional cloning.

Detection of phenotype
Determination of inheritance
Association with molecular markers using restriction fragment length polymorphism (RFLP) or multiple arbitrary amplicon profiling (MAAP)
Determination of genetic distance
Determination of physical distance
Isolation of large DNA segments by cloning into special vectors such as yeast artificial chromosomes (YACs) and bacterial artificial chromosomes (BACs)
Ordering of clones relative to each other
Functional complementation by transformation into mutant cells or embryos to assign function to cloned DNA sequences
Molecular characterization of DNA sequences with a known function

---

number per plant. The genetic analysis of these mutations, induced by chemical mutagenesis, suggested that they were located at separate unlinked loci. They were inherited as single recessive Mendelian genes (Delves *et al.*, 1988; Mathews *et al.*, 1989b, 1990, 1992). Most research has focused on the supernodulation locus (see Gresshoff, 1993, for a review).

Many plant characteristics, such as disease resistance, are understood at the genetic but not at the molecular level. This process of locating genes by genome analysis is termed positional cloning or map-directed cloning and involves techniques such as chromosome walking or gene landing. Table 5.1 outlines the steps required for the positional cloning of a gene of interest. Variations in the strategy may occur depending on the biological system used. Other molecular markers such as microsatellites or retrotransposons may also be applied (Wicking and Williamson, 1991; Bond *et al.*, 1992).

5.2.1 Genetic analysis of nodulation

A total of 12 alleles have been detected at the *nts* locus (Delves *et al.*, 1988). These vary in the degree of supernodulation. For example, mutant nts1116 forms only 2 to 4 times the nodule number compared to parent cultivar Bragg. In contrast, nts382 and nts1007 may form as many as 5 to 20 times the nodule number. Other *nts* mutants of soyabean have been isolated (Gremaud and Harper, 1989; Buzzell *et al.*, 1990; Akao and Kouchi, 1992). Mutant En6500 is at the same genetic locus as the Bragg mutant nts382 (S. Akao, personal communication). All nts mutants share the shoot control of autoregulation shown first by Delves *et al.* (1986).

Mutations conditioning either supernodulation or non-nodulation were also isolated in pea (Jacobsen and Feenstra, 1984; Kneen and LaRue, 1984; Engvild, 1987; Duc and Messager, 1989; Pracht *et al.*, 1993) and *Phaseolus vulgaris* (Park and Buttery, 1988). The pea mutants used by the Cornell laboratory (*nod3*) have been mapped in part. Interestingly, many symbiotic loci and nodulin genes mapped to chromosome 1, some forming a cluster. The co-segregation of an RFLP locus defining a nodulin and a symbiotic mutation suggests some functional association. However, as yet, none of the symbiotic mutations in all legumes has been causally associated with a nodulin or other gene product.

Recently we were able to map the early nodulin gene *enod2* on to linkage group 3 of the RIL (recombinant inbred lines; Lark *et al.*, 1993) map generated from a cross between soyabean cultivars Noir 1 and Minsoy (F. Ghasemmi and Gresshoff, unpublished data). Interestingly, this region of the RIL map is equivalent to the central part of linkage group A of the USDA RFLP map (Shoemaker *et al.*, 1992). This region is characterized by coding for a soyabean cyst nematode resistance gene and the black colour locus (*I*). The *nts* locus segregated in good fit to Mendelian segregation both in crosses with its parent Bragg, but also with the wild-type progenitor of cultivated soyabean, *Glycine soja* (Sieb. and Zucc.). It is thus assumed to be a single nuclear gene.

All *nts* mutants isolated in the Bragg background failed to complement suggesting that they were all situated in the same complementation group. The question arose whether the *nts* locus is the only locus controlling autoregulation. In the absence of evidence this is difficult to answer. However, three considerations may be taken into account. First, the *nts* locus was shown (see later discussion; and Funke *et al.*, 1993) to be located in a section of linkage group H, which is known to be single copy. Other autoregulation genes may be in duplicated regions of the genome and are not detectable by mutagenesis as recessive mutations. Second, in two EMS mutagenesis experiments (Carroll *et al.*, 1986), the frequency of *nts* mutations was similar to that determined for non-nodulation, nitrate reductase, and chlorophyll-deficiency mutations. Thus there is no apparent hot spot of mutation. Third, in an attempt to generate larger deletions in the *nts* gene, fast neutron bombardment was carried out in collaboration with Dr H. Brunner (IAEA, Seibersdorf, Austria); both *G. max* cv. Bragg and *G. soja* seeds were treated. The subsequent M_1 generation was grown to maturity in Knoxville greenhouses, bulk-harvested, and M_2 seedlings were screened for mutation. Chlorophyll deficiency ranged from 1 to 1.4%. If the same ratio of nts mutants to chlorophyll-deficiency mutants was to be found, a total of 8 to 12 supernodulating plants was expected. However, none was found. Since we know that the supernodulation phenotype exists in soyabean, and since we measured that fast neutron bombardment was effective, one has to conclude that the unlikely detection of no supernodulation mutants may be caused by the possible lethality of deletions in the *nts* gene. This suggests that the EMS-defined *nts* gene was detected as a mutant, because the phenotype was marginal and that the extreme phenotype caused by a deletion is lethal. This implies that the *nts* gene (or a nearby gene) controls a biological function that is not symbiosis specific. Other genes in the autoregulation pathway may be more stringently required, and thus their mutation may also lead to lethality.

The supernodulation locus segregated independently from the two non-nodulation loci (*nod49* and *nod139*) displaying negative epistasis of non-nodulation over supernodulation (Mathews *et al.*, 1990). F_2 segregation ratios of such crosses were 9 : 3 : 4. Double recessive mutants of *nts382* and *nod49* (or *nod139*) exhibited a non-nodulation phenotype.

The *nod49* non-nodulation locus segregated as a single Mendelian recessive in crosses with both Bragg and *G. soja* PI468.397. Earlier complementation tests demonstrated that *nod49* is allelic with the naturally occurring non-nodulation allele *rj1* (Mathews *et al.*, 1989b). The same study showed that the *nod139* mutation defines a separate gene controlling non-nodulation. Interestingly, mutant *nod139* when crossed to Bragg gave a Mendelian segregation for non-nodulation. However, when crossed to *G. soja* PI468 397, a 15 : 1 ratio was observed; this cross (M_{70}) was tested on several occasions; Landau-Ellis and Gresshoff (unpublished data). We proposed the genetic symbol *rj6* for the newly discovered nodulation gene. Since the original isolation of the *nod139* mutant, Pracht *et al.* (1993) have isolated the NN5 nodulation mutant of Williams soyabean. Genetic analysis suggests that this mutant carries a deficiency at two loci, one of which is allelic to *nod139*. The other locus is called *rj5*, which presumably is already in the mutant form or absent in the Bragg genetic background.

These genetic relationships hint at the ancestral tetraploidy of soyabean. Many American soyabean varieties have variety Peking as an ancestor. This variety shares many characteristics with *G. soja* lines, and it is possible that some commercial lines are dominant at both *rj5* and *rj6* loci (like *G. soja*), while others are semi-dominant (like Bragg), facilitating the isolating of *nod139*. Mutant *nod139* may be very early in the signal recognition chain which links the bacterium and its lipo-oligosaccharide nodulation factor to the cellular response (cell division, root hair curling, infection) in the plant.

5.3 DNA Markers for the Genetic Mapping of Soyabean

Restriction fragment length polymorphisms (RFLPs) were the first means by which DNA markers were used to produce genetic maps. Several RFLP maps have been generated for soyabean. DuPont Company produced one of the most extensive but the information and clones are not publicly available. Gordon Lark, in collaboration with others, has started to produce a *G. max* × *G. max* map (Lark *et al.*, 1993). The map is based on an F_2 population derived from a cross of cultivars Noir I and Minsoy. The RIL population and map (now at F_{11}) are very valuable as the map is characterized for many agronomic properties. Furthermore, it is immortal, allowing sharing of representative seeds from families. This, in turn, permits more complicated molecular analyses which require the isolation of high molecular weight DNA from living protoplasts.

The most extensive, but mortal, public map so far comes from USDA, ARS in Ames, Iowa, who crossed *G. max* and *G. soja* to evaluate the co-segregation of RFLPs of random *Pst*I clones (Keim *et al.*, 1989, 1990; Shoemaker *et al.*, 1992). The mapped F_2 population consisted of 66 plants, delineating the accuracy of any genetic distance to about 2cM (centimorgans). At present about 500 markers are distributed over 23 linkage groups spanning a total of 3000cM (Shoemaker *et al.*, 1994). Unfortunately, the Iowa F_2 population was not maintained as recombinant inbred lines, restricting the sharing of plant material and limiting the extent of future DNA analysis as DNA will become exhausted. This problem is further increased, if the need exists for the isolation of large molecular weight DNA. Such DNA is essential for YAC cloning as well as physical mapping (see Funke and Kolchinsky, 1994), but can only be isolated from living plants.

The genome size of soyabean is 1090 megabases (Gurley *et al.*, 1979; Blackhall *et al.*, 1991), about seven times that of the model plant *Arabidopsis thaliana*, and about twice that of the rice genome. Soyabean has 20 chromosomes which are distinguishable at the pachytene stage of meiosis (Singh and Hymowitz, 1988). All of the soyabean chromosomes now exist in trisomic lines, permitting the coupling of RFLP linkage groups to chromosomes (T. Hymowitz, unpublished results). Recently, Kolchinsky and Gresshoff (1995) discovered a 92 bp satellite, which represents about 0.7% of the total DNA. The satellite exists in large clusters (as large a 1 Mb), but is not located in telomeric regions. Presumably this satellite is centro-metric. *In situ* hybridization data of P. Klein (Arizona, unpublished data) point to this chromosomal location on several soyabean chromosomes.

Table 5.2. Soyabean genome characteristics relevant to genome analysis and gene isolation (as of September 1994).

One billion base pairs
20 chromosomes ($2n = 40$)
23 RFLP linkage groups
500 linked RFLP, morphological and enzyme markers
Genetic length is approximately 3000 centimorgans
Approximately 35% of DNA is highly repeated
A single nucleolar organizer region is present
Duplicated regions of the genome occur
Behaves as a normal diploid at meiosis
Recessive mutations have been identified
Trisomic stocks (all 20) are available
0.7% of the genome consists of a 92 bp satellite DNA sequence
Large segments of DNA have been cloned in YACs
RILs and NILs are available
cDNA and genomic libraries are available

General features of the soyabean genome are summarized in Table 5.2.

F_2 populations from crosses of nts lines and *G. soja* (carrying the wild-type allele of the *nts* locus) were analysed to reveal close co-segregation of an RFLP marker with the mutant gene. Probe pUTG-132a (Landau-Ellis *et al.*, 1991) and PCR primers defining a SCAR (Sequence Characterized Amplified Region; Paran and Michelmore, 1993) within the probe were used to map the *nts* locus (Kolchinsky *et al.*, 1995). At present over 171 plants have been analysed (representing over 342 chromosomes), and only one recombination event has been observed. This places the SCAR/RFLP within about 0.3 cM of the *nts* locus. The one recombinant is very informative as it confirms the gene order in the region of the *nts* locus to be: pA-36 – pUTG132a – nts – pA-381.

The pUTG-132a region segregated in an expected 1 : 2 : 1 ratio in the population as did other closely linked markers, such as pA-36. The same pUTG-132a-SCAR/RFLP site was mapped to two separate alleles (namely

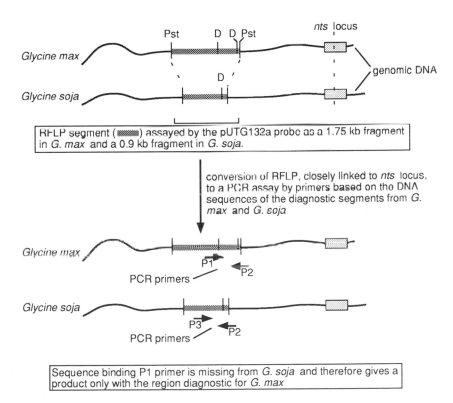

Fig. 5.1. The molecular region close to the supernodulation gene of soyabean as detected by RFLP polymorphism and PCR primers. P3 primer site was duplicated. D = *Dra*I site, Pst = *Pst*I site.

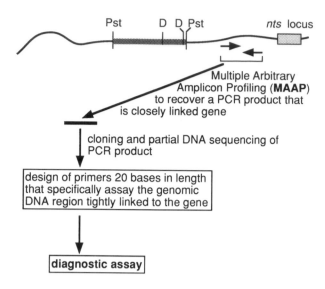

Fig. 5.2. Molecular markers close to a gene of interest. Either an RFLP site or an arbitrary primer amplification site can be detected by tight co-segregation in F$_2$ populations. DNA sequencing will direct the synthesis of specific PCR primers. The PCR product may be polymorphic to generate a SCAR. See Fig. 5.1 legend for abbreviations.

nts382 and *nts1007*) of the *nts* locus, supporting previous complementation date (Delves *et al.*, 1988; Landau-Ellis and Gresshoff, 1992). The RFLP between nts382 and *Glycine soja* was characterized by an 877 bp deletion which removed an internal *Dra*I site (Kolchinsky *et al.*, 1994; Gresshoff, 1993). Figure 5.1 shows the nature of the RFLP probe, the genomic regions and the positioning of PCR primer, which were used to detect the same region as a SCAR. The pUTG-132a probe detected a single copy region of the soyabean genome, supporting the genetic data placing the recessive *nts* locus in a single copy region of the genome (Landau-Ellis and Gresshoff, 1994). This makes pUTG-132a a useful starting point for a chromosome walk or a gene landing, provided that the pA-381 marker is close enough. Recently it was possible to place additional RFLP markers within the pA-36 to pA-381 region.

The scarcity of additional RFLP markers close to *nts* suggested that other molecular markers were required. Instead of using the specificity of a restriction nuclease site and the length polymorphism between these, amplification polymorphisms were sought. Figure 5.2 shows how a RFLP marker, like pUTG-132a, was converted to a SCAR. The same is possible for amplification markers such as those found closely linked to *nts* using DNA amplification fingerprinting (see below; also Caetano-Anollés *et al.*, 1993).

5.4 Physical Mapping

Leaf protoplasts of soyabean were embedded in low melting agarose and digested *in situ* to yield high molecular weight DNA (Funke and Gresshoff, 1992). The digested DNA was separated by pulse field gel electrophoresis and probed with closely linked RFLP probes derived from linkage group H (the *nts* locus is located in the same region of linkage group H). Co-migration and co-hybridization were interpreted as evidence that sequences homologous to RFLP probes were contained on the same DNA fragment generated by rare cutting restriction nucleases (such as *Not*I and *Sfi*I; Funke *et al.*, 1993). Complexities arose as many probes detected more than one genomic region. Some of the additional regions were not polymorphic and these were not mapped on the RFLP map. Using CHEF electrophoresis in one dimension, followed by lane excision and *in situ* restriction nuclease digestion and subsequent agarose gel electrophoresis at right angles to the original CHEF gel, revealed which PFGE band harboured the monomorphic and the polymorphic (mapped) homologue of the probe. Figure 5.3 details the steps used in the two-dimensional electrophoresis. It was possible to predict the physical arrangement of the duplicated genome pK-9 and pA-89 region on linkage group I for which only one polymorphic marker was detected (Funke *et al.*, 1993). The most significant finding from the physical mapping study was the calculation that one centimorgan in this region of the soyabean genome was equivalent to about 500 kb of DNA. The accuracy of this conversion factor is not very high, as the genetic distance was determined from just 66 plants of the USDA/ISU map. It is likely that errors of ± 0.5 cM may occur. Taken at face value, the *nts* locus may be as close as 150 kb from the pUTG 132a marker and the PCR primer defined SCAR. The other flanking marker may be pA-381, which in our material appears to be within less than 4 cM (perhaps 600 kb).

The ability to make high molecular weight DNA from soyabean allowed the construction of the first yeast artificial chromosomes (YACs) carrying soyabean genomic DNA (Funke and Kolchinsky, 1994; Funke *et al.*, 1994; Fig. 5.4). Agarose embedded leaf protoplasts were partially digested with *Eco*RI, then exposed to a size selection by pulse field gel electrophoresis to eliminate the majority of small *Eco*RI fragments. Size selected DNA was cloned into pYAC4 and transformed into yeast. The YACs were confirmed by ethidium bromide staining in pulse field gels, as well as by Southern hybridization with total soyabean genomic DNA or bacterial vector sequences. Hybridization with genomic DNA gave signals of different intensity, suggesting the presence of repeated DNA sequences in several YAC clones. At present the YACs range in size from about 50 to 960 kb.

YAC cloning and chromosome walking represent some technical problems, because YACs in other organisms are often chimeric. This is especially

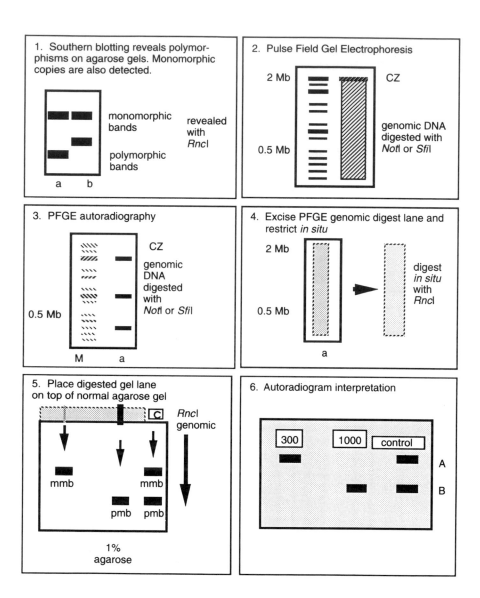

Fig. 5.3. Diagram of two-dimensional gel electrophoresis and its application in physical mapping of soyabean. CZ = compression zone of around 2 Mb DNA; *Rnc*I = restriction nuclease I used to reveal the original RFLP in genomic digests; pmb = polymorphic band; mmb = monomorphic band; A = the monomorphic band; B = the polymorphic band; C = control lane; sizes in panel 6 are in base pairs; M = yeast chromosome size markers; a (box 3) = bands revealed by probing.

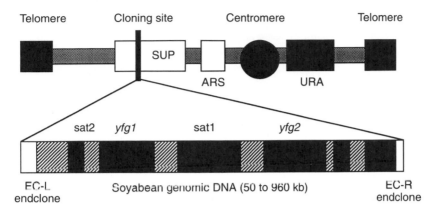

Fig. 5.4. General scheme for yeast artificial chromosome construction. The insertion of foreign DNA into the sup gene causes an alteration of the adenine metabolism of the yeast cell allowing the colony to develop a red colour. This colour is used as a selection step. *yfg* = your favourite gene; sat = putative satellite sequence; EC-L = end-clone left; EC-R = end-clone right. Unique and polymorphic end clone sequences can be mapped and used to find neighbouring YACs.

the case when large (> 500 kb) inserts are selected. To ascertain whether a YAC is chimeric, end clones can be generated using ligation of vectorettes, followed by amplification from PCR primers annealed to the cassette. If end clones are polymorphic, then their position can be mapped. Co-segregation is a strong argument for their common origin from one genomic region, suggesting that the YAC is not chimeric. Such end clones are presumed to be useful for the determination of physical distance between sites that are genetically mapped. For this type of study a large F_2 population is needed. The same approach may be used to assign selected YAC clones to a mapped morphological marker. While the presence of flanking RFLPs on a YAC provides strong evidence, the further mapping of polymorphic YAC end clones to the mapped locus strengthens any conclusion. End clones also permit the detection of a neighbouring clone, therefore laying the basis for chromosome walking. This approach is hindered by the fact that most plant DNA contains repeated sequences. Thus the end clones are likely to carry these repeats as well. Thus multiple sites are detected in the genome making these YACs unsuitable for positional cloning.

A high frequency of linked and ordered molecular markers, some knowledge regarding physical distances and large stable YACs are a prerequisite for a successful chromosome walk or landing. However, there exists the chance that the gene of interest is flanked by too much repeated DNA, eliminating the chance to approach the gene through a contiguous region. In that case, Funke and Kolchinsky (unpublished data) isolated the ends of soyabean DNA cloned into YACs and found that about half contained repeated DNA, however, unlike the pattern obtained for telomeric or subtelomeric regions. It is therefore paramount that cloning strategies do not rely exclusively on the chance to walk to a gene using YAC cloning. It seems to be more profitable to detect as many molecular markers closely linked to the gene of interest, then order the markers genetically, prior to isolating one large YAC, which contains both the flanking markers. Thus touching down on the gene may be more efficient than walking to it. Even relatively simple and low complexity genomes like those of *Arabidopsis thaliana* and rice experience this problem with YACs. Since RFLP markers often are absent close to the gene of interest, alternative methods for generating high marker density are needed. This may be able to be accomplished using novel DNA amplification profiling techniques as outlined below.

5.5 New Approaches for Identifying DNA Markers

While the information base of nucleotide sequences is exponentially growing, methods are needed to investigate plant genomes at a level of complexity *above* the primary sequence, but *below* the cytogenetic, karyotypic arrangement.

Over the last few years, single, short primer-based DNA amplification techniques were developed and applied to plant genomes. These techniques were collectively labelled as MAAP (Multiple Arbitrary Amplicon Profiling; Caetano-Anollés *et al.*, 1992c, 1993, 1994). In essence, MAAP involves the use of a short, arbitrarily chosen oligonucleotide primer, which, annealed to DNA, will direct DNA amplification of multiple genome regions (amplicons; Mullis, 1991). Temperature cycling and the use of a thermostable DNA polymerase are common components with more specific and targeted PCR (polymerase chain reaction). In contrast to standard PCR, MAAP procedures use a single primer which is of arbitrary sequence. MAAP intentionally generates multiple products and a primer used for one species can be used repeatedly for others, even if evolutionary distances between the template DNA are large. For example, bacterial, soyabean, and human DNA have been analysed with the same primer (generating different type profiles; Caetano-Anollés *et al.*, 1991a).

Amplification products are separated and recorded by a variety of detection methods; in all cases, a linear array of signals generates a profile, which is representative for the target DNA and specified by the DNA sequence of the primer. Variations in primer sites on the target DNA, length variations between primer sites, and possibly changes in the secondary structure of target DNA between or flanking the primer recognition sites generate molecular polymorphisms. These amplification polymorphisms define molecular regions of the plant genome and thus were used as: (i) potential sequence tagged sites for positional cloning approaches; or (ii) components of profile used in DNA profiling and diagnostics.

Amplification polymorphisms were termed AFLP (amplification fragment length polymorphism) in analogy to RFLPs (Caetano-Anollés *et al.*, 1991a,b). Recently, the same term has been applied to polymorphisms generated by the yet unpublished Key-PCR technique, involving DNA digestion, cassette ligation and primer amplification (M. Zabeau, personal communication). The use of the same term may be warranted if the products indeed are similar. MAAP procedures were developed independently and apparently concurrently in three laboratories. Welsh and McClelland (1990) developed AP-PCR, which uses PCR-length primers (18 to 32 nucleotides (nt)) of arbitrary sequence to amplify target DNA under low stringency annealing conditions for two amplification cycles. This allows abundant mismatching and the generation of multiple amplification products. Increased stringency of annealing at later amplification cycles generated reproducible products which were resolved on polyacrylamide gels and detected by autoradiography (Gresshoff and MacKenzie, 1994).

Williams *et al.* (1990) invented the RAPD procedure, in which arbitrary primers of nine or ten nucleotides produced amplification products after temperature cycling. Products are routinely resolved on agarose gels and visualized by ethidium bromide. This provides a rapid (hence the acronym) method of scanning a genome. Alternative methods of detection, such as polyacrylamide and silver staining, coupled with careful optimization of amplification parameters (Collins and Symons, 1993) improved the utility of the approach.

5.5.1 The DAF procedure

Caetano-Anollés *et al.* (1991a,b) discovered DNA amplification fingerprinting (DAF). This MAAP method utilizes very short primers, 5 to 15 nucleotides in length (for a review of parameters affecting DAF reactions, see Caetano-Anollés, 1993). The optimal length was found to be seven or eight nucleotides, a length which does not produce efficient amplification in the RAPD system. Informative amplification profiles were generated with five-nucleotide primers (5-mers) using soyabean DNA as a template (Caetano-Anollés *et al.*, 1993). DAF products are separated by thin polyacrylamide gels, backed on to plastic Gel-Bond film run in a commercially available vertical gel chamber (such as the Mini-Protean II unit marketed by BioRad. Inc., CA). This gel-plastic support is stained by an improved silver staining method (Bassam *et al.*, 1991), which allows detection of DNA at about $1 \, pg \, mm^{-2}$. Resultant gels are air-dried and kept for permanent record and evaluation. A pattern of bands, reminiscent of the Universal Product Code (UPC) is generated, permitting the easy distinction of genotypes (Fig. 5.5).

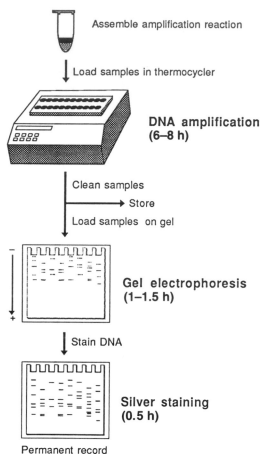

Assemble amplification reaction

Load samples in thermocycler

**DNA amplification
(6–8 h)**

Clean samples
→ Store
Load samples on gel

**Gel electrophoresis
(1–1.5 h)**

Stain DNA

**Silver staining
(0.5 h)**

Permanent record

Fig. 5.5. Flow diagram showing DAF amplification, in a thermocycler, followed by polyacrylamide gel electrophoresis, and silver staining.

The silver staining procedure has found additional utility in other applications; for example, Promega Inc. markets a kit based on Bassam *et al.* (1991) used for silver staining of DNA sequencing gels. Figure 5.5 shows the overall DAF procedure from amplification to gel viewing.

5.5.2 DAF pattern detection

The PAGE/silver staining technique provides a low cost, high throughput analytical method of DAF products. Profiles are easily stored in photo albums, without the need for photography. The gel itself offers a depository of the actual DNA, facilitating subsequent reanalysis or fragment cloning. PAGE gels dry easily in air and were scanned by laser densitometry (see Jayarao *et al.*, 1992; Prabhu and Gresshoff, 1994).

DAF products were also resolved by alternative methods. Agarose gels give clear resolution, but fewer products (Prabhu and Gresshoff, 1994). Kolchinsky *et al.* (1993) amplified isolated yeast chromosomes with DAF primers, separated the products on agarose, isolated individual bands by cloning, and rehybridized the cloned fragments on to pulsed field gel (PFG) karyotypes of yeast. Cloned fragments detected only their appropriate template chromosome. Similar cloning, but following reamplification prior to cloning of picked bands from dried, silver stained polyacrylamide gels was achieved by Weaver *et al.* (1994).

Caetano-Anollés *et al.* (1992a) collaborated with Applied Biosystems Inc. (Foster City, CA) by generating fluorochrome-labelled octamer primers, which then directed amplification of plant DNA. The resultant amplification products were separated on an ABI Gene Scanner. Single nucleotide resolution was obtained for lower sized amplification products. Preliminary tests using capillary electrophoresis have been positive (Pat Williams, Gaithersburg, MD, personal communication). DNA profiles can be generated within 30 min using automatic loading and data collection. High signal-to-noise ratios were obtained for amplification products between 100 and 1000 bp; however, additional research is needed to sieve amplification mixes to remove higher molecular weight products.

In general, the DAF procedure generates scoreable polymorphisms in the molecular size range from 100 to 800 bp. Recently, we have used the precast and automated PhastGel system (Pharmacia Inc.) to generate profiles for pathogenic nematodes on soyabean (Braum *et al.*, 1994). The pre-cast gradient PAGE gels and the automated silver staining permitted rapid throughput (160 samples per day). Bands at higher molecular weight (up to 1500 bp) were scoreable and species as well as race-specific polymorphisms were detected. Denaturing gradient gel electrophoresis (DGGE) is another method which would help to distinguish polymorphic products of wheat (He *et al.*, 1992). Recently, we developed novel mini-hairpin primers (Caetano-Anollés and Gresshoff, 1994). These primers are about ten nucleotides long, but their 5' region contains a stable mini-hairpin. This pin closes before the annealing temperature to the template DNA is reached. Thus as few as three 3'-terminal nucleotides govern the amplification site. These primers are useful for low complexity genomes, such as viruses, plasmids, YACs, or isolated chromosomes.

In summary, detection can be carried out by several procedures, offering different costs, convenience and resolution. Automation is possible at present but may require lower cost machines and procedures to make MAAP of great utility to the plant breeder.

5.5.3 *Parameters affecting DAF*

DAF generates multiple amplification products which are seen on silver stained PAGE gels as bands of different intensity. Theoretical analysis showed that in complex genomes such as soyabean not all existing amplicons are amplified effectively enough to be scored by PAGE and silver staining. In general, DAF profiles contain about 20 to 40 scoreable bands. The number varies with primer and template DNA. Co-migration is likely and requires cloning to generate verified single molecular markers.

Many physical and chemical parameters are also critical and were analysed for the DAF reaction (Bassam *et al.*, 1992b). We found that magnesium ion levels are critical, being optimal around 2mM $MgCl_2$ for complex genomes such as soyabean, or 6mM $MgCl_2$ for less complex genomes like those of bacteria (Bassam *et al.*, 1992b; Jayarao *et al.*, 1992). while *Taq* polymerase (such as Amplitaq, Perkin-Elmer-Cetus, Emeryville, CA) produces excellent results, we found that the truncated Stoffel fragment (from the same source) provided stronger amplification products at lower sizes (50 to 200 bp).

Primer composition and length as well as concentration are critical. Amplification was never detected with four nucleotide long primers, although five nucleotide primers work efficiently (Caetano-Anollés *et al.*, 1991a, 1993). Clear reiteration of the same nucleotide seems to block amplification. For example, primer 8.10a (eight G residues) did not generate amplification profiles in soyabean (Prabhu and Gresshoff, 1994). Attention also needs to be paid to intrinsic amplification products. These may arise from low level DNA contamination of the *Taq* polymerase or the primer. Addition of template DNA in excess usually removes the impact of the intrinsic amplification products. However, when working with extremely low DNA concentration, dummy runs need to be included to test for artefacts of this kind, revealed only because of the high resolution gel and staining procedures used for DAF.

Primer engineering, which involved taking a core sequence and adding additional nucleotides to it, showed that the 3' end of the primer is most critical in determining the specificity of the target interaction (Caetano-Anollés *et al.*, 1992b). Substitutions on the 5' end in the number 7 or 8 position influenced profiles, but only in a minor way. Addition of extra bases to the 5' end extending the octamer to a 9-mer or 10-mer had little influence on the amplification profile, suggesting that the optimal primer length is seven to eight nucleotides, and that additional bases increase synthesis cost and may increase the chance for stochastic annealing under suboptimal conditions. This may arise when: (i) primer concentration is too low (below 1μM); (ii) magnesium ion concentration is not optimal for the genome complexity; (iii) template DNA concentration is too high (greater than 20ng per 25μl reaction mix); or (iv) variations in temperature profiles occur in the thermocycler.

Primer concentration for DAF reactions needs to be around 3μM, when using an octamer. This is about ten times the concentration recommended for RAPD reactions and may be one of the explanations why shorter primers and higher annealing temperatures (above 45°C) fail to produce RAPD products. The detection of amplified fragment length polymorphisms between *Glycine max* (the commercial soyabean) and *Glycine soja* (an ancestral soyabean, fully fertile with *G. max*) is independent of GC content (over the 50 to 100% range) as is the number of amplification products (Prabhu and Gresshoff, 1994). DAF worked with an octamer of 37% GC content, showing that presumed limitations of the RAPD reaction do not hold true for DAF. The occurrence of non-functional primers is very low (perhaps 1 in 20). In contrast, the frequency of non-functional primers in RAPD reactions is as high as 50% (J. Carlson, University of British Columbia, Vancouver, personal communication).

5.5.4 Genetic uses of DAF

The ability to detect molecular markers closely associated with genes of agricultural importance, makes marker-based breeding (also called backcross conversion) an attractive proposition. The need for maintaining large plant populations through advanced breeding cycles can be reduced by detecting heterozygotes. MAAP markers converted through cloning, partial sequence analysis and specific PCR primer synthesis may provide SCARs (sequence characterized amplified regions) which are diagnostic for either a gene region in a plant or a pathogen. Figure 5.2 illustrates the utility of RFLPs and MAAP markers in generating diagnostic tools. For example, with further analysis, it may be possible to find high certainty markers specific for a soyabean nematode race. It may be possible to convert this marker to a SCAR, then use a diagnostic, proactive test on agricultural soil to predict which nematode race is predominant in the field prior to planting. Thus, many samples either from the breeder or the pathologist may need to be tested.

DAF markers detect both nuclear and cytoplasmic DNA. Prabhu and Gresshoff (1994) found one 8 nucleotide primer which generated two AFLPs. One was inherited in a Mendelian fashion, the other in a maternal way. Several DAF polymorphisms were mapped on to the RIL map (Lark *et al.*, 1993; Prabhu and Gresshoff, 1994). Rafalski *et al.* (1994) described the use of automated machines that analyse RAPD-generated MAAP samples. Plant genome analysis for agricultural purposes as compared to medical or forensic diagnoses in humans is different in many of its costing and volume parameters. Plant analysis generates high numbers of samples, which need to be analysed at an intermediate resolution and accuracy. Costs need to be low per analysis because the final product is of relatively low unit value, and large numbers of analyses need to be carried out to achieve the marketable product.

DAF applications in genotyping

The ability to generate many amplification products means that DAF is very efficient in scanning the genome of an organism for variable sites. In a survey of 25 primers (all octamers), Prabhu and Gresshoff (1994), working with *G. max* and *G. soja*, detected an average of 1.5 amplified fragment length polymorphisms per primer. This compares to 0.5 amplified fragment length polymorphisms per primer claimed by Williams *et al.* (1990) working with the same *Glycine* species. Interestingly, RAPD gels produce an average of five to seven scoreable bands, while DAF in soyabean produced an average of 20 to 25 bands. Accordingly, since the ratio of scored polymorphism to scoreable band is nearly the same, DAF is not picking up more amplified fragment length polymorphisms because of the shorter primer length but because of the detection method.

The large number of products allow a high density genotyping and genotype differentiation (Gresshoff, 1992; Fig. 5.6). Reliable exclusion is obtained when one or more bands differ between samples. Inclusion is more difficult, as many primers need to be tested, frequency of variation within the sampled species needs to be known, and careful statistical statements need to be generated. It is impossible to declare with 100% certainty that two things are the same; it always needs to be a probabilistic statement. It is up to the

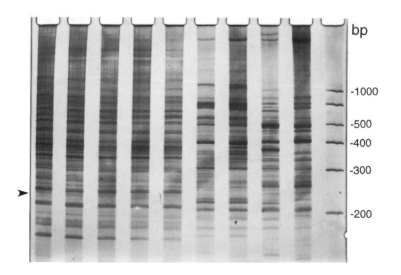

Fig. 5.6. DNA amplification fingerprint gel derived from several soyabean nematode species and races. Primer 8.6d (62.5% GC content) and Stoffel fragment of Amplitaq were used in a DAF reaction. Molecular size markers range from 100 bp to 200 bp. The polyacrylamide gel was stained with silver. (Photograph kindly supplied by Dr B. Bassam.)

user (society, courts, scientists) to concur on acceptable levels of confidence for such probabilities.

DAF allowed the easy distinction of variant turfgrass material in commercial plots (Callahan *et al.*, 1993). For example, foundation stock from several geographic locations gave identical profiles for Bermudagrass Tifway 419, while samples analysed from golf courses repeatedly showed major variation. Sunflower material provided by a seed company was categorized into several groups. Some common bands permitted the suggestion of a possible pedigree. This type of analysis has utility for product verification and plant variety rights. DAF analysis may also reveal the putative parents of a commercially protected hybrid (say maize or sunflower), permitting the reconstruction of the hybrid by renewed breeding from unprotected parental stock.

The determination of genetic identity is also essential for the determination of plant product quality as many food manufacturers use processes directly optimized for a specific biological feedstock. The food industry relies on biological material, and it is essential that the correct biological feedstock is entered into the manufacturing process. All maize varieties, for example, have the same flowers morphologically and yet the products of different varieties can differ significantly for starch or sugar content. Moreover, often it is impossible to inspect the source plant, as one looks at a harvested product. It is for these industrial and related horticultural applications that a new technology was needed. The advent of DNA analysis has provided the means by which closely related organisms are distinguished for industrial, manufacturing and retailing purposes.

As yet there are no guidelines for commercial DNA fingerprinting practices. How are samples to be deposited? Which techniques are acceptable? Who will (if at all) certify laboratories? What controls need to run on each gel? So far the human forensic, paternity and criminological applications of DNA profiling have received much legal and commercial interest. Plant genotyping for agricultural, legal and ecological purposes is still in its infancy.

DAF application in positional cloning

DAF markers were shown to be repeatable polymorphisms in different DNA isolations, operators, time periods and amplifications. They are heritable, as about 75% of the differences between *G. max* and *G. soja* segregated as dominant Mendelian markers in F_2 populations (Prabhu and Gresshoff, 1994; Caetano-Anollés *et al.*, 1993). Interestingly, the other 25% segregated in a uniparental way, being either maternal or paternal. Maternal inheritance presumably stems from amplification of cytoplasmic replicons. As yet, paternal replication is unexplained, and may represent either highly repeated chromosomal replicons or possibly alterations from normal cytoplasmic inheritance patterns in soyabean.

DAF markers are useful in defining closely linked regions in bulked segregant analysis (Michelmore *et al.*, 1991). Preliminary studies by Blauenfeldt and Gresshoff (1993) and Abbit and Caetano-Anollés (personal communication) used F_2 bulks of either nodulating or non-nodulating plants derived from a cross between *G. soja* (wild type) and *G. max* (non-nodulating, *nod49* allele). Nearly 100 DAF primers were screened to detect amplification length polymorphisms between the pooled DNA samples. Several were found that repeatedly give band differences between the Nod$^+$ and the Nod$^-$ pools. Segregation analysis on large-scale individual plant DNA samples is in progress as is cloning of the diagnostic band.

Induced plant mutations have the advantage that they are in near-isogenic background as the genetic difference between parent and mutant in minimal. Using 25 DAF primers Caetano-Anollés *et al.* (1993) showed that the induced supernodulation mutant nts382 and its wild-type parent cv. 'Bragg' did not show polymorphisms despite the pairwise comparison of nearly 500 amplification products. Only by the use of a new tec-MAAP method, in which the target DNA was predigested with two restriction nucleases (four base cutters) and then amplified with a single octamer, could polymorphisms be detected between mutant and wild-type parent. Only 19 primers were needed to reveal 42 putative AFLPs. Fourteen of these segregated at 100% with the supernodulation phenotype in *G. soja* (wild type) and *G. max* (mutant) derived F_2 populations. Some polymorphisms appeared to distinguish between the *nts382* and *nts1007* alleles. It is likely that these are valuable markers close to the *nts* locus and their cloning and further characterization will facilitate the isolation and ordering of yeast artificial chromosomes in that region.

5.6 Summary

These findings show that DAF analysis of plant genomes adds a further genetic tool to construct the high density maps needed for positional cloning and marker-based breeding approaches. Taken together, DAF and RFLP analysis have provided several confirmed or putative DNA markers closely linked to *nts*. Many of these need to be converted into molecular probes to permit the isolation of large DNA fragments homologous to them. Genetic mapping and ordering will go hand-in-hand with the physical analysis of the region.

The detailed molecular knowledge about plant genomes using a variety of techniques is generating new approaches to tracing specific genes in plant breeding and agriculture. Marker-assisted breeding and plant variety diagnosis, as well as pathogen typing and detection, may be the natural overlapping areas of classical plant genetics and molecular analysis of plant genomes.

Acknowledgements

Research was supported by the United Soybean Board, the Human Frontiers Science Programme, the Tennessee Soybean Promotion Board, CSRS project TEN0174 and the Ivan Racheff Endowment. The help of the Racheff staff and the supply of unpublished data is acknowledged.

References

Akao, S. and Kouchi, H. (1992) A supernodulating mutant isolated from soybean cultivar Enrei. *Soil Science and Plant Nutrition* 38, 183–187.

Bassam, B.J., Caetano-Anollés, G. and Gresshoff, P.M. (1991) A fast and sensitive silver-staining for DNA in polyacrylamide gels. *Analytical Biochemistry* 196, 80–83.

Bassam, B.J., Caetano-Anollés, G. and Gresshoff, P.M. (1992a) DNA amplification fingerprinting and its potential for genome analysis. In: Gresshoff, P.M. (ed.), *Plant Biotechnology and Development. Current Topics in Plant Molecular Biology*, vol. 1. CRC press, Boca Raton, FL, pp. 1–9.

Bassam, B.J., Caetano-Anollés, G. and Gresshoff, P.M. (1992b) DNA amplification fingerprinting of bacteria. *Applied Microbiology and Biotechnology* 38, 70–76.

Baum, T.J., Gresshoff, P.M., Lewis, S.A. and Dean, R.A. (1994) Characterization and phylogenetic analysis of four root-knot nematode species using DNA amplification fingerprinting and automated polyacrylamide gel electrophoresis. *Molecular Plant-Microbe Interactions* 7, 39–47.

Blackhall, N.W., Hammatt, N. and Davey, M.R. (1991) Analysis of variation in the DNA content of *Glycine* species: a flow cytometric study. *Soybean Genetics Newsletter* 18, 194–200.

Blauenfeldt, J. and Gresshoff, P.M. (1993) DAF analysis of non-nodulation in soybean. abstract, 4th Gatlinburg Symposium. *Plant Genome Analysis*. p. 4.

Bond, J., McDonnell, R., Finer, J. and Gresshoff, P.M. (1992) Construction and use of a low cost micro-projectile gene gun for gene transfer in plants. *Tennessee Farm and Home Science* 162, 4–14.

Buzzell, R.I., Buttery, B.R. and Ablett, G. (1990) In: Gresshoff, P.M., Roth, L.E. Stacey, G. and Newton, W.E. (eds), *Nitrogen Fixation: Achievements and Objectives*. Routledge, Chapman & Hall, New York, NY, p. 726.

Caetano-Anollés, G. (1993) Amplifying DNA with arbitrary oligonucleotide primers. *PCR Methods and Applications* 3, 85–94.

Caetano-Anollés, G. and Gresshoff, P.M. (1991) Plant genetic control of nodulation. *Annual Review of Microbiology* 45, 345–382.

Caetano-Anollés, G. and Gresshoff, P.M. (1993) DNA amplification fingerprinting: a general tool with applications in breeding, identification and phylogenetic analysis of plants. In: Schierwater, B., Streit, B., Wagner, G.P. and DeSalle, R. (eds), *Molecular Ecology and Evolution: Approaches and Applications*. Birkhauser Verlag, Frankfurt, pp. 17–31.

Caetano-Anollés, G. and Gresshoff, P.M. (1994) DNA amplification fingerprinting using arbitrary mini-hairpin oligonucleotide primers. *Bio/Technology* 12, 619–623.

Caetano-Anollés, G., Bassam, B.J. and Gresshoff, P.M. (1991a) DNA amplification fingerprinting using very short arbitrary oligonucleotide primers. *Bio/Technology* 9, 553–557.

Caetano-Anollés, G., Bassam, B.J. and Gresshoff, P.M. (1991b) DNA amplification fingerprinting: a strategy for genome analysis. *Plant Molecular Biology Reporter* 9, 292–305.

Caetano-Anollés, G., Bassam, B.J. and Gresshoff, P.M. (1992a) DNA amplification fingerprinting with very short primers. In: Neff, M. (ed.), *Application of RAPD Technology to Plant Breeding*. ASHS, St Paul, MN, pp. 18–25.

Caetano-Anollés, G., Bassam, B.J. and Gresshoff, P.M. (1992b) Printer-template interactions during *in vitro* amplification with short oligonucleotides. *Molecular and General Genetics* 235, 157–165.

Caetano-Anollés, G., Bassam, B.J. and Gresshoff, P.M. (1992c) DNA fingerprinting: MAAPing out a RAPD redefinition. *Bio/Technology* 10, 937.

Caetano-Anollés, G., Bassam, B.J. and Gresshoff, P.M. (1993) Enhanced detection of polymorphic DNA by multiple arbitrary amplicon profiling of endonuclease digested DNA: identification of markers linked to the supernodulation locus of soybean. *Molecular and General Genetics* 241, 57–64.

Caetano-Anollés, G., Bassam, B.J. and Gresshoff, P.M. (1994) Multiple arbitrary amplicon profiling using short oligonucleotide primers. In: Gresshoff, P.M. (ed.), *Plant Genome Analysis. Current topics in Plant Molecular Biology*, vol. 3, chapter 4. CRC Press, Boca Raton, FL.

Callahan, L.M., Caetano-Anollés, G., Bassam, B.J., Weaver, K., MacKenzie, A. and Gresshoff, P.M. (1993) DNA fingerprinting of turfgrass. *Golf Course Management*, June, 80–86.

Carroll, B.J., McNeil, D.L. and Gresshoff, P.M. (1985a) A supernodulation and nitrate tolerant symbiotic (*nts*) soybean mutant. *Plant Physiology* 78, 34–40.

Carroll, B.J., McNeil, D.L. and Gresshoff, P.M. (1985b) Isolation and properties of soybean mutants which nodulate in the presence of high nitrate concentrations. *Proceedings of the National Academy of Sciences, USA* 82, 4162–4166.

Carroll, B.J., McNeil, D.L. and Gresshoff, P.M. (1986) Mutagenesis of soybean (*Glycine max* (L.) Merr.) and the isolation of non-nodulating mutants. *Plant Science* 47, 109–114.

Collins, G.G. and Symons, R.H. (1993) Polymorphisms in grapevine DNA detected by the RAPD PCR technique. *Plant Molecular Biology Reporter* 11, 105–112.

Delves, A.C., Carter, A., Carroll, B.J., Mathews, A. and Gresshoff, P.M. (1986) Regulation of the soybean–*Rhizobium* symbiosis by shoot and root factors. *Plant Physiology* 82, 588–590.

Delves, A.C., Carroll, B.J., and Gresshoff, P.M. (1988) Genetic analysis and complementation studies on a number of mutant supernodulating soybean lines. *Journal of Genetics* 67, 1–8.

Duc, G. and Messager, A. (1989) Mutagenesis of pea (*Pisum sativum* L.) and the isolation of mutants for nodulation and nitrogen fixation. *Plant Science* 60, 207–213.

Engvild, K.C. (1987) Nodulation and nitrogen fixation mutants of pea, *Pisum sativum. Theoretical and Applied Genetics* 74, 711–713.

Funke, R. and Gresshoff, P.M. (1992) Soybean genome analysis using pulsed field gel electrophoresis. In: Gresshoff, P.M. (ed.), *Plant Biotechnology and Development. Current Topics in Plant Molecular Biology*, vol. 1. CRC Press, Boca Raton, FL, pp. 111–116.

Funke, R.P. and Kolchinsky, A. (1994) Plant yeast artificial chromosome libraries and their use: status and some strategic considerations. In: Gresshoff, P.M. (ed.), *Plant Genome Analysis. Current Topics in Plant Molecular Biology*, vol. 3, Chapter 11. CRC Press, Boca Raton, FL, USA.

Funke, R. and Kolchinsky, A. and Gresshoff, P.M. (1993) Physical mapping of a region in the soybean (*Glycine max*) genome containing duplicated sequences. *Plant Molecular Biology* 22, 437–446.

Funke, R.P., Kolchinsky, A. and Gresshoff, P.M. (1994) High EDTA concentrations cause entrapment of small DNA molecules in the compression zone of pulsed field gels, resulting in smaller than expected insert sizes in YACs prepared from size selected DNA. *Nucleic Acids Research* 22, 2708 2709.

Gremaud, M.F. and Harper, J.E. (1989) Selection and initial characterization of partially nitrate tolerant nodulation mutants of soybean. *Plant Physiology* 89, 169–173.

Gresshoff, P.M. (1992) DNA fingerprinting brings high-tech genetics into commercial greenhouses. *Grower Talks*, July, 119–127.

Gresshoff, P.M. (1993a) Molecular genetic analysis of nodulation genes in soybean. *Plant Breeding Reviews* 11, 275–318.

Gresshoff, P.M. (1993b) Genome analysis of soybean; coupling structure and function. *Plant Genetics Newsletter* 10, 2–9.

Gresshoff, P.M. and MacKenzie, A. (1994) Low experimental variability of DNA profiles generated by arbitrary primer based amplification (DAF) of soybean. *Chinese Journal of Botany* 6, 1–6.

Gresshoff, P.M., Sayavedra-Soto, L.A., Landau-Ellis, D.R., Caetano-Anollés, G., Culpepper, J.H., Angermüller, S.A. and Bassam, B.J. (1990) DNA fingerprinting – the powerful new tool for the genetic detective. *Tennessee Farm and Home Science* 155, 4–10.

Gurley, W.B., Hepburn, A.G. and Key, J.L. (1979) Sequence organization of the soybean genome. *Biochimica et Biophysica Acta* 561, 167–183.

He, S., Ohm, H. and McKenzie, S. (1992) Detection of DNA sequence polymorphisms among wheat varieties. *Theoretical and Applied Genetics* 84, 573–578.

Jacobsen, E. and Feenstra, W.J. (1984) A new pea mutant with efficient nodulation in the presence of nitrate. *Plant Science Letters* 33, 337–344.

Jayarao, B.M., Bassam, B.J., Caetano-Anollés, G., Gresshoff, P.M. and Oliver, S.P. (1992) Subtyping of *Streptococcus uberis* by DNA amplification fingerprinting. *Journal of Clinical Microbiology* 30, 1347–1350.

Keim, P., Schoemaker, R.C. and Palmer, R.G. (1989) Restriction fragment length polymorphism diversity in soybean. *Theoretical and Applied Genetics* 77, 786–792.

Keim, P., Diers, B.W., Olson, T.C. and Schoemaker, R.C. (1990) RFLP mapping in soybean: Association between molecular markers and variation in quantitative traits. *Genetics* 126, 735–742.

Kneen, B.E. and LaRue, T.A. (1984) Nodulation resistant mutant of *Pisum sativum* (L.). *Journal of Heredity* 75, 238–240.

Kolchinsky, A. and Gresshoff, P.M. (1995) A major satellite DNA of soybean is a

124 P.M. Gresshoff

92-base pairs tandem repeat. *Theoretical and Applied Genetics* 5, 621–626.

Kolchinsky, A., Funke, R.P. and Gresshoff, P.M. (1993). DAF-amplified fragments can be used as markers for DNA from pulse field gels. *BioTechniques* 14, 400–403.

Kolchinsky, A., Landau-Ellis, D. and Gresshoff, P.M. (1995) Molecular analysis of a polymorphic DNA region tightly linked to the supernodulation (*nts*) locus of soybean. *Molecular and General Genetics* (in press).

Landau-Ellis, D. and Gresshoff, P.M. (1992) Supernodulating soybean mutant alleles *nts382* and *nts1007* show no recombination with the same RFLP marker supporting complementation data. *Molecular Plant-Microbe Interactions* 5,428–429.

Landau-Ellis, D. and Gresshoff, P.M. (1994) The RFLP molecular marker closely linked to the supernodulation locus of soybean contains three inserts. *Molecular Plant–Microbe Interactions* 7, 432–433.

Landau-Ellis, D., Angermüller, S.A., Shoemaker, R. and Gresshoff, P.M. (1991) The genetic locus controlling supernodulation co-segregates tightly with a cloned molecular marker. *Molecular and General Genetics* 228, 221–226.

Lark, K.G., Wisemann, J.M., Mathews, B.F., Palmer, R., Chase, K. and Macalma, T. (1993) A genetic map of soybean (*Glycine max* L.) using an interspecific cross of two cultivars Minsoy and Noir 1. *Theoretical and Applied Genetics* 86, 901–906.

Marcker, A., Lund, M., Jensen, E.Ø. and Marcker, K.a. (1984) Transcription of the soybean leghemoglobin genes during nodule development. *EMBO Journal* 3, 1691–1695.

Mathews, A., Carroll, B.J. and Gresshoff, P.M. (1987) Characterisation of non-nodulation mutants of soybean (Glycine max (L) Merr.): Bradyrhizobium effects and absence of root hair curling. *Journal of Plant Physiology* 131, 349–361.

Mathews, A., Carroll, B.J. and Gresshoff, P.M. (1989a) Biological characterization of root exudates and extracts from nonnodulating and supernodulating soybean mutants. *Protoplasma* 150, 40–47.

Mathews, A., Carroll, B.J. and Gresshoff, P.M. (1989b) A new recessive gene conditioning non-nodulation in soybean. *Journal of Heredity* 80, 357–360.

Mathews, A., Carroll, B.J. and Gresshoff, P.M. (1990) The genetic interaction between nonnodulation and supernodulation in soybean: an example of developmental epistasis. *Theoretical and Applied Genetics* 79, 125–130.

Mathews, A., Carroll, B.J. and Gresshoff, P.M. (1992) Studies on the root control of non-nodulation and plant growth of non-nodulating mutants and a super-nodulating mutant of soybean (*Glycine max* (L.) Merr.). *Plant Science* 83, 35–43.

Michelmore, R.W., Paran, I. and Kesseli, R.V. (1991) Identification of markers linked to disease resistance genes by bulked segregant analysis: a rapid method to detect markers in specific genomic regions using segregating populations. *Proceedings of the National Academy of Sciences, USA* 88, 9828–9832.

Mullis, K.B. (1991) The polymerase chain reaction in an anemic mode: how to avoid cold oligodioxyribonuclear fusions. *PCR Methods and Applications* 1, 1–4.

Nutman, P.S. (1952) Studies on the physiology of nodule formation. III. Experiments on the excision of root-tip and nodules. *Annals of Botany N.S.* 16, 81–102.

Paran, I. and Michelmore, R.W. (1993) Development of reliable PCR markers linked to downy mildew resistance genes in lettuce. *Theoretical and Applied Genetics* 85, 985–993.

Park, S.J. and Buttery, B.R. (1988) Nodulation mutants of white bean (*Phaseolus vulgaris* L.) induced by ethyl-methane sulphonate. *Canadian Journal of Plant Science* 68, 199–202.

Prabhu, R.R. and Gresshoff, P.M. (1994) Inheritance of polymorphic markers generated by short single oligonucleotides using DNA amplification fingerprinting in soybean. *Plant Molecular Biology* 26, 105–116.

Pracht, J.E., Nickell, C.D. and Harper, J.E. (1993) Rj_5 and Rj_6 genes controlling nodulation in soybean. *Crop Science* 33, 711–713.

Rafalski, J.A., Hanafey, M.K., Tingey, S.V. and Williams, J.G.K. (1994) Technology for molecular breeding: RAPD markers, microsatellites and machines. In: Gresshoff, P.M. (ed.), *Plant Genome Analysis. Current Topics in Plant Molecular Biology*, vol. 3, chapter 3. CRC Press, Boca Raton, FL, USA.

Sanchez, F., Padilla, J.E., Pérez, H. and Lara, M. (1991) Control of nodulin genes in root-nodule development and metabolism. *Annual Review of Plant Physiology and Plant Molecular Biology* 42, 507–528.

Shoemaker, R.C., Guffy, R.D., Lorenzen, L. and Specht, J.E. (1992) Molecular genetic mapping of soybean: map utilization. *Crop Science* 32, 1091–1098.

Shoemaker, R.C., Lorenzen, L.L., Diers, B.W. and Olson, T.C. (1994) Genome mapping and agriculture. In: Gresshoff, P.M. (ed.), *Plant Genome Analysis. Current Topics in Plant Molecular Biology*, vol. 3, chapter 1. CRC Press, Boca Raton, FL, pp. 1–10.

Singh, R.J. and Hymowitz, T. (1988) The genomic relationship between *Glycine max* (L.) Merr. and *Glycine soja* (Sieb and Zucc.) as revealed by pachytene chromosomes analysis. *Theoretical and Applied Genetics* 76, 705–711.

Stacey, G., Burris, R.H. and Evans, H.J. (1992) *Biological Nitrogen Fixation.* Chapman & Hall, New York, NY.

Verma, D.P.S. (1992) Signals in root nodule organogenesis and endocytosis of *Rhizobium. Plant Cell* 4, 373–382.

Weaver, K., Caetano-Anollés, G., Gresshoff, P.M. and Callahan, L.M. (1994) Isolation and cloning of DNA amplification products from silver stained polyacrylamide gels. *BioTechniques* 16, 226–227.

Welsh, J. and McClelland, M. (1990) Fingerprinting genomes using PCR with arbitrary primers. *Nucleic Acids Research* 18, 7213–7218.

Wicking, C. and Williamson, B. (1991) From linked marker to gene. *Trends in Genetics* 7, 288–290.

Williams, J.G.K., Kubelik, A.R., Livak, K.J., Rafalski, J.A. and Tingey, S.V. (1990) DNA polymorphisms amplified by arbitrary primers are useful genetic markers. *Nucleic Acids Research* 18, 6531–6535.

Nucleic Acid Techniques in Testing for Seedborne Diseases

J.C. Reeves

Molecular Biology and Diagnostics Section, National Institute of Agricultural Botany, Huntingdon Road, Cambridge CB3 0LE, UK.

6.1 Introduction

The importance of seed as a vector of a range of diseases is well established (Neergaard, 1977; Agarwal and Sinclair, 1987). Pathogens can be carried on, in and with seed and disease can begin when the host crop is sown. Subsequent disease development is dependent on a number of factors and transmission rates from the seed to the crop can vary according to the particular pathogen involved. Nevertheless, even where transmission rates are low, there can be sufficient infected seed in commercial seedlots for epidemics to begin (Hewlett, 1983). Because of this and with the growth of commercial seed trading, particularly with the lifting of trade barriers, there is potential for the proliferation of seedborne pathogens within and between regions, countries and continents.

The spread of seedborne diseases can be restricted in a number of ways. Many, but not all, fungal pathogens can be adequately controlled with modern seed treatments, many of which are systemic and can kill spores or mycelium carried beneath the seedcoat. However, seed treatments can be expensive and often ineffective against bacteria, viruses and some fungi so that other control strategies are required. Often these strategies include seed health testing as part of seed certification schemes, either statutory or voluntary within a federal or nation state, or as part of quarantine control imposed at national borders. Seed certification seeks to ensure seed quality through field inspections of seed crops and seed testing. Seed stocks which fail to meet quality criteria are not certified and are thereby made unavailable for use as seed by their removal from the seed multiplication cycle. Quarantine laws restrict the entry of infected seed across national borders where a particular phytopathogen is known not to occur. In these ways seedborne diseases can be limited through controls over the movement and use of infected seed.

6.2 Seed Health Testing

Field inspections of seed crops for the presence of diseases which could be transmitted on the seed can be effective in the control of certain diseases, for example loose smut in barley (Reeves and Wray, 1994). However, there are disadvantages: the disease must show unambiguous field symptoms not masked by other factors such as the presence of another disease, perhaps non-seedborne, for a correct diagnosis to be made. These symptoms must also be manifest at the time of the inspection. The fact that no disease is identified during a field inspection cannot be taken to indicate freedom from infection on the seed and the most effective way to establish the disease status of a seedlot is to conduct an appropriate seed health test.

Seed health tests seek:

1. to establish the presence of a pathogen on the seed;
2. to make an estimate of the extent of this infection within the seedlot.

This second objective is not always required if the pathogen is known to cause epidemics from very low levels of inoculum which is often the case with phytopathogenic bacteria. Additionally, with some pathogens there are insufficient epidemiological data to fix tolerance levels of infection on the seed.

The general requirements of a seed health test depend in part on the purpose for which the test is being conducted. Apart from seed certification and quarantine, seed tests can be done to establish the commercial value of seed, to help to make rational seed treatment decisions, to investigate reasons for poor germination often caused by seedborne fungi, for example *Fusarium* sp. in wheat, and for research. In routine seed testing there is a requirement for speed when large numbers of samples are being processed before certification and entry into commerce. This is often the case with crops sown before winter in European areas and places considerable demands on the nature of the test employed. As well as a need for speed, a test should be relatively simple with few stages to minimize handling and the possibility of error. A test should be reliable giving repeatable, easily interpreted results, taking into account sampling variation, and should be independent of the individual operator. In addition, the cost of the test should not prejudice the profitability of the seed production enterprise thus affecting the availability of suitably priced seed for the farmer.

6.3 Current Test Methods

The technology used in large-scale seed health testing at present is well established and some seed health examinations can be complex and others very simple. An example of a superficially simple method is the growing on test where the seed for test is germinated and the disease in the seedlings is assessed. This relies on testing a large number of seeds which have to be viable and which do not become infected during the test procedure. This test is not very sensitive, requires that disease symptoms are unambiguous and is slow and unsuited to large-scale testing. Most fungal pathogens are detected in seed by variations on a direct plating procedure whereby seeds are surface sterilized and plated on to the surface of an agar plate or incubated under moist conditions on paper blotters. The growth of the fungus out of the seed allows its visual identification by observation of its mycelium or the morphology of its reproductive structures. This kind of technique can also be used for bacteria and often provides a presumptive identification for subsequent verification by host inoculation or other tests (Schaad and Kendrick, 1975; Lelliot and Stead, 1987). Host inoculation is often used as the final confirmatory stage of seedborne disease testing, for example in the test for *Septoria apiicola* in celery seeds (Maude, 1963; Hewett, 1968) but it can also be used as a test method in its own right. An indicator plant can be sprayed with a seed extract in conjunction with an abrasive compound to help the disease organism gain entry to the plant or the extract can be directly injected. An example of this is the test for lettuce mosaic virus (Marrou and Messiaen, 1967).

Most of the recent advances in seed health testing have been in the improvement of semi-selective media to help with the isolation of the target pathogen and with the introduction of immunodetection techniques. The latter involve the use of polyclonal antisera or, more recently, monoclonal antibodies to detect phytopathogens in seed extracts or as pure cultures isolated in some way from seed. Despite some major disadvantages arising from their lack of specificity for some organisms, the use of variants of the ELISA technique and immunofluorescence microscopy has become widespread in routine seed testing (van Vuurde and Maat, 1983, 1985; van Vuurde *et al.*, 1983).

Many tests for seedborne pathogens (particularly bacteria) can be divided into three stages: an extraction stage where the organism is separated from the seed; an isolation stage where the organism is obtained in pure culture; an identification stage where a final diagnosis is made. In practice it is possible for two of these stages to be combined, for example an isolation stage can be omitted and an identification attempted directly in a seed extract, or identification can take place at the same time as isolation. Examples of both these cases have been given earlier for viral tests and for fungal direct plating tests respectively.

There are a number of general disadvantages with these standard tests for some, but not necessarily all, seedborne pathogens. Firstly, they can be slow since they often rely on an incubation period for the organism to grow and may also require development of symptoms in a host plant. A test can take anywhere between a few days and up to 2 weeks or more. Secondly, accuracy of diagnosis can be a problem if the pathogen is difficult to distinguish visually from other organisms. Thirdly, sensitivity of detection may be prejudiced if the organism is present only at a low level or the background microflora of the seed is high or grows more aggressively on the test medium. For bacteria the development of semi-selective media has helped with this latter problem but for most tests for seedborne fungi it still remains a difficulty. Because of these disadvantages alternative approaches to the testing of seed for infection by various pathogens have been taken.

6.4 Nucleic Acid Techniques

6.4.1 Nucleic acid hybridization probes

A DNA probe is a section of single-stranded DNA having a sequence homologous to a specific portion of the genome of a particular organism. Under the appropriate conditions a probe can be made to hybridize to its homologous target DNA sequence in this organism and the presence of the hybrid can be detected through the incorporation of a label or reporter molecule on the probe. If the target sequence for the probe is unique then this reaction is the basis of a highly specific means of identification for the organism. DNA probes can be used both as means of identification and for detection of a particular target organism and their use with phytopathogenic bacteria has recently been reviewed (Rasmussen and Reeves, 1992).

Bacteria

The initial development of DNA probes for bacterial diagnostic purposes took place in the early 1980s with the use by Moseley *et al.* (1980) of DNA sequences specific for enterotoxigenic strains of *Escherichia coli* and much of their subsequent application has been in the medical and food microbiological fields. However, many of the detection and identification problems associated with plant pathogenic bacteria are similar to those in these other areas and consequently there has been much interest in the use of DNA probes in plant pathology. Unfortunately in the more specific field of seed pathology there has been relatively less development of these techniques and almost no reports of their use in routine seed testing programmes. Schaad *et al.* (1986) published work on the use of a cloned phaseolotoxin gene for the identification of the seedborne bacterium *Pseudomonas syringae* pv. *phaseoli-*

cola and in a subsequent report (Schaad *et al.*, 1989) referred to its utility in routine seed testing. In this study the pathogen was detected in naturally infected bean seeds by dot-blotting washings from isolation plates containing a semi-selective medium and hybridizing the blots with the probe. Results were variable when there were low numbers of colony-forming units on the plates and were in part dependent on the quantity of saprophytes co-isolated with the pathogen. It was also found to be important that sufficient time for the growth of the pathogen on the isolation plate was allowed so that enough target DNA was available for the probe to detect it.

Denny (1988) was able to differentiate between *P. syringae* pv. *tomato*, a seedborne pathogen, and *P. syringae* pv. *syringae*, a ubiquitous and less important phytopathogen, using a DNA probe. The objective of this work was to develop a test which could identify the target pathogen from leaf lesions rather than in seed. Unfortunately this probe also hybridized to other *P. syringae* pathovars although these were not usually found on tomato and so did not necessarily compromise the use of the probe. The probe was produced using random genomic clones but work by Cuppels *et al.* (1990), also on *P. syringae* pv. *tomato*, used probes derived from the chromosomal region controlling the production of the phytotoxin coronatine. In principle this is similar to the probe used by Schaad *et al.* (1989) which was also derived from a gene coding for the phytotoxic phaseolotoxin. Again Cuppels *et al.* (1990) did not attempt to develop a seed test but used their probe to screen field tomato leaves using colony hybridization.

Thompson *et al.* (1989) also developed DNA probes for the identification of a seedborne phytopathogenic bacterium, in this case *Clavibacter michiganense* subsp. *michiganense* the causal agent of bacterial canker in tomato. They did not attempt to develop a test to detect this pathogen in seed but again used dot-blot hybridization with plant sap to test their probes, presumably as a preliminary stage to further development. Johansen *et al.* (1989) found DNA fragments which allowed specific identification of *Clavibacter michiganense* subsp. *sepedonicus* which causes bacterial ringrot in potatoes and which can be spread on seed tubers. These workers also did not develop a routine test for this organism but showed that the sensitivity of detection could be affected both by the nature of the label used with the probe (they compared radioactive and non-radioactive labels and showed they could confer comparable sensitivities) and by the presence of contaminating DNA. A plasmid DNA probe for two closely related seedborne bacterial pathogens of beans, *Xanthomonas campestris* pv. *phaseoli* and *X. c. p.* var. *fuscans*, was developed by Gilbertson *et al.* (1989). These workers also did not detect these organisms directly in seed but concentrated on detection by colony hybridization in bean leaves and debris and by squash and dot-blotting of leaves. Nevertheless colony hybridization is a technique which could be incorporated into a seed test. In fact Rasmussen and Wulff (1990) evaluated the use of both colony hybridization and dot-blotting as part of an improved seed test, using their probe for *Pseudomonas syringae* pv. *pisi*, a seedborne bacterial pathogen of peas. They suggested that by using colony hybridization after the standard extraction and isolation stages, the time taken for a seed test could be reduced from about 2 weeks to 5 days. These workers indicated that the dot-blotting and probing of pure cultures obtained from the standard seed test could replace the final pathogenicity test as a confirmation of diagnosis. Schaad *et al.* (1989) also found that out of 173 strains closely or distantly related to *Pseudomonas syringae* pv. *phaseolicola* only one gave a positive test result when tested with the probe whereas all 34 strains of the target pathogen tested gave a positive result. Investigation of the false positive indicated that the isolate involved also produced phaseolotoxin and characteristic halo symptoms in beans.

The main ways that DNA probes could be used in a seed test for bacteria are either by dot-blotting and probing seed extracts of some kind, in a way analogous to dot-blotting plant extracts, or by dot-blotting and probing pure cultures obtained from isolation plates. Alternatively, these isolation plates could be used in a colony hybridization technique. Dot-blotting of seed extracts involves using a DNA probe overtly to detect the target pathogen directly whereas colony hybridization and the dot-blotting of pure cultures imply that some isolation of the target pathogen has taken place. Consequently, in principle the use of seed extracts offers a more efficient seed test although it would not necessarily provide a culture of the pathogen for any further testing required, for instance race typing. There is therefore an important distinction here between the use of a probe for detection, where both sensitivity and accuracy are required, and its use after an isolation stage has taken place where the probe is being used primarily for identification and sensitivity of detection is less critical because of the relative abundance of the target. The detection limit of dot-blotting is influenced by the presence of contaminating DNA from other microorganisms and by the amount of plant DNA present. This is a major disadvantage of DNA probes as part of a detection system (Schaad *et al.*, 1989; Johansen *et al.*, 1989). Although there are several ways to ameliorate the effects of this non-target material they are less than satisfactory and it is doubtful whether they would contribute to an efficient seed test. Another factor which can affect the sensitivity of DNA probes used as a means of detection is the method by which they are labelled. In early work ^{32}P was frequently used to label probes but this would not be suitable in routine testing because of the quantities needed and because, through radioactive decay, probes would continually need to be produced. Modern non-radioactive labelling methods have now largely overcome this difficulty.

It seems therefore that the most likely function which a DNA probe could usefully fulfil in seed testing is in the identification of colonies of pure cultures either on an isolation plate using colony hybridization or by dot-blotting cells lifted from single colonies from an isolation plate. Such a test format would therefore still require an extraction stage and an isolation stage.

Fungi

Despite the importance of seedborne fungal pathogens there appear to be fewer reports of their detection and identification in seed using DNA probes than for seedborne bacteria. Much of the work with nucleic acid techniques on fungal phytopathogens has concentrated on identifying DNA polymorphisms, relating these to pathotype characterization and the investigation of genetic diversity within fungal species and to their application in fungal taxonomy (Manicom *et al.*, 1987, 1990; Braithwaite and Manners, 1989; Panabières *et al.*, 1989). Although it is often acknowledged that this work may find an application in fungal identification it is not clear how this approach would be used in large-scale routine testing or how it could be suited to sensitive detection of the pathogen in seed or other plant tissue, particularly in the absence of disease symptoms.

Most of the work on detecting fungal phytopathogens using DNA probes has concentrated on establishing infection in plant material other than seeds or in soil. Early work by Rollo *et al.* (1987) identified a probe for *Phoma tracheiphila*, a pathogen of lemons. In this case the probe was used to detect the pathogen at different periods of time after inoculation of healthy lemon seedlings or in lignified adult tree branches. These workers found that the symptom independence of this method of detection was one of its most important advantages since diagnosis by visual inspection of over 100 branch samples detected infection in 29% but using the probe in a dot-blot format detected 84% infection.

There have been a number of other similar reports on the detection of various pathogens. Goodwin *et al.* (1990) developed a probe for *Phytophthora parasitica*, a pathogen of tomatoes, which they used in colony hybridization to detect the pathogen in crushed leaf disks and crushed fungal colonies. The probe was also used successfully to detect the pathogen in soil samples but attempts to quantify *P. parasitica* infection in this medium gave variable results. Tisserat *et al.* (1991) used a DNA probe to identify *Leptosphaeria korrae* in total DNA extracted from turfgrass roots, while Yao *et al.* (1991) were able to detect *Peronosclerospora sacchari* in maize. There have been others more recently reporting on the detection and identification of fungal phytopathogens (Bateman *et al.*, 1992; Marmeisse *et al.*, 1992; Thomas *et al.*, 1992; Sauer *et al.*, 1993). All this work has concentrated on pathogens which are difficult to detect or identify using traditional techniques but none involved seedborne fungal pathogens.

Viruses

Some of the earliest work on the use of DNA probes in plant pathology was on the detection of potato viruses and some of these viruses can be transmitted in tubers and also in true seed, for example potato spindle tuber viroid (PSTV). The testing of potato planting material is analogous to the testing of seed samples because it is routinely done on a large scale as part of the process of producing large quantities of virus-free material for planting and in breeding programmes. Owens and Diener (1981) showed that a cDNA probe could detect PSTV in sap from sprouts of infected tubers and in sap from the eyes and skin from between the eyes spotted on to a nitrocellulose membrane. Harris *et al.* (1986) and Harris and James (1987) also used this method in comparison with traditional techniques for screening seminal material for breeding and distribution to the UK seed potato industry and found it to be rapid and reliable. These early reports used radioactively labelled probes but Heling *et al.* (1988) used a proprietary non-radioactive labelling method to detect PSTV in crude nucleic acid extracts from potato and tomato plant tissue. Baulcombe *et al.* (1984) successfully used a similar technique for the detection of potato virus X (PVX) in crude sap samples and their view was that the use of the probe was as reliable and sensitive as ELISA but required less labour. This was confirmed for potato virus Y and potato leafroll virus in a report by Boulton *et al.* (1988) where cDNA probes were routinely used to detect these viruses in up to 5000 samples per year and by work in Russia (Nikolaeva *et al.*, 1990) on a range of potato viruses. Potato virus S has also been detected in crude potato extracts using non-radioactively labelled (biotinylated) cDNA or RNA probes (Eweida *et al.*, 1989).

Other plant viruses have been detected using nucleic acid hybridization techniques with cDNA, DNA and RNA probes depending on the virus type (Varveri *et al.*, 1987, 1988; Navot *et al.*, 1989). These generally use sap, squashes or crude extracts of the plant material (or insects where this is a vector) blotted on to a membrane or in a dot-blot format.

6.4.2 Polymerase chain reaction

Assuming that an extraction stage is unavoidable, as is usually the case in seed tests for bacteria, then for optimum efficiency detection should take place in the medium which has been used to extract the target bacterium from the seed. Any test that attempts detection directly *in planta* or in an extraction medium needs to be extremely sensitive to detect the low numbers of target cells that may be present when the inoculum level is low because these would not be multiplied through growth in an isolation stage. The polymerase chain reaction (PCR) (Saiki *et al.*, 1985) is a technique by which a specific DNA sequence is enzymatically amplified many times to a level at which it can be easily detected, either electrophoretically or by other means. Theoretically it requires very little template DNA from which to initiate the reaction and therefore could offer the necessary level of sensitivity to allow direct detection. Consequently PCR has attracted a great deal of attention both in research and in diagnostics of all kinds.

Bacteria

One of the earliest reports of the use of PCR to detect a seedborne phytopathogenic bacterium in seed was by Rasmussen and Wulff (1990) who were able to detect 10^2 cells of *Pseudomonas syringae* pv. *pisi* per millilitre of ground pea seed extract. These workers also immediately identified one of the problems of using PCR for direct detection which is the inhibition of the enzyme used in the reaction by compounds in the extract. To make the PCR reliable an eight-step nucleic acid extraction stage was required and although this inevitably makes the test procedure more cumbersome it was possible for a single person to do 20 extractions in 1.5 h. It was also found that detection was possible if pea seeds were soaked rather than ground and that there appeared to be no inhibition of the reaction for soak times of less than 6 h but complete inhibition after 16 h of soaking. This was confirmed by Reeves *et al.* (1994) (Table 6.1) and indicates that inhibitory compounds are released by long soak times. The PCR product was observed electrophoretically and by dot-blotting and probing with a probe internal to the amplified DNA sequence. False negative (2–4%) results were obtained, which could be overcome by using duplicate amplifications from each seed sample. False positives were also obtained and presumably arose from contamination problems for which PCR is notorious. A related seedborne phytopathogenic bacterium, *Pseudomonas syringae* pv. *phaseolicola* causing

Table 6.1. The results of PCR detection of *Pseudomonas syringae* pv. *pisi* in pea seed soak liquors sampled after 6 h and 24 h compared with the standard seed test.

Seed sample	PCR		Standard test
	6 h	24 h	
445	+	−	+
618	+	−	+
646	+	+	+
652	+	+	+
657	−	−	+
1009	−	−	−
1023	−	−	−
1026	−	−	−
1027	−	−	−
1028	−	−	−

halo blight of beans, has also been detected using PCR (Prosen *et al.*, 1991; Tourte and Manceau, 1991). Similar sensitivity of 10^2 cells per millilitre was reported by Tourte and Manceau (1991) with similar inhibition problems (personal communication). A recent report (Prosen *et al.*, 1993), also on the PCR detection of *P. syringae* pv. *phaseolicola*, did not find any indication of inhibition of the PCR until the seed extracts were concentrated 100-fold in a test based on amplification from extracted DNA rather than directly from seed extracts. Interestingly this work also showed that the presence of heterologous DNA improved amplification, possibly because this DNA sequestrated the inhibitors of Taq polymerase.

A test for the detection of *Erwinia stewartii* in maize seed has been developed (Blakemore *et al.*, 1992; Blakemore and Reeves, 1993). In this test a DNA fragment found to be specific for the pathogen was identified using random amplified polymorphic DNA (RAPDs) analysis. The fragment was cloned and sequenced, and specific primers were designed for PCR amplification. This test is being developed to use nested primers to increase

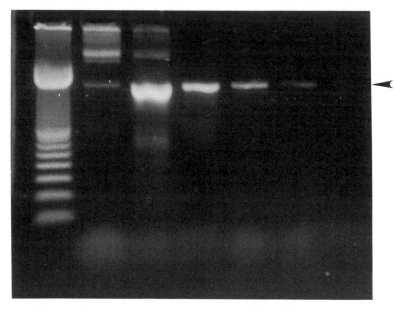

Fig. 6.1. A 1% agarose gel stained with ethidium bromide showing PCR-amplified *Erwinia stewartii* DNA. The amplifications are from a series of tenfold dilutions of template DNA from 500 ng (lane 2) down to 5 pg (lane 7). Lane 1 shows a standard set of DNA markers. Details of the PCR reaction have been described in Blakemore and Reeves (1993). The low recovery of product in lane 2, containing the highest amount of template DNA, may be explained by inhibition of the PCR reaction owing to excessive amounts of template DNA and/or associated impurities. The arrow indicates the PCR product that is of particular interest.

specificity and to allow colorimetric observation of the PCR product. Although this test is sensitive enough to detect 5 pg of bacterial DNA (Fig. 6.1), the presence of seed material also causes inhibition of the reaction.

The problem of inhibition of PCR used in direct detection in the presence

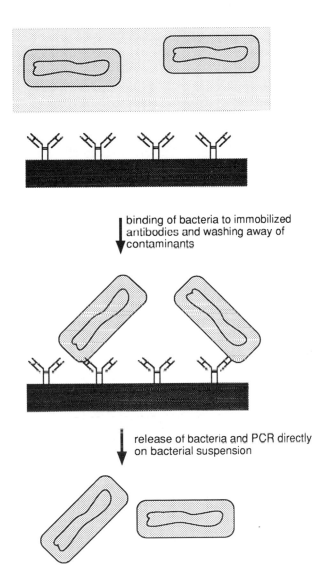

binding of bacteria to immobilized antibodies and washing away of contaminants

release of bacteria and PCR directly on bacterial suspension

Fig. 6.2 Summary of procedure to pre-purify bacteria by immunoabsorption prior to PCR. The PCR includes a 95°C step and this generally releases sufficient DNA from the cells to allow the reaction to proceed.

of contaminating biological material has also been noted in food microbiology (Rossen *et al.*, 1992) and these workers tested a large range of compounds for their inhibitory effects. In addition Hartung *et al.* (1993) found that in using PCR for the detection of *Xanthomonas campestris* pv. *citri*, the reaction buffer composition was critical to the success of the amplification and that sensitivity of detection was affected by the length of the denaturation step in the PCR cycle. Unlike the interference with hybridization techniques noted above it does not seem that PCR inhibition is primarily caused by the presence of non-homologous DNA since Rasmussen and Wulff (1991) found no effect on their method for detection of *Pseudomonas syringae* pv. *pisi* of the addition of up to 10^6 *Erwinia carotovora* cells.

PCR can be made to work in the presence of biological matrices. Bessessen *et al.* (1990) were able to detect *Listeria monocytogenes* in artificially infected milk and in cerebrospinal fluid and Koch *et al.* (1993) were able to detect *Vibrio cholerae* in a number of artificially infected foods. In other cases (Tsai and Olson, 1992; Tsai *et al.*, 1993) the template DNA was separated from inhibitory compounds using Sephadex G-200 spun columns which allowed the optimal PCR detection of *Escherichia coli* in sewage and sludge. An interesting technique which has been used to separate the target cells from PCR inhibitors is immunocapture (Fig. 6.2) in which the bacteria are specifically separated from matrix in which they occur using polyclonal antisera or monoclonal antibodies. This has been used successfully in conjunction with magnetic particles as a support for the antiserum or antibody which allows captured bacterial cells to be separated from the inhibitory matrix using magnetism before being subject to PCR (Widjo-joatmodjo *et al.*, 1992). This kind of assay has been called magnetic immuno PCR assay or MIPA.

Fungi

Although there has been much interest in PCR detection and character-ization of phytopathogenic fungi in general, the detection of seedborne fungal pathogens has received very little attention. Nevertheless some of the techniques used would be applicable in a seed test and some of the contexts in which PCR detection has been used are analogous to seed testing. In some cases PCR methods have been developed as refinements of already existing DNA probe-based tests. For example Rollo *et al.* (1990) used the probe they developed to detect *Phoma tracheiphila* in lemon trees (Rollo *et al.*, 1987) as a basis on which to improve the sensitivity and speed of detection by using PCR. They subcloned the probe and sequenced a fragment which they used to design primers for the PCR which repeatably gave a 102bp product from only and all *P. tracheiphila* isolates tested. This test was 5 days quicker than using the probe to detect the pathogen in lignified tissues. It is interesting that no inhibition of the PCR by the presence of lignin was reported but this is

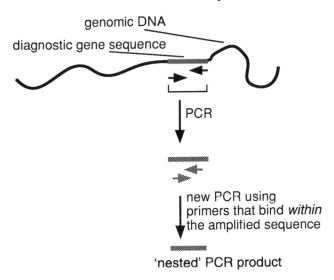

Fig. 6.3. Summary of the nested PCR technique that is used to eliminate non-specific amplification products that may arise in the first reaction.

probably because crude nucleic acid extractions were made from the lignified tissue. These extracts were then used as the templates for PCR and this simple purification may have prevented any inhibition.

Schesser *et al.* (1991) used PCR to detect *Gaeumannomyces graminis* in wheat plants but had to use nested PCR whereby a second round of amplification was done using primers internal to those used in the first round, in order to overcome non-specific amplification products from other fungi and also from uninfected wheat material. These products were not of the expected molecular weight but were removed, to avoid confusion, by the nested PCR technique (Fig. 6.3). Again the primer sets were designed from a DNA probe previously developed but which was unable to detect the pathogen directly in plant tissue. In further development of this work (Henson *et al.*, 1993) PCR was able to detect the pathogen in naturally and artificially infected host plants without prior extraction of DNA simply by boiling the tissues. However, there was an indication that this technique was not always successful with some naturally infected grasses, possibly because of the presence of inhibitors. Soil was also found to inhibit the PCR.

Work in the UK, on the same pathogen complex, by Ward and Gray (1992) used PCR to amplify a fragment of ribosomal DNA (rDNA) to take advantage of the highly variable sequences between conserved regions to which the primers anneal. This fragment was used as a probe in RFLP studies but these workers suggested that it could have some use in identification, its attraction being that probe production is very rapid. Using a related principle, Lee *et al.* (1993) were able to distinguish different *Phytophthora* species by hybridizing oligonucleotide probes to PCR-amplified internally transcribed spacer sequences (ITS) in the rDNA. Johanson and Jeger (1993) also used these ITS regions as a basis to design a PCR detection method for *Mycosphaerella fijiensis* and *M. musicola* in bananas.

Random amplified polymorphic DNA markers (RAPDs) were first reported by Welsh and McClelland (1990) and Williams *et al.* (1990). This variant of the PCR technique has been used by Goodwin and Annis (1991) to produce DNA profiles of *Leptosphaeria maculans* for genetic analysis and to divide isolates into groups of phenotypes correlating with pathogenicity. A similar approach has been adopted has been adopted by Guthrie *et al.* (1992) working on *Colletotrichum graminicola*, a seedborne disease of sorghum. Reeves and Ball (1991) also used RAPDs to group isolates of *Pyrenophora* spp., a seedborne barley pathogen, according to their pathogenicity. However, RAPDs are in themselves unlikely to be of any direct use in seedborne disease testing. This is because of the non-specific nature of the amplification produced where the template is contaminated with other DNA. RAPDs can contribute as a means of rapidly identifying DNA fragments which can be evaluated for their specificity as probes and if suitable can be sequenced to provide longer specific primers for conventional PCR. This technique was used by Blakemore *et al.* (1992) and Blakemore and Reeves (1993) in the development of a PCR detection method for *Erwinia stewartii* and is in progress for the detection and identification of members of the *Diaporthe/Phomopsis* complex in soyabean seeds (Jaccoud Filho and Reeves, 1993).

Viruses

With the accumulation of sequence data for viral genomes it would be surprising if there had been no attempts to detect viruses in plants and seeds using these data for PCR primer design. Most groups of plant viruses have RNA genomes which implies that before PCR can be done the RNA must be transcribed into cDNA using a reverse transcriptase enzyme and this step must be incorporated into the test method. This need not normally present any problems and Kohnen *et al.* (1992) were able to do this immediately prior to PCR in the same tube in their pathotype-specific detection method for pea seedborne mosaic virus (PSbMV). Viral RNA was detectable in whole nucleic acid extracts from seed parts (embryo axes), and other plant parts but not in seedcoats. This may be because of inhibition of the reactions by the presence of polyphenols derived from the seedcoat or because of the absence of functional RNA in this plant fraction. Reverse transcriptase PCR (RT-PCR) detection of cucumber mosaic virus has been achieved in lupin seed by Wylie *et al.* (1993) and again the reverse transcriptase step was done immediately before the PCR and in the same tube. As before there were indications of inhibition of the reactions since dilution of the lupin seed extract improved amplification, presumably because inhibitors were also diluted. This test was more sensitive than the existing ELISA test for this pathogen which was able to detect 1 infected seed in 100. RT-PCR with a rapid RNA extraction method detected 1 infected seed in 200 and with a more rigorous extraction detected 0.1% infection.

Care must be exercised with RT-PCR because it has been shown that reverse transcriptase can inhibit Taq polymerase activity and block PCR (Sellner *et al.*, 1992; Fehlmann *et al.*, 1993) although no indications of this appeared in the two examples above. In the case of Wylie *et al.* (1993) the reaction conditions were optimized using a range of reverse transcriptase concentrations.

6.5 Future Developments

6.5.1 Quantification

One of the major disadvantages of nucleic acid methods in seed health testing becomes evident when it is necessary to determine the level of infection in the seed sample. More specifically, this is the number of seeds in the test sample in which pathogen can be detected rather than the amount of inoculum carried on individual seeds. Unless individual seeds are examined, as they frequently are with traditional techniques, the level of infection is difficult to determine and nucleic acid methods could not realistically be used on each of the individual seeds comprising a credible seed health test.

Colony hybridization could be used to give an estimate of the number of colony-forming units (CFUs) per unit volume of seed extract plated on to the isolation plates, but relating this to the number of seeds infected in the sample tested is difficult unless the mean level of CFU/seed is known. Frequently these data are not available but Taylor (1970) showed for *Pseudomonas syringae* pv. *phaseolicola*, a seedborne disease of beans, that although there was considerable variation, over 80% of single infected seeds contained more than 10^5 bacteria. This work also showed that it was possible to determine the level of infection in seed by using the method of maximum likelihood or the most probable number (MPN) technique as it is alternatively known. This method involves a series of tests on samples containing decreasing numbers of seeds and the series of positive and negative results are interpreted using statistical tables (Swaroop, 1951), an approach which has recently been refined (Roberts *et al.*, 1993; Taylor *et al.*, 1993).

Methods of quantitative PCR have been developed which take into account the intrinsic unsuitability of this technique for quantitative studies. Because PCR is an amplification procedure the small variations in the reactions, to which the method is prone, could give rise to large differences in the yield of product, making estimates of the quantity of template material uncertain. Gilliland *et al.* (1990) devised a method whereby a competitive DNA template, effectively an internal control, is co-amplified with the target DNA from which it can be distinguished by size if additional DNA sequences are inserted in the competitor DNA, or by restriction analysis if an additional restriction site is present. The target DNA is co-amplified with a dilution series of the competitor DNA present in known concentrations and any variation in the PCR reaction is assumed to affect the amplification of both components equally. From the relative amounts of both target and competitor DNA after amplification, the starting concentration of the target DNA can be calculated. There have been other PCR methods developed to determine the concentration of the initial template DNA (Lundeberg *et al.*, 1991; Chevrier *et al.*, 1993) and Hu *et al.* (1993) used a method based on the use of internal control templates successfully to quantify *Verticillium* infection in alfalfa and sunflower plants. Although these methods can give an estimate of the amount of target DNA in the sample, the problem of relating this to numbers of infected seeds in the sample is the same as for the use of colony hybridization. If results were to be presented differently, perhaps as CFU/unit of seed or DNA/unit of seed, these could be interpreted using a calibration of these units against percentage infection of seed. Such calibrations would need to be produced for each host/pathogen species but would have to assume that the level of infection of each seed, i.e. its loading of pathogen, was constant. Perhaps the easiest approach would be to do multiple tests as part of a MPN analysis. The problem of providing quantitative data from seed tests based on nucleic acid technique will require further work in the future before the methods can be fully exploited.

6.5.2 Multiplex PCR

Often a seed sample is tested for more than one pathogen and using traditional techniques it may be possible to conduct these tests simultaneously. For example in testing linseed for seed certification in the UK, the seed is examined for up to five fungal pathogens as part of a single test. This is only possible if the test conditions are appropriate for all the organisms, otherwise separate tests are required as might be the case if testing for bacterial and fungal pathogens in the same seed sample. Using PCR, simultaneous testing for more than one organism may be feasible through the use of mixtures of primers specific for each of the pathogens in question. This technique is known as multiplex PCR and as yet its use is rare in diagnostics. However, in a recent report Cadieux *et al.* (1993) were able to detect the mollicutes *Mycoplasma pneumoniae* and *M. genitalium*, two related human pathogens, in a triplex PCR reaction in a single tube. It was found to be critically important to the success of the reaction to optimize the PCR conditions carefully and in particular the relative concentrations of the various primers.

This is because amplification of the larger fragments in the mixture is inhibited in some way by the presence of the smaller fragments or their primers. These data suggest that although at present there are no reports of the use of multiplex PCR for diagnostic purposes in plant pathology, the approach could have great potential for cost-effective seed testing.

6.5.3 Ligase chain reaction

The ligase chain reaction (LCR), like PCR, is a DNA amplification technique and is capable of high sensitivity and specificity (Lee, 1993). In this technique double stranded DNA is melted at high temperature and then cooled so that two oligonucleotides can hybridize to the single stranded DNA such that the 3' end of one oligonucleotide is immediately adjacent to the 5' end of the other. Using a thermostable DNA ligase the oligonucleotides are then ligated to form a product which acts as a template for subsequent cycles of the reaction. LCR is highly specific since even a single base mismatch at the junction of the oligonucleotides will prevent ligation and this feature has been used to detect human genetic disorders (Barany, 1991). There have also been a number of recent reports in which LCR has been used to detect clinically important organisms, particularly *Chlamydia tracomatis* (Dille *et al.*, 1993) and *Mycobacterium tuberculosis* (Iovannisci and Winn-Deen, 1993: Winn-Deen *et al.*, 1993). As yet there do not appear to be any reports of its use in plant pathology but because it is well suited to automation and multiplexing, LCR warrants further attention.

6.6 Conclusions

The potential of nucleic acid techniques for diagnostic work in plant and seed pathology is now well established and it seems likely that their use in routine seed testing programmes will increase. This use will remain limited in the short term by difficulties with inhibition of PCR and by the problems associated with quantification but as these are overcome the full benefits of the new test methods, especially with multiplexing and automation, may be realized. At present it is only possible to speculate whether or not the economic advantages of the new technology will match the technical benefits but it is probably that large-scale testing, maybe in some central facility, will be the best means of restricting unit costs.

Acknowledgements

The members of the Molecular Biology and Diagnostics Section at NIAB deserve thanks for their support and forbearance during the writing of this chapter and possibly at other times. Other friends and colleagues have also been a great encouragement.

References

Agarwal, V.K. and Sinclair, J.B. (1987) *Principles of Seed Pathology.* CRC Press, Boca Raton, FL, USA.

Barany, F. (1991) Genetic disease detection and DNA amplification using cloned thermostable ligase. *Proceedings of the National Academy of Sciences, USA* 88, 189–193.

Bateman, G.L., Ward, E. and Antoniw, J.F. (1992) Identification of *Gaeumannomyces graminis* var. *tritici* and *G. graminis* var. *avenae* using a DNA probe and non-molecular methods. *Mycological Research* 96, 737–742.

Baulcombe, D., Flavell, R.B., Boulton, R.E. and Jellis, G.J. (1984) The sensitivity and specificity of a rapid nucleic acid hybridization method for the detection of potato virus X in crude sap samples. *Plant Pathology* 33, 361–370.

Bessessen, M.T., Luo, Q., Rotbart, H.A., Blaser, M.J. and Ellison III, R.T. (1990) Detection of *Listeria monocytogenes* by using the polymerase chain reaction. *Applied and Environmental Microbiology* 56, 2930–2932.

Blakemore, E.J.A. and Reeves, J.C. (1993) PCR used in the development of a new seed health test to identify *Erwinia stewartii*, a bacterial pathogen of maize. *Proceedings of the 1st ISTA Plant Disease Committee Symposium on Seed Health Testing.* Ottawa, Canada, pp. 19–22.

Blakemore, E.J.A., Reeves, J.C. and Ball, S.F.L. (1992) Research note: polymerase chain reaction used in the development of a DNA probe to identify *Erwinia stewartii*, a bacterial pathogen of maize. *Seed Science and Technology* 20, 331–335.

Boulton, R.E., Jellis, G.J. and Squire, A.M. (1988) An improved method of evaluating resistance to potato virus Y and potato leafroll virus in a breeding programme. *Plant Varieties and Seeds* 1, 109–115.

Braithwaite, K.S. and Manners, J.M. (1989) Human hypervariable minisatellite probes detect DNA polymorphisms in the fungus *Colletotrichum gloeosporioides*. *Current Genetics* 16, 473–475.

Cadieux, N., Lebel, P. and Brousseau, R. (1993) Use of a triplex polymerase chain reaction for the detection and differentiation of *Mycoplasma pneumoniae* and *Mycoplasma genitalium* in the presence of human DNA. *Journal of General Microbiology* 139, 2431–2437.

Chevrier, D., Rasmussen, S.R. and Guesdon, J. (1993) PCR product quantification by non-radioactive hybridization procedures using an oligonucleotide covalently bound to microwells. *Molecular and Cellular Probes* 7, 187–197.

Cuppels, D.A., Moore, R.A. and Morris, V.L. (1990) Construction and use of a nonradioactive DNA hybridization probe for detection of *Pseudomonas syringae* pv. *Tomato* on tomato plants. *Applied and Environmental Microbiology* 56, 1743–1749.

Denny, T.P. (1988) Differentiation of *Pseudomonas syringae* pv. *tomato* from *P. s. syringae* with a DNA hybridization probe. *Phytopathology* 78, 1186–1193.

Dille, B.J., Butzen, C.C. and Birkenmeyer, L.G. (1993) Amplification of *Chlamydia trachomatis* DNA by ligase chain reaction. *Journal of Clinical Microbiology* 31, 729–731.

Eweida, M., Sit, T.L. and Abouhaidar, M.G. (1989) Molecular cloning of the genome of the carlavirus potato virus S: biotinylated RNA transcripts for virus detection in crude potato extracts. *Annals of Applied Biology* 115, 253–261.

Fehlmann, C., Krapf, R. and Solioz, M. (1993) Reverse transcriptase can block polymerase chain reaction. *Clinical Chemistry* 39, 368–369.

Gilbertson, R.L., Maxwell, D.P., Hagedorn, D.J. and Leong, S.A. (1989) Development and application of a plasmid DNA probe for detection of bacteria causing common bacterial blight of bean. *Phytopathology* 79, 518–525.

Gilliland, G., Perrin, S. and Bunn, H.F. (1990) Competitive PCR for quantitation of mRNA. In: Innis, M.A., Gelfand, D.H., Sninsky, J.J. and White, T.J. (eds), *PCR Protocols: A Guide to Methods and Applications*, chapter 8. Academic Press, San Diego, CA, USA.

Goodwin, P.H. and Annis, S.L. (1991) Rapid identification of genetic variation and pathotype of *Leptosphaeria maculans* by random amplified polymorphic DNA assay. *Applied and Environmental Microbiology* 57, 2482–2486.

Goodwin, P.H., English, J.T., Neher, D.A., Duniway, J.M. and Kirkpatrick, B.C. (1990) Detection of *Phytophthora parasitica* from soil and host tissue with a species-specific DNA probe. *Phytopathology* 80, 277–281.

Guthrie, P.A.I., Magill, C.W., Frederiksen, R.A. and Odvody, G.N. (1992) Random amplified polymorphic DNA markers: a system for identifying and differentiating isolates of *Colletotrichum graminicola*. *Phytopathology* 82, 832–835.

Harris, P.S. and James, C.M. (1987) Exclusion of viroids from potato resources and the modified use of a cDNA probe. *Bulletin OEPP/EPPO* Bulletin 17, 51–60.

Harris, P.S., James, C.M. and Kelly, P. (1986) Use of a cDNA probe for sensitive detection of potato spindle tuber viroid in potato quarantine. *Viroids of Plants and their Detection International Seminar*, 12–20 August, Warsaw Agricultural University.

Hartung, J.S., Daniel, J.F. and Pruvost, O.P. (1993) Detection of *Xanthomonas campestris* pv. *citri* by the polymerase chain reaction method. *Applied and Environmental Microbiology* 59, 1143–1148.

Heling, Z., Balbo, I. and Salazar, L.F. (1988) Nonradioactive detection of PSTV by nucleic acid hybridization with pAV 401 probe. *Acta Scientiarum Naturalium Universitatis Intramongolicae* 19, 515–521.

Henson, J.M., Goins, T., Grey, W., Mathre, D.E. and Eliot, M.L. (1993) Use of polymerase chain reaction to detect *Gaeumannomyces graminis* DNA in plants grown in artificially and naturally infested soil. *Phytopathology* 83, 283–287.

Hewett, P.D. (1968) Viable *Septoria* spp. in celery seed samples. *Annals of Applied Biology* 61, 89–98.

Hewett, P.D. (1983) Epidemiology – fundamental for disease and control. *Seed Science and Technology* 11, 697–706.

Hu, X., Nazar, R.N. and Robb, J. (1993) Quantification of *Verticillium* biomass in wilt disease development. *Physiological and Molecular Plant Pathology* 42, 23–36.

Iovannisci, D.M. and Winn-Deen, E.S. (1993) Ligation amplification and fluorescence detection of *Mycobacterium tuberculosis* DNA. *Molecular and Cellular Probes* 7, 35–43.

Jaccoud Filho, D.S. and Reeves, J.C. (1993) The detection and identification of species in the *Phomopsis/Diaporthe* complex from soya bean seed using PCR. *Proceedings of the 1st ISTA Plant Disease Committee Symposium on Seed Health Testing*, Ottawa, Canada, pp. 34–43.

Johansen, I.E., Ramussen, O.F. and Heide, M. (1989) Specific identification of *Clavibacter michiganense* subsp. *sepedonicum* by DNA-hybridization probes. *Phytopathology* 79, 1019–1023.

Johanson, A. and Jeger, M.J. (1993) Use of PCR for detection of *Mycosphaerella fijiensis* and *M. musicola*, the causal agents of Sigatoka leaf spots in banana and plantain. *Mycological Research* 97, 670–674.

Koch, W.H., Payne, W.L., Wenz, B.A. and Cebula, T.A. (1993) Rapid polymerase chain reaction method of detection of *Vibrio cholerae* in foods. *Applied and Environmental Microbiology*, 59, 556–560.

Kohnen, P.D., Doughterty, W.G. and Hampton, R.O. (1992) Detection of pea seedborne mosaic potyvirus by sequence specific enzymatic amplification. *Journal of Virological Methods* 37, 253–258.

Lee, H. (1993) Infectious disease testing by ligase chain reaction. *Clinical Chemistry* 39, 729–730.

Lee, S.B., White, T.J. and Taylor, J.W. (1993) Detection of *Phytophthora* species by oligonucleotide hybridization to amplified ribosomal DNA spacers. *Phytopathology* 83, 177–181.

Lelliot, R.A. and Stead, D.E. (1987) *Methods for the Diagnosis of Bacterial Diseases of Plants. Methods in Plant Pathology 2.* Blackwell Scientific Publications, Oxford, UK.

Lundeberg, J., Wahlberg, J. and Uhlén, M. (1991) Rapid colorimetric quantification of PCR-amplified DNA. *BioTechniques* 10, 68–75.

Manicom, B.Q., Bar-Joseph, M., Rosner, A., Vogodsky-Haas, H. and Kotzé, J.M. (1987) Potential applications of random DNA probes and restriction fragment length polymorphisms in the taxonomy of the fusaria. *Phytopathology* 77, 669–672.

Manicom, B.Q., Bar-Joseph, M. and Kotzé, J.M. (1990) Molecular methods of potential use in the identification and taxonomy of filamentous fungi, particularly *Fusarium oxysporum*. *Phytophylactica* 22, 233–239.

Marmeisse, R., Debaud, J.C. and Casselton, L.A. (1992) DNA probes for species and strain identification in the ectomycorrhizal fungus *Hebeloma*. *Mycological Research* 96, 161–165.

Marrou, J. and Messiaen, C.M. (1967) The *Chenopodium quinoa* test: a critical method for detecting seed transmission of lettuce mosaic virus. *Proceedings of the International Seed Testing Association* 32, 49–57.

Maude, R.B. (1963) Testing the viability of *Septoria* on celery seed. *Plant Pathology* 12, 15–17.

Moseley, S.L., Huq, L., Alim., A.R.M.A., So, M., Samadpour-Motalebi, M. and Falkow, S. (1980) Detection of enterotoxigenic *Escherichia coli* by DNA colony hybridization. *Journal of Infectious Diseases* 142, 892–895.

Navot, N., Ber, R. and Czosnek, H. (1989) Rapid detection of tomato yellow leaf curl virus in squashes of plants and insect vectors. *Phytopathology* 79, 562–568.

Neergaard, P. (1977) *Seed Pathology*. Macmillan Press, London.

Nikolaeva, O.V., Morozov, S.Yu., Zakhariev, V.M. and Skryabin, K.G. (1990)

Improved dot-blot hybridization assay for large-scale detection of potato viruses in crude potato tuber extracts. *Journal of Phytopathology* 129, 283–290.

Owens, R.A. and Diener, T.O. (1981) Sensitive and rapid diagnosis of potato spindle tuber viroid disease by nucleic acid hybridization. *Science* 213, 670–671.

Panabières, F., Marais, A., Trentin, F., Bonnet, P. and Rucci, P. (1989) Repetitive DNA polymorphism analysis as a tool for identifying *Phytophthora* species. *Phytopathology* 79, 1105–1109.

Prosen, D., Hatziloukas, E., Panopoulos, N.J. and Schaad, N.W. (1991) Direct detection of the halo blight pathogen *Pseudomonas syringae* pv. *phaseolicola* in bean seeds by DNA amplification. (Abstr.) *Phytopathology* 81, 1159.

Prosen, D., Hatziloukass, E., Schaad, N.W. and Panopoulos, N.J. (1993) Specific detection of *Pseudomonas syringae* pv. *phaseolicola* DNA in bean seed by polymerase chain reaction-based amplification of a phaseolotoxin gene region. *Phytopathology* 83, 965–970.

Rasmussen, O.F. and Reeves, J.C. (1992) DNA probes for the detection of plant pathogenic bacteria. *Journal of Biotechnology* 25, 203–220.

Rasmussen, O.F. and Wulff, B.S. (1990) Identification and use of DNA probes for plant pathogenic bacteria. In: Christiansen, C., Munck, L. and Villadsen, J. (eds), *Proceedings 5th European Congress on Biotechnology*. Munksgaard, Copenhagen, pp. 693–698.

Rasmussen, O.F. and Wulff, B.S. (1991) Detection of *Ps.* pv. *pisi* using PCR. In: Durbin, R.D., Surico, G. and Mugnai, L. (eds), *Proceedings 4th International Working Group on* Pseudomonas syringae *pathovars*. Stamperia Granducale, Florence, Italy, pp. 369–376.

Reeves, J.C. and Ball, S.F.L. (1991) Research note: preliminary results on the identification of *Pyrenophora* species using DNA polymorphisms amplified from arbitrary primers. *Plant Varieties and Seeds* 4, 185–189.

Reeves, J.C. and Wray, M.W. (1994) Seed testing, seed certification and seed treatment in the control of cereal seed-borne disease. *Seed Treatment – Progress and Prospects*. BCPC Monograph No. 57 (Martin, T. (ed.)), pp. 39–46.

Reeves, J.C., Rasmussen, O.F. and Simpkins, S.A. (1994) The use of a DNA probe and PCR for the detection of *Pseudomonas syringae* pv. *pisi* in pea seed. *Proceedings of the 8th ICPPB*. Les Colloques de INRA, pp. 383–390.

Roberts, S.J., Phelps, K., Taylor, J.D. and Ridout, M.S. (1993) Design and interpretation of seed health assays. *Proceedings of the 1st ISTA Plant Disease Committee Symposium on Seed Health Testing*. Ottawa, Canada, pp. 115–125.

Rollo, F., Salvi, R. and Torchia, P. (1990) Highly sensitive and fast detection of *Phoma tracheiphila* by polymerase chain reaction. *Applied Microbiology and Biotechnology* 32, 572–576.

Rollo, F., Amici, A., Foresi, F. and di Silvestro, I. (1987) Construction and characterization of a cloned probe for the detection of *Phoma tracheiphila* in plant tissues. *Applied Microbiology and Biotechnology* 26, 352–357.

Rossen, L., Norskov, P., Holmstrom, K. and Rasmussen, O.F. (1992) Inhibition of PCR by components of food samples, microbial diagnostic assays and DNA-extraction solutions. *International Journal of Food Microbiology* 17, 37–45.

Saiki, R.K., Scharf, S., Faloona, F., Mullis, K.B., Horn, G.T., Erloch, H.A. and Arnheim, N. (1985) Enzymatic amplification of β-globin genomic sequences and restriction site analysis for diagnosis of sicklecell anaemia. *Science* 230, 1350–1354.

Sauer, K.M., Hulbert, S.H. and Tisserat, N.A. (1993) Identification of *Ophiosphaerella herpotricha* by cloned DNA probes. *Phytopathology* 83, 97–102.

Schaad, N.W. and Kendrick, R.(1975) A qualitative method for detecting *Xanthomonas campestris* in crucifer seed. *Phytopathology* 65, 1034–1036.

Schaad, N.W., Azad, H., Peet, R.C. and Panopoulos, N.J. (1986) Cloned phaseolotoxin gene as a hybridization probe for identification of *Pseudomonas syringae* pv. *phaseolicola*. (abstr.) *Phytopathology* 76, 846.

Schaad, N.W., Azad, H., Peet, R.C. and Panopoulos, N.J. (1989) Identification of *Pseudomonas syringae* pv. *phaseolicola* by a DNA hybridization probe. (Abstr.) *Phytopathology* 79, 903–907.

Schesser, K., Luder, A. and Henson, J.M. (1991) Use of polymerase chain reaction to detect the take-all fungus, *Gaeumannomyces graminis*, in infected wheat plants. *Applied and Environmental Microbiology* 57, 553–556.

Sellner, L.N., Coelen, R.J. and Mackenzie, J.S. (1992) Reverse transcriptase inhibits Taq polymerase activity. *Nucleic Acids Research* 20, 1487–1490.

Swaroop, S. (1951) The range of variation of the most probable number of organisms estimated by the dilution method. *Indiana Journal of Medical Research* 39, 107–134.

Taylor, J.D. (1970) The quantitative estimation of the infection of bean seed with *Pseudomonas phaseolicola* (Burkh.) Dowson. *Annals of Applied Biology* 66, 29–36.

Taylor, J.D., Phelps, K. and Roberts, S.J. (1993) Most probable number (MPN) method: origin and application. *Proceedings of the 1st ISTA Plant Disease Committee Symposium on Seed Health Testing*. Ottawa, Canada, pp. 106–114.

Thomas, D., Maraite, H. and Boutry, M. (1992) Identification of rye- and wheat-types of *Pseudocercosporella herpotrichoides* with DNA probes. *Journal of General Microbiology* 138, 2305–2309.

Thompson, E., Leary, J.V. and Chun, W.W.C. (1989) Specific detection of *Clavibacter michiganense* subsp. *michiganense* by a homologous DNA probe. *Phytopathology* 79, 311–314.

Tisserat, N.A., Hulbert, S.H. and Nus, A. (1991) Identification of *Leptosphaeria korrae* by cloned DNA probes. *Phytopathology* 81, 917–921.

Tourte, C. and Manceau, C. (1991) Direct detection of *Pseudomonas syringae* pathovar *phaseolicola* using the polymerase chain reaction (PCR). In: Durbin, R.D., Surico, G. and Mugnai, L. (eds), *Proceedings of the 4th International Working Group on* Pseudomonas syringae *Pathovars*. Stamperia Granducale, Florence, Italy, pp. 377–379.

Tsai, Y. and Olson, B.H. (1992) Rapid method for separation of bacterial DNA from humic substances in sediments for polymerase chain reaction. *Applied and Environmental Microbiology* 58, 2292–2295.

Tsai, Y., Palmer, C.J. and Sangermano, L.R. (1993) Detection of *Escherichia coli* in sewage and sludge by polymerase chain reaction. *Applied and Environmental Microbiology* 59, 353–357.

van Vuurde, J.W.L. and Maat, D.Z. (1983) Routine application of ELISA for the detection of lettuce mosaic virus in lettuce seeds. *Seed Science and Technology* 11, 505–513.

van Vuurde, J.W.L. and Maat, D.Z. (1985) Enzyme-linked immunosorbent assay

(ELISA) and disperse dye immunoassay (DIA): comparison of simultaneous and separate incubation of sample and conjugate for the routine detection of lettuce mosaic virus and pea early browning virus. *Netherlands Journal of Plant Pathology* 91, 3–13.

van Vuurde, J.W.L., van den Bovenkamp, G.W. and Birnbaum, Y. (1983) Immunofluorescence microscopy and enzyme-linked immunosorbent assay as potential routine tests for the detection of *Pseudomonas syringae* pv. *phaseolicola* and *Xanthomonas campestris* pv. *phaseoli* in bean seed. *Seed Science and Technology* 11, 547–559.

Varveri, C., Ravelonandro, M. and Dunez, J. (1987) Construction and use of a cloned cDNA probe for the detection of plum pox virus in plants. *Phytopathology* 77, 1221–1224.

Varveri, C., Candresse, T., Cugusi, M., Ravelonandro, M. and Dunez, J. (1988) Use of a ^{32}P-labelled transcribed RNA probe for dot hybridization detection of plum pox virus. *Phytopathology* 78, 1280–1283.

Ward, E. and Gray, R.M. (1992) Generation of a ribosomal DNA probe by PCR and its use in identification of fungi within the *Gaeumannomyces–Phialophora* complex. *Plant Pathology* 41, 730–736.

Welsh, J. and McClelland, M. (1990) Fingerprinting genomes using PCR with arbitrary primers. *Nucleic Acids Research* 18, 7213–7218.

Widjojoatmodjo, M.N., Fluit, A.D.C., Torensma, R., Verdonk, G.P.H.T. and Verhoef, J. (1992) The magnetic immuno polymerase chain reaction assay for direct detection of Salmonellae in fecal samples. *Journal of Clinical Microbiology* 30, 3195–3199.

Williams, J.G.K., Kubelik, A.R., Livak, K.J., Rafalski, J.A. and Tingey, S.V. (1990) DNA polymorphisms amplified by arbitrary primers are useful as genetic markers. *Nucleic Acids Research* 18, 6531–6535.

Winn-Deen, E.S., Batt, C.A. and Wiedmann, M. (1993) Non-radioactive detection of *Mycobacterium tuberculosis* LCR products in a microtitre plate format. *Molecular and Cellular Probes* 7, 179–186.

Wylie, S., Wilson, C.R., Jones, R.A.C. and Jones, M.G.K. (1993) A polymerase chain reaction assay for cucumber mosaic virus in lupin seeds. *Australian Journal of Agricultural Research* 44, 41–51.

Yao, C., Magill, C.W., Frederiksen, R.A., Bonde, M.R., Wang, Y. and Wu, P. (1991) Detection and identification of *Peronosclerospora sacchari* in maize by DNA hybridization. *Phytopathology* 81, 901–905.

Fungal Immunodiagnostics in Plant Agriculture

7

F.M. Dewey[1] and C.R. Thornton[2]

[1]Department of Plant Sciences, University of Oxford,
South Parks Road, Oxford OX1 3RB, UK;
[2]Department of Plant Sciences, University of Cambridge,
Downing Street, Cambridge CB2 3EA, UK.

7.1 Introduction

Early detection and correct identification of fungal pathogens is of primary importance in determining the most effective course of treatment to prevent the spread of fungi causing plant diseases and postharvest storage rots. Correct diagnosis, on the basis of symptoms alone, is difficult for many diseases. Confirmation of a given pathogen by classical methods generally involves the plating out of infected, surface sterilized, plant tissue on appropriate growth media. Identification of the pathogen is based either on the characteristic morphology of spores produced *in vitro*, which can be determined only by microscopy, or by the morphological features of mycelial growth on selective media. Both methods are time consuming, require mycological expertise and are not suitable for the following: obligate pathogens, pathogens on the surfaces of plants or seeds, pathogens that only grow slowly in culture because the growth is often masked by faster growing saprophytic fungi associated with the pathogen and soilborne pathogens. Identification and enumeration of spores of airborne fungi is also important especially for disease forecasting and the development of control strategies. Microscopical methods used to identify and count airborne spores are slow and tedious and confirmation, which involves the plating out of trapped spores, is often needed to validate visual identification. Thus, in many aspects of plant pathology, there is a need for assay systems such as immunoassays that are specific, sensitive and quick to confirm visual symptoms and to detect early and latent infections before symptoms are visible. Immunodetection assays will never entirely replace classical methods but they do enable advisers and growers, who may have little mycological expertise, to screen large numbers of samples either on site or with minimal facilities.

In this chapter, we concentrate on the development of immunological assays for detection of fungal pathogens of crop plants, fruits and vegetables and soilborne pathogens. Detection of airborne spores of fungal pathogens is discussed briefly. The difficulties involved in raising antibodies to fungi that are taxonomically specific, either antibodies in antiserum (i.e. polyclonal pAb) or monoclonal antibodies (mAb) and the different assay formats are addressed generally, as are methods on antigen extraction from plant tissues.

7.2 Difficulties in Raising Fungal Antibodies that are Specific

Immunological assays have been used very successfully for a number of years for the detection of viruses in diseased plants but development of immuno-assays for the detection of fungal pathogens has been slow. The difficulties encountered in producing antisera to fungi that have the required specificity has been one of the main reasons that has retarded the development of fungal immunodiagnostic techniques. When antisera raised to mycelial fragments, extracts from lyophilized mycelia, surface washings of solid cultures or culture filtrates are tested by enzyme-linked immunosorbent assays (ELISA) or immunofluorescence (IMF), they generally cross-react with both related and unrelated fungi and host tissues of their extracts (El-Nashaar *et al.*, 1986; Bolik *et al.*, 1987; Mohan, 1988; Dewey *et al.*, 1989a; Xia *et al.*, 1992; Brill *et al.*, 1994).

Most of the antibodies in fungal antisera that are non-specific are of the immunoglobulin class IgM, whereas many of the Abs that are species specific are IgG antibodies (Dewey *et al.*, 1989a,b, 1990; Mitchell *et al.*, 1994; Cahill and Hardham, 1994; Thornton *et al.*, 1993, 1994; Bermingham *et al.*, Oxford, UK, personal communication). Preliminary results (Wakeham, HRI, Well-esbourne, Oxon, UK, personal communication) indicate that the specificity of fungal antisera can be improved by replacing the secondary antibody enzyme conjugate (i.e. reporter molecule) with a protein A enzyme conjugate (see Fig. 7.1 for format) that binds only to IgG antibodies. A number of workers have obtained reasonable levels of specificity by using, as their immunogen, proteins precipitated from either culture filtrates or mycelial extracts (Gleason *et al.*, 1987; Gerik *et al.*, 1987; Barker and Pitt, 1988; Mohan, 1988). Antisera raised against specific fungal fractions, such as species-specific bands on SDS gels (Poupard *et al.*, 1991; Sundaram *et al.*, 1991; Priestley and Dewey, 1993) or soluble carbohydrates (Notermans *et al.*, 1987; Notermans and Kamphuis, 1990; De Ruiter *et al.*, 1994), generally have a higher degree of specificity.

The advent of hybridoma technology has enabled highly specific fungal immunoassays to be developed and an increasing number of these are being used for diagnostic purposes. The level of difficulty in raising hybridomas

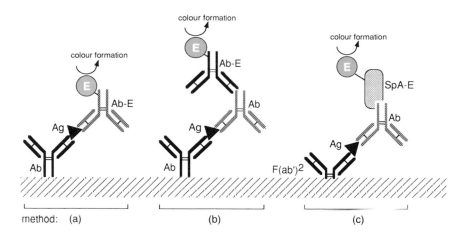

Fig. 7.1. Various double-antibody-sandwich ELISA formats that have been used to detect fungal pathogens. Ab, antibody; Ab-E, antibody-enzyme conjugate; Ag, antigen; E, enzyme; F(Ab')², antibody fragment; SpA, *Staphylococcus aureus* Protein A conjugate.

Examples of methods: (a) Miller *et al.* (1992), detection of *Phytophthora*, *Pythium* and *Rhizoctonia* species in turfgrass; (b) Priestley and Dewey (1993), detection of *Pseudocercosporella herpotrichoides* in stem-base region of cereals; (c) Harrison *et al.* (1990), detection of *Phytophthora infestans* in potato leaves.

that secrete mAbs which recognize a specific pathogen appears to vary considerably from fungus to fungus. On the whole we have found that it is more difficult to raise mAbs to pathogens than to non-pathogens. The reasons for this are not clear but we have found that the water-soluble components from surface washings of cultures of saprophytic fungi grown on solid media are more immunogenic and contain more proteinaceous material than do the water-soluble surface washings of pathogenic fungi (F.M. Dewey, unpublished). We have also found that it is generally more difficult to raise isolate- and species-specific mAbs to fungi than it is to raise genus-specific mAbs and we, like De Ruiter *et al.* (1994), have found that assays employing mAbs that recognize carbohydrate epitopes are more sensitive and robust than those employing mAbs that recognize proteins (Bossi and Dewey, 1992; Dewey *et al.*, 1993).

7.3 Types of Immunodetection Assays

Several different assay formats exist but the most common are the ELISA-based systems. These assays have many advantages: they are highly sensitive and can easily be replicated, automated and quantified. The only disadvantage is that they require laboratory facilities. A number of 'user-friendly',

membrane-bound, dot-blot and dipstick assays, which do not require laboratory facilities, have been developed and are being used increasingly as 'on site'
screening tests. Very few agglutination assays have been developed despite the
obvious ease and rapidity of such assays. Some immunofluorescence assays,
which were developed for the immunolocalization of fungal antigens, have
been adapted for detection purposes. Such assays are useful but they are never
likely to become widely used for mass screening because they involve
microscopy and a UV light source. None the less, they may prove useful in
epidemiology for the enumeration of airborne fungal spores settling on petals
(Salinas and Schots, 1994) or leaves or trapped on tapes as, for example, in a
Berkhard spore trap (F.M. Dewey et al., unpublished).

7.3.1 ELISA

Most of the ELISA tests developed for fungi are, with the exception of the
commercial assays, all simple indirect assays in which the microtitre wells are
directly coated with the test sample or fungal antigens (i.e. plate trapped
antigens; Fig. 7.2). Fungal antigens, particularly glycoproteins, which appear
to be the immunodominant molecules, bind strongly to microtitre wells.
Once the wells have been coated with fungal carbohydrates or glycoproteins,
washed and dried, they can be stored dry at 4°C for several years (Dewey,
1992). The disadvantage of using antigen-coated wells is the relatively long
binding step, 5h or more (generally an overnight step), needed to ensure
maximum binding of the antigens to the wells but the simplicity of such
assays is attractive. Fungal carbohydrates and most glycoproteins are heat
stable, which means that many of the cross-reactive proteinaceous antigens
can be precipitated from the test sample by heat treatment before the antigen
mixture is used to coat wells. Heat treatment of soil extracts has proved very
useful in the detection of Rhizoctonia solani antigens (Thornton et al., 1993).
We have found, as did Gleason et al. (1987), that assays involving antigen-
coated wells work particularly well where the fungus is present on or near
the surface of the infected tissue and where passive release by overnight
soaking is sufficient to enable detection at very low infection levels. This
method has proved invaluable in the detection of two fungi involved in
postharvest spoilage of rice grains, Humicola lanuginosa and Penicillium
islandicum (Dewey et al., 1989b and 1990 respectively). Such methods should
prove useful in the detection of seedborne pathogens that are present at or
near the surface of seedcoats.

Most of the commercial enzyme immunoassays involve double-
antibody-sandwich tests in which one of the antibodies is generally a specific
mAb and the other polyclonal antisera (see Fig. 7.2). Sensitivity in these
assays can be improved, as it can with antigen-coated wells, by the use of
reporter molecules such as biotinylated secondary antibodies or biotinylated
protein-A conjugates.

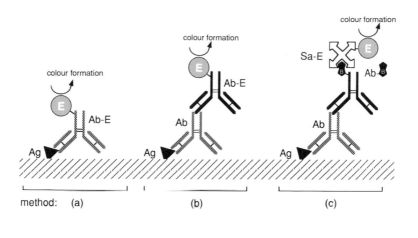

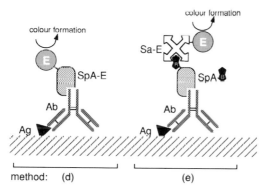

Fig. 7.2. Various ELISA formats employing plate trapped antigens that have been used to detect fungal pathogens. SpA, *Staphylococcus aureus* Protein A-conjugate; Ab, antibody; Ag, antigen; E, enzyme; B, biotin; Sa, streptavidin.

Examples of the methods: (a) Beyer *et al.* (1993), detection of *Fusarium* species in stem-base region of rye; (b) Nameth *et al.* (1990), detection of *Leptosphaeria korrae* in stems, crowns and roots of turf grass; (c) Dewey *et al.* (unpublished), detection of latent infections of *Botrytis cinerea* in apple fruits; (d) Harrison *et al.* (1993), detection of *Spongospora subterranea* in skins of potato tubers; (e) Thornton *et al.* (1994), detection of *Trichoderma harzianum* in soils.

Magnetic beads provide a promising alternative solid support system for capturing fungal antigens. Immunoassays incorporating magnetized beads have been used for some time in the fields of medicine and plant virology (e.g. Banttari *et al.*, 1991) but their use in the diagnosis of fungal plant pathogens has yet to be fully exploited. Enzyme immunoassays that use small beads have a number of advantages over conventional assays that use microtitre plates or membranes as the immunosorbent surface. They are less constrained by slow binding kinetics and the small surface areas of plates and membranes and are ideally suited to efficiently extracting antigens from samples that contain particulate debris. Thornton *et al.* (unpublished) have developed an indirect-magnetic microsphere enzyme immunoassay for the detection and enumeration of conidia of *Trichoderma harzianum* in complex environments. Samples containing conidia are incubated with a mixture of IgM and IgG1 mouse monoclonal antibodies specific to *T. harzianum*. Water-soluble antigens that diffuse from the conidia are bound by the antibodies and the complex extracted using commercially available beads (Dynabeads, Dynal, Oslo, Norway) precoated with rat monoclonal antibodies raised against mouse antibodies of the immunoglobulin subclass IgG1. The bound complex is visualized by sequential exposure to commercial goat anti-mouse IgM (μ-chain specific) alkaline phosphatase conjugate and PNP substrate.

Another type of assay system which is rarely used at present for fungal diagnostics but which could prove to be attractive, because of the apparent increase in sensitivity and specificity, is the competition assay (Kitagawa *et al.*, 1989; Lyons and White, 1992; Yuen *et al.*, 1993).

7.3.2 Dot-blot and dipsticks assays and tissue printing

The mechanism of the dot-blot or dipstick assay is essentially the same as that in ELISA tests. Some test systems use nitrocellulose membranes or nitrocellulose-coated plastic tags or cards and others use polyvinylidene difluoride (Immobilon P, Millipore). Cahill and Hardham (1994) have found that positively charged nylon membranes are useful for the detection of *Phytophthora cinnamomi* because their charged surfaces attract zoospores that are within the immediate vicinity. The reporter conjugate in these assays is generally an enzyme conjugate but gold conjugates, which can be silver enhanced, have been used by a number of workers and are thought to be more sensitive (Dewey *et al.*, 1989b; Cahill and Hardham, 1994). Some dot-blot type commercial assays have been developed. The most notable are those sold by Neogen (Lansing, MI, originally developed by Agridiagnostics, Cinnaminson, NJ, USA) for the detection of turfgrass pathogens (see later for details). These assays are designed for use in the field and can be completed in 10 min.

Tissue printing or squash blot systems hold considerable promise. They have been used successfully for the detection and immunolocalization of the pathogen *Fusarium oxysorum* f. sp. *narcissi* that causes basal rot of daffodil bulbs (Linfield and Lyons, HRI, Wellesbourne, Oxon, UK, personal communication) and *Fusarium* spp. in stems of seedlings (Arie *et al.*, 1993). Gwinn *et al.* (1991) report that their tissue immunoprint-blot method for the detection of an endophyte, *Acremonium coenophialum*, in tall fescue, has the same level of sensitivity as their protein-A sandwich ELISA. Cahill and

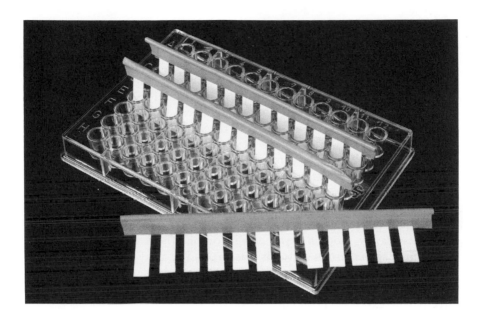

Fig. 7.3. Multiple dipstick device ('toothcomb') used by Cahill and Hardham (1994) for the detection of *Phytophthora cinnamoni*. (Photograph courtesy of Dr A.R. Hardham, Australian National University.)

Hardham (1994) have developed a novel dipstick immunoassay for the detection of *Phytophthora cinnamomi* in soils. Zoospores of the pathogen are attracted to squares of nylon membranes that are precoated with the chemoattractant aspartic acid and attached to plastic dipsticks. Several dipsticks can be attached to a support, as in the teeth of a toothcomb, and processed simultaneously in microtitre wells (Fig. 7.3). The minimum detection limit of these dipsticks was found to be 40 zoospores per millilitre. Arie *et al.* (1993) have developed a combined culture/ immunoblot assay for the detection of pathogens in soils. A membrane is overlaid in a Petri dish with a nutrient gel on which a dilute soil sample is spread. After 20h, the gel is peeled off and the membrane is probed with the specific antibody. Fungal propagules in the soil, which have grown during the incubation period and released antigens which diffuse through the gel on to the membrane, can then be immunolabelled. This method is not dissimilar from that used by Gleason *et al.* (1987), for the detection of *Phomopsis longicolla* in soyabean seed. Here, seeds are left to germinate directly on a membrane overlying moist tissue and the seedborne pathogen, if present, grows out over the membrane. This method has the additional advantage over ELISA tests in that it determines only the percentage of infection in viable seeds.

7.3.3 Agglutination assays

There are two types of agglutination assay: those that employ immunoglobu-lins bound to latex particles and those in which the antibodies are bound to protein-A on the surface of formalin treated cells of *Staphylococcus aureus*. The assays developed by Notermans and Kamphuis (1990) for the detection of aspergilli and penicillia employ latex-sensitized beads that recognize the extracellular polysaccharide produced by these fungi under some, but not all, conditions. The sensitivity of these latex agglutination tests has been found to correspond well with the Howard mould colony count for detection of the fungi in seeds, animal feeds and spices. However, no correlation has been found between these tests on fruit juices or cheese products.

Lyons *et al.* (HRI, Wellesbourne, UK, personal communication), have developed *S. aureus* agglutination assays for the detection of *Pythium violae* in carrots (cavity spot), *Botrytis allii* in onions (neck rot) and *Fusarium oxysporum* f.sp. *narcissi* in daffodils (basal rot). The assay system is the same as that developed by Lyons and Taylor (1990) for the detection of bacterial plant pathogens except that tests for the fungal antigens are performed directly with expressed plant sap.

7.4 Extraction of Antigens from Plant Tissues and Soils

Many workers have experienced difficulties in extracting fungal antigens from infected tissues and it appears that antigen extraction is one of the most limiting steps in the development of sensitive immunoassays for systemic pathogens. Many systems such as those for the detection of turfgrass pathogens involve the use of carborundum paper to mechanically shred the infected tissues and expose the pathogen to the buffer system (Miller *et al.*, 1992; Priestley and Dewey, 1993). Other systems involve grinding the sample in a mortar and pestle, or polytron (Beckman *et al.*, 1994) or grinding the tissue in liquid nitrogen (Newton and Reglinski, 1993). Some involve freeze drying the samples (Beyer *et al.*, 1993) or their extracts to increase the concentration of the antigens (Wakeham, HRI, Wellesbourne, Oxon, UK, personal communication).

Buffers used for extraction of antigens vary; phosphate buffered saline (PBS) buffers (pH 7.2) are commonly used for antigen coating of wells where carbohydrate epitopes are involved and bicarbonate buffers (pH 9.6) for systems employing antibodies recognizing proteinaceous epitopes. Many buffer systems include azide or thimerosol to prevent growth of con-taminant microorganisms. PVP is commonly added to sample buffers to reduce interference from plant phenolic compounds (e.g. Beckman *et al.*, 1994). In sandwich type immunoassay systems sample buffers generally include 0.05% Tween 20 and reconstituted non-fat dried milk, casein or BSA as blockers to reduce non-specific binding. In the development of a quantitative assay for the detection of the eyespot pathogen (*Pseudocercosporella herpotrichoides*) in very young seedlings (8–15 days), Priestley and Dewey (1993) found it necessary to use an acetate buffer system at pH 4.5 to prevent interference from wheat germ agglutinin, which, at higher pH levels, binds non-specifically to the rabbit polyclonal antiserum.

7.5 Immunoassays for Different Kinds of Fungal Pathogens

7.5.1 Soilborne and root-infecting fungi

While a number of immunoassays have been developed for the detection of soilborne pathogens *in planta* (for example, *Verticilium dahliae* (Gerik *et al.*, 1987), *Phytophthora fragariae* (Werres, 1988) and *Pythium ultimum* (Yuen *et al.*, 1993)), few attempts have been made to specifically detect pathogens resident in the soil *per se* using immunological techniques. Of the assays developed, only a limited number have been developed to detect live mycelium alone, for example *Rhizoctonia solani* (Thornton *et al.*, 1993). The majority rely on the detection of spores in the form of oospores (*Pythium* spp. (Miller *et al.*, 1989)), zoospores (*Phytophthora* spp. (Jones and Shew, 1988; Klopmeyer *et al.*, 1988; Cahill and Hardham, 1994), *Pythium aphanidermatum* (Mitchell *et al.*, 1994)), cystosori (*Spongospora subterranea* (Harrison *et al.*, 1993)) and resting spores (*Plasmodiophora brassicae* (Wakeham, HRI, Wellesbourne, Oxon, UK, personal communication)). This reflects not only the life histories of the target organisms but also the difficulties inherent in the extraction of the organisms and, hence, their antigens from a complex medium. Many of the immunoassays employ a period of biological amplification, which usually involves baiting with host tissue (Klopmeyer *et al.*, 1988) or enrichment on or in solid or aqueous semi-selective media (MacDonald and Duniway, 1979; Thornton *et al.*, 1993) to allow germination and growth of viable propagules or, alternatively, lengthy extraction and concentration procedures involving sieving, flotation, centrifugation and filtration (e.g. Harrison *et al.*, 1993; Wakeham, HRI, Wellesbourne, Oxon, UK, personal communication).

The assays themselves are invariably simple direct or indirect ELISA (Miller *et al.*, 1989; Harrison *et al.*, 1993; Thornton *et al.*, 1993) or double-antibody-sandwich ELISA (Jones and Shew, 1988; Klopmeyer *et al.*, 1988) formats although a number of workers have developed immunofluorescence, dipstick and immunobinding assays (Malajczuk *et al.*, 1975; Arie *et al.*, 1993; Thornton *et al.*, 1993). We have developed an *in situ* immunofluorescence plate culture assay that has proved useful in differentiating mycelia of *Rhizoctonia solani* and mycelia from other *Rhizoctonia* species growing out from soil particles on a solid semi-selective medium (Thornton *et al.*, 1993 and Fig. 7.4).

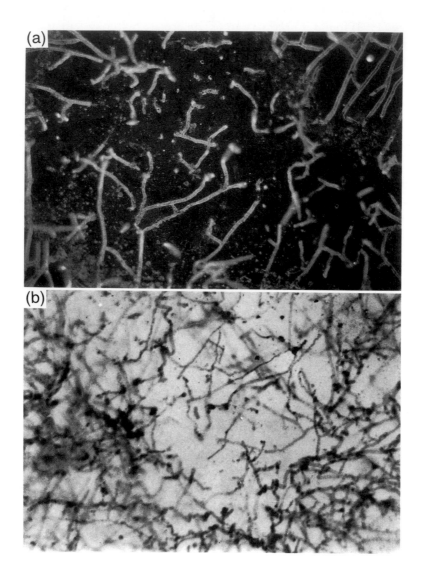

Fig. 7.4. (a) Photomicrograph of species of *Rhizoctonia*, and other fungi isolated on a solid semi-selective medium from soil artificially infested with *R. solani* and immunolabelled with *R. solani*-specific mAb and goat anti-mouse- FITC, viewed with a Zeiss Axiophot epifluorescent microscope (× 125) using UV light; (b) same viewed under white light with DIC optics. *Note*: only hyphae of *R. solani* become immunolabelled and fluoresce with UV light.

At present, no kits are available in the agricultural market for the detection of soilborne pathogens *in situ*. However, Neogen Corporation (Lansing, MI, USA) now supply ALERT™ On-Site Crop Disease Detection Kits (originally developed by Agri-Diagnostics) for the diagnosis of *Phytophthora*, *Pythium* and *Rhizoctonia* spp. in root, stem and leaf samples. These kits detect a wide range of important species within the target genus and each test takes approximately 10 min. Other kits (e.g. Agri-Diagnostics laboratory-based *Phytophthora* E and *Pythium* C multiwell ELISA kits) have been used by a number of workers for the detection of the pathogens in the soil or in irrigation water (Schmitthenner and Miller, 1988; Ali-Shtayeh *et al.*, 1991; Dunsuncell and Fox, 1992; Pscheidt *et al.*, 1992; Timmer *et al.*, 1993).

It is clear that the development of specific, sensitive and user-friendly kit assays for the specific detection of pathogens in the soil has been hindered by technical issues such as the elimination of effects by the soil, collection and concentration of propagules, and sample preparation. However, these problems can be solved and should pave the way to the development of rapid 'on site' detection assays. These will enable the grower to determine which soils are infested with damaging organisms prior to sowing, thereby evading infection of susceptible crop plants and eliminating the need for costly pesticide applications.

7.5.2 Stem-based and vascular pathogens

A number of immunoassays, both pAb and mAb based, have been developed for the detection and differentiation of stem-based pathogens. Differentiation and quantification of the eyespot pathogen of cereals, *Pseudocercosporella herpotrichoides*, from the other stem-base pathogens such as *Rhizoctonia cerealis*, *Microdochium nivale*, *Fusarium avanaceum* and *F. culmorum* is important because it is the only pathogen that is known both to cause a significant yield loss and to respond to fungicide treatment. Some of the pAbs raised to *P. herpotrichoides*, for the detection of eyespot, are known to cross-react with one or more of the other stem-base pathogens (Bolik *et al.*, 1987) but the assays developed by Du Pont (see Cagnieul and Lefebre, 1991), Poupard *et al.* (1991), and Priestley and Dewey (1993) are specific. As with many pathogens, early detection of *P. herpotrichoides* before symptoms are apparent is invaluable. Characteristic eyespot lesions are not seen normally until 3–4 weeks after infection but Priestley and Dewey (1993) using a double-antibody-sandwich ELISA have shown that it is possible to detect the pathogen in symptomless seedlings 5 to 8 days after inoculation. Similarly, Velicheti (1993), using antisera to *Phomopsis longicolla* in a double-antibody-sandwich ELISA format has detected the pathogen in extracts from inoculated asymptomatic stems of soyabean plants. Beyer *et al.* (1993) have raised an antiserum to *F. culmorum* that recognizes all species of *Fusarium* but does not cross-react with the other stem-based pathogens and so could be used to determine the total biomass of species of *Fusarium* attacking the stem-base region.

Immunological differentiation of the species and *forma speciales* of the vascular wilt pathogens, *Verticillium albo-atrum,V. dahliae* and *Fusarium oxysporum*, has proved difficult (Mohan and Ride, 1982; Ianelli *et al.*, 1983). Van de Koppel and Schots (1994) have developed a mAb-based double-antibody-sandwich ELISA for the detection of *V. dahliae* in stems of infected roses and chrysanthemums. The antibodies used in this assay system, which also recognize *V. albo-atrum*, could be used potentially for the detection of *V. dahliae* in crop plants such as potato and in high value horticultural plants such as pelargonium. Sanofi-phytodiagnostics sell a pAb-based assay for the detection of *V. dahliae* in pelargonium stems. Del Sorbo *et al.* (1993), using an immunological tolerance technique, have produced antisera that will distinguish, by ELISA, between *F. oxysporum* f. sp. *lycopersici* and *F. oxysporum* f. sp. *dianthi*. Using competition assays, Kitagawa *et al.* (1989) were able to develop two tests, one that allowed the specific detection of *F. oxysporum* f.sp. *cucumerinum* and another that detected all species of *Fusarium*. Curiously, Wong *et al.* (1988) have been able to differentiate, by IMF, thick-walled chlamydospores of race 4 of *F. oxysporum* f. sp. *cubense* from races 1 and 2 using a mAb that they raised to mycelial fragments. None of the above *Fusarium* assays appears to have been adopted for routine diagnostic purposes.

Stace-Smith *et al.* (1995), using mAbs raised to *Leptosphaeria maculans* (imperfect state of *Phoma lignam*), the causal agent of stem canker of *Brassica napus* and *B. rapa*, have developed an ELISA test that will differentiate virulent from weakly virulent stains of the pathogen. To our knowledge, this is the first time that anyone has developed an immunoassay that will distinguish between virulent forms of a pathogenic fungus. Identification of highly virulent strains of *L. maculans* is important in Canada where the climate prevents the cultivation of resistant cultivars.

The commercial immunoassays developed by Agri-Diagnostics for the detection of turfgrass pathogens (see Section 7.5.1 for details) have been evaluated by a number of workers including Baldwin (oral presentation at BSPP meeting, Norwich, UK, July 1989), Shane (1991) and Holmes (1993). Most workers agree that such assays are useful in confirming suspected infections, provided that the instructions, included with the kits, are followed correctly. Nameth *et al.* (1990) have developed a monoclonal-antibody-based assay for the detection of *Leptosphaeria korae*, the causal agent of the necrotic ringspot disease of turfgrass that attacks the stem-base, crown and root. They found that ELISA and conventional plate assays give comparable results.

7.5.3 Foliar pathogens

Immunoassays have proved useful in the identification of foliar pathogens in mixed infections and in the estimation of mycelial biomass of a particular pathogen in leaf tissue for breeding and research purposes. Commercial immunoassays have been developed by both Du Pont (Newark, DE, USA) and CIBA Agriculture (Whittlesford, Cambridge, UK) for the detection and differentiation of *Septoria nodorum* and *S. tritici* and causal agents of leaf and glume blotch and leaf blotch respectively of cereals, particularly wheat. Differentiation of the two pathogens, which have different teleomorphic stages (*Leptosphaeria nodorum* and *Mycosphaerella graminicola*) is important because they respond to different fungicides. Both detection assays are similar in that they are laboratory-based, double-antibody-sandwich ELISA tests employing pAb and specific mAbs.

Quantitative pAb-ELISA tests have been developed by Newton and Reglinski (1993) for *Erysiphe graminis* f. sp. *hordei* in green or senescent barley leaves and by Harrison *et al.* (1990), for *Phytophthora infestans* in potato leaf tissue.

Barker *et al.* (1994) have developed a mAb ELISA for the detection of *Colletotrichum acutatum*, the causal agent of blackspot disease of strawberry fruits. Latent infections can be detected in petioles of the planting material provided that a short biological amplification method is included in the assay.

7.5.4 Seedborne pathogens

Immunodetection of seedborne pathogens is frequently difficult because of the small amount of fungal material present and high background interference encountered with extracts from whole seeds. Gleason *et al.* (1987), in developing an assay for *Phomopsis longicola* in soyabean seeds, have shown that one method of overcoming this problem is to allow the seeds to germinate on a membrane laid on moist tissues. Once the pathogen has grown out from the infected seeds on to the membrane, it can be immunolabelled. This method as well as ELISA tests have been applied to the detection of *Pyrenophora graminae* and *Pyrenophora terres*, two seedborne pathogens of barley that cause leaf stripe and net blotch respectively, using a monoclonal antibody that detects both species (Burns *et al.*, 1994).

7.5.5 Spores of airborne pathogens

Surprisingly, few immunoassays have been developed for the detection of airborne spores despite the clear need both in the medical field for the detection of specific allergenic fungal spores and for disease forecasting and spray management programmes. Xia *et al.* (1992), have raised mAbs to

Pyricularia grisea with the intention of using these to develop an IMF assay to detect conidia in rice fields. Recently, we have developed a system of immunostaining spores of *Botrytis cinerea* caught on a modified tape in a Berkhard spore trap (F.M. Dewey *et al.*, unpublished). Other mechanisms of trapping and immunodetecting spores, for example directly on to rotating microtitre wells which are then processed by ELISA, are also being explored (Schmechel *et al.*, 1994).

Chaparral Diagnostics (Burlington, VT, USA) have developed a blot immunoassay for the detection of ascospores of *Venturia inaequalis*, which causes apple scab. Leaves with visible lesions are overwintered in special traps and at appropriate intervals during the season are soaked in water to initiate the release of spores from the leaf surface (SporeCard[TM]) which are then fixed, blocked and immunostained.

7.5.6 *Fungi causing postharvest spoilage*

Fungi are responsible for a diverse range of postharvest problems. Severe losses can be incurred immediately postharvest or later during shipping and storage of fruits, vegetables and grains. Rapid detection of latent infections before the produce is packaged or stored is critical to the industry. It is surprising therefore that, to date, very few immunoassays have been developed for the detection of latent infections. A few assays using pAb have been developed for the detection of *B. cinerea* in grapes (Ricker *et al.*, 1991) and foods generally (Cousin *et al.*, 1990).

Monoclonal antibodies to *B. cinerea* were raised by Salinas and Schots (1994) and used for the detection, by immunofluorescence, of spores of the fungus on the petals of gerbera flowers prior to marketing. We have also raised mAbs to *B. cinerea* and have developed an ELISA test for the detection of *B. cinerea* in strawberries (Bossi and Dewey, 1992) and symptomless grapes, pears and apples (F.M. Dewey *et al.*, unpublished). Determining the threshold for such a ubiquitous pathogen as *B. cinerea*, at which infection levels result in significant losses, is a slow process. It can be done only by comparison of the immunassay data with the level of disease after storage.

Detection of mycotoxins produced by fungi growing on or in stored grains, nuts and other produce is important to both the food and agriculture industry (see Chapter 10) and will not be discussed further here. However, a quick screening test for the detection of the fungi themselves could be useful (see Dewey *et al.*, 1993). We have developed mAb-based assays for individual fungi notably *Penicillium islandicum* (Dewey *et al.*, 1990) and *Humicola lanuginosa* (Dewey *et al.*, 1989b) on rice grains and *Aspergillus flavus* on peanuts (F.M. Dewey, unpublished). Notermans and his group have raised both antisera and mAbs to heat-stable extracellular poly-saccharides (EPS) from several fungi involved in spoilage of foods (to mucoraceous fungi (De Ruiter *et al.*, 1994) and to ascomycete fungi

(Notermans *et al.*, 1987; Notermans and Kamphuis, 1990)). These assays, in the ELISA format, can be used, quantitatively, to detect fungal antigens in processed as well as fresh produce. They have also developed agglutination assays but care has to be taken in both types of assay formats to eliminate false positives (Notermans and Kamphuis, 1990). Most of these assays are 'genus' specific but those employing pAbs are not able to distinguish between species of *Aspergillus* and *Penicillium*. MAbs that recognize species of *Aspergillus* but not *Penicillium* have now been raised and could prove useful in a negative screen to demonstrate the absence of aflotoxin-producing fungi (Dewey *et al.*, 1993).

7.6 Commercially Available Antisera, Monoclonal Antibodies and Kits and Future Potential

Relatively few kits are sold for the detection of fungal pathogens. Most have been developed by Agri-Diagnostics for the detection of turfgrass diseases (Miller *et al.*, 1992). They produced both field usable dot-blot and laboratory-based assays for the detection of species of *Phytophthora*, *Pythium*, *Sclerotinia* and *Rhizoctonia* (see Section 7.5.1). These kits are now available through Neogen, Lansing, Michigan in the USA and ADGEN, Auchincruive, in the UK. CIBA Agriculture have developed commercial kits for the detection and differentiation of *Septoria nodorum* and *S. tritici*. Du Pont have also developed assay kits to detect and differentiate the two *Septoria* species and kits for the detection of *Pseudocercosporella herpotrichoides* on cereals (Smith *et al.*, 1990), and *Botrytis cinerea* on grapes. The kits developed by CIBA Agriculture and Du Pont are, at present, marketed only with their agrochemical products.

The commercial potential of fungal immunodiagnostics is, as yet, unclear. Two key factors that will determine the viability of such assays are environmental legislation and the amount by which they increase the profit margin of a particular crop. It is fairly certain that fungal immunodetection assays will be commercially viable for high value crops where detection of the pathogen by conventional means is difficult and early treatment is effective. Immunodetection of soilborne pathogens with long survival rates may also be commercially viable particularly where the land is rented under short-term contracts. However, it is doubtful whether immunoassays for foliar pathogens and stem-based pathogens of cereals and other arable crops will be commercially viable unless legislation is introduced to limit the widespread prophylactic used of fungicides. It is probable that for vegetables and fruits legislation will be introduced to limit application of fungicides to the preharvest stages. In this event, there will be a considerable demand for the detection of latent infections before shipping and storage.

References

Ali-Shtayeh, M.S., McDonald, J.D. and Kabashima, J. (1991) A method for using commercial ELISA tests to detect zoospores of *Phytophthora* and *Pythium* species in irrigation water. *Plant Disease* 75(3), 305–311.

Arie, T., Hayashi, S., Schimazaki, K., Yoneyama, I. and Yamaguchi, I. (1993) Novel diagnosis of *Fusarium* infestation of seedlings and soils by immunoassay methods. In: *Proceedings of the Sixth International Congress of Plant Pathology*, Montreal, Canada, July–August 1993. Abstract 2.1.15, p. 40.

Banttari, E.E., Clapper, D.L., Sheau-Ping, Hu., Daws, K.M. and Khurana, S.M.P. (1991) Rapid magnetic microsphere enzyme-linked immunoassay for potato virus X and potato leafroll virus. *Phytopathology*, 81, 1039–1042.

Barker, I. and Pitt, D. (1988) Detection of the leaf curl pathogen of anemones in corms by enzyme-linked immunosorbent assay (ELISA). *Plant Pathology* 37, 417–422.

Barker, I., Brewer, G., Cook, R.T.A., Crossley, S. and Freeman, S. (1994) Strawberry blackspot disease (*Colletotrichum acutatum*). In: Schots, A., Dewey, F.M. and Oliver, R. (eds), *Modern Assays for Plant Pathogenic fungi: Identification, Detection and Quantification*. CAB International, Wallingford, UK, pp. 179–189.

Beckman, K.B., Harrison, J.G. and Ingram, D.S. (1994) Optimization of a polyclonal enzyme-linked immunosorbent assay (ELISA) of fungal biomass for use in studies of plant defence responses. *Physiological and Molecular Plant Pathology* 44, 19–32.

Beyer, W., Hoxter, H., Miedaner, T., Sander, E. and Geiger, H.H. (1993) Indirect ELISA for quantitative assessment of *Fusarium* spp. in rye. *Journal of Plant Disease and Detection* 100(3), 278–284.

Bolik, M., Casper, R. and Lind, V. (1987) Einsatz serolgischer und gelektrophor-etischer Verfahren zum Nachweis von *Pseudocercosporella herpotrichoides*. *Journal of Plant Disease and Protection* 94, 449–456.

Bossi, R. and Dewey, F.M. (1992) Development of a monoclonal antibody-based immunodetection assay for *Botrytis cinerea*. *Plant Pathology* 41, 472–482.

Brill, L.M., McClary, R.D. and Sinclair, J.B. (1994) Analysis of two ELISA formats and antigen preparations using polyclonal antibodies directed against *Phomopsis longicolla*. *Phytopathology* 84, 173–179.

Burns, R., Vernon, M.L. and George, E.L. (1994) Monoclonal antibodies for the detection of *Pyrenophora graminea*. In: Schots, A., Dewey, F.M. and Oliver, R. (eds), *Modern Assays for Plant Pathogenic Fungi: Identification, Detection and Quantification*. CAB International, Wallingford, UK, pp. 199–204.

Cagnieul, P. and Lefebvre, A.D. (1991) Diagnostic kit (diagnolab) for cereal eyespot. *Proceedings of the Third International Conference on Plant Diseases*, Bordeaux, pp. 547–553.

Cahill, D.M. and Hardham, A.R. (1994) Exploitation of zoospore taxis in the development of a novel dipstick immunoassay for the detection of *Phytophthora cinnamomi*. *Phytopathology* 84, 193–200.

Cousin, M.A., Dufrenne, J., Rombouts, F.M. and Notermans, S. (1990) Immuno-logical detection of *Botrytis* and *Monascus* species in food. *Food Microbiology* 7, 227–235.

De Ruiter, G.A., Bos, W., Van Bruggen-Van der Lugt, A.W., Hofstra, H. and Rombouts, F.M. (1994) Development of an ELISA and dot-blot assay to detect *Mucor racemosus* and related species of the Order Mucorales. In: Schots, A., Dewey, F.M. and Oliver, R. (eds), *Modern Assays for Plant Pathogenic Fungi: Identification, Detection and Quantification*. CAB International, Wallingford, UK, pp. 157–164.

Del Sorbo, G., Scala, F., Capparelli, R., Ianelli, D. and Noviello, C. (1993) Differentiation between two *forma speciales* of *Fusarium oxysporum* by antisera produced in mice immunologically tolerized at birth through lactation. *Phytopathology*, 83, 1178–1182.

Dewey, F.M. (1992) Detection of plant invading fungi by monoclonal antibodies. In: Duncan, J.M. and Torrance, L. (eds), *Techniques for the Rapid Detection of Plant Pathogens*. Blackwell Scientific Publications, Oxford, pp. 47–62.

Dewey, F.M., Munday, C.J. and Brasier, C.M. (1989a) Monoclonal antibodies to specific components on the Dutch elm pathogen *Ophiostoma ulmi*. *Plant Pathology* 38, 9–20.

Dewey, F.M., McDonald, M.M. and Phillips, S.I. (1989b) Development of monoclonal-antibody-ELISA, -DOT-BLOT and -DIP-STICK immunoassays for *Humicola lanuginosa* in rice. *Journal of General Microbiology* 135, 361–374.

Dewey, F.M., McDonald, M.M., Phillips, S.I. and Priestley, R.A. (1990) Development of monoclonal-antibody-ELISA and -DIP-STICK immunoassays for *Penicillium islandicum* in rice grains. *Journal of General Microbiology* 136, 753–760.

Dewey, F.M., Banham, A.H., Priestley, R.A., Martin, B., Hawes, C., Phillips, S.I. and Wareing, P.W. (1993) Monoclonal antibodies for the detection of spoilage fungi. *International Biodeterioration and Biodegradation* 32, 127–136.

Dunsuncdi, F. and Fox, R.T.V. (1992) Accuracy of methods for estimating the size of *Thanatephorus cucumeris* populations in soil. *Soil Use and Management* 8(1), 21–26.

El-Nashaar, H.M., Moore, L.W. and George, R.A. (1986) Enzyme-linked immunosorbent assay for the quantification of initial infection of wheat by *Gaeumannomyces graminis* var. *tritici* as moderated by biocontrol agents. *Phytopathology* 76, 1319–1322.

Gerik, J.S., Lommel, S.A. and Huisman, O.C. (1987) A specific serological staining procedure for *Verticillium dahliae* in cotton root tissue. *Phytopathology* 77, 261–266.

Gleason, M.L., Ghabrial, S.A. and Ferris, R.S. (1987) Serological detection of *Phomopsis longicolla* in soybean seeds. *Phytopathology* 77, 371–375.

Gwinn, K.D., Collins-Shepard, M.H. and Reddick, B.B. (1991) Tissue print-immunoblot, an accurate method for the detection of *Acremonium coenophialum* in tall fescue. *Phytopathology* 81(7), 747–748.

Harrison, J.G., Barker, H., Lowe, R. and Rees, E.A. (1990) Estimation of amounts of *Phytophthora infestans* in leaf tissue by enzyme-linked immunosorbent assay. *Plant Pathology* 39, 274–277.

Harrison, J.G., Rees, E.A., Barker, H. and Lowe, R. (1993) Detection of spore balls of *Spongospora subterranea* on potato tubers by enzyme-linked immunosorbent assay. *Plant Pathology* 42, 181–186.

Holmes, S.J. (1993) Rapid serological tests for the detection of root and stem base pathogens of HNS and protected crops. In: *Proceedings of the Sixth*

International Congress of Plant Pathology, Montreal, Canada, July–August 1993. Abstract 2.1.27, p. 42.

Iannelli, D., Capparelli, R., Mariziano, F., Scala, F. and Noviello, C. (1983) Production of hybridoma secreting monoclonal antibodies to the genus *Fusarium*. *Mycotaxon* 17, 523–532.

Jones, K. and Shew, H.D. (1988) Immunoassay procedure for the detection of *Phytophthora parasitica* var. *nicotianae* in soil. *Phytopathology* 78(12), 1577 (Abs.).

Kitagawa, T., Sakamoto, Y., Furumi, K. and Ogwra, H. (1989) Novel enzyme immunoassays for specific detection of *Fusarium oxysporum* f.sp. *cucumerinum* and for the general detection of various *Fusarium* species. *Phytopathology* 79, 162–165.

Klopmeyer, M.J., Miller, S.A., Rittenburg, J.H., Petersen, F.P. and Grothaus, G.D. (1988) Detection of *Phytophthora* in soybean soil by immunoassay analysis of infected bait. *Phytopathology* 52, 1576 (Abs.).

Lyons, N.F. and Taylor, J.D. (1990) Serological detection and identification of bacteria from plants by the conjugated *Staphylococcus aureus* slide agglutination test. *Plant Pathology* 39, 584–590.

Lyons, N.F. and White, J.G. (1992) Detection of *Pythium violae* and *Pythium sulcatum* in carrots with cavity spot using competition ELISA. *Annals of Applied Biology* 120, 235–244.

MacDonald, J.D. and Duniway, J.M. (1979) Use of fluorescent antibodies to study the survival of *Phytophthora megasperma* and *P. cinnamomi* zoospores in soil. *Phytopathology* 69, 436–441.

Malajczuk, N., McComb, A.J. and Parker, C. (1975) An immunofluorescence technique for detecting *Phytophthora cinnamomi* Rands. *Australian Journal of Botany* 23, 289–309.

Miller, S.M., Petersen, F.P., Miller, S.A., Rittenburg, J.H., Wood, S.C. and Grothaus, G.D. (1989) Development of a direct immunoassay to detect *Phytophthora megasperma* f.sp. *glycinea* in soil. *Phytopathology* 79, 1139 (Abs.).

Miller, S.M., Rittenburg, J.H., Petersen, F.P., and Grothaus, G.D. (1992) From the research bench to the market place: development of commercial diagnostic kits. In: Duncan, J.M. and Torrance, L. (eds), *Techniques for the Rapid Detection of Plant Pathogens*. Blackwell Scientific Publications, Oxford, pp. 208–221.

Mitchell, A.J., Mackie, A.J., Roberts, A.M., Hutchison, K.A., Estrada-Garcia, M.T., Callow, J.A. and Green, J.R. (1994) Specificity of monoclonal antibodies raised to *Pythium aphanidermatum* and *Erysiphe pisi*. In: Schots, A., Dewey, F.M. and Oliver, R. (eds), *Modern Assays for Plant Pathogenic Fungi: Identification, Detection and Quantification*. CAB International, Wallingford, UK, pp. 231–238.

Mohan, S.B. (1988) Evaluation of antisera raised against *Phytophthora fragariae* for detecting the red core disease of strawberries by enzyme-linked immunosorbent assay (ELISA). *Plant Pathology* 38, 352–363.

Mohan, S.B. and Ride, J.P. (1982) An immunoelectrophoretic approach to the identification of progressive and fluctuating isolates of the hop wilt fungus *Verticillium albo-atrum*. *Journal of General Microbiology* 128, 255–265.

Nameth, S.T., Shane, W.W. and Stier, J.C. (1990) Development of a monoclonal antibody for detection of *Leptosphaeria korrae*, the causal agent of necrotic ringspot disease of turfgrass. *Phytopathology* 80, 1208–1211.

Newton, A.C. and Reglinski, T. (1993) An enzyme-linked immunosorbent assay for quantifying mildew biomass. *Journal of Plant Disease and Detection* 100, 176–179.

Notermans, S. and Kamphuis, H. (1990) Detection of moulds in food by latex agglutination: a collaborative study. *Food and Agriculture Immunology*, 2, 37–47.

Notermans, S., Wieten, G., Engel, H.W.B., Rombouts, R.M., Hoogerhout, P. and Van Boom, J.H. (1987) Purification and properties of extracellular polysaccharide (EPS) antigens produced by different mould species. *Journal of Applied Bacteriology* 62, 157–166.

Poupard, P., Grare, S. and Cavalier, N. (1991) Using ELISA method for assessment of *in vitro* development of strains of *Pseudocercosporella herpotrichoides,* the cereal eyespot fungus. *Proceedings of the Third International Conference of Plant Diseases*, Bordeaux, pp. 655–662.

Priestley, R.A. and Dewey, F.M. (1993) Development of a monoclonal antibody immunoassay for the eyespot pathogen *Pseudocercosporella herpotrichoides.* *Plant Pathology* 42, 403–412.

Pscheidt, J.W., Burket, J.Z., Fischer, S.L. and Hamm, P.B. (1992) Sensitivity and clinical use of *Phytophthora*-specific immunoassay kit. *Plant Disease* 76, 928–932.

Ricker, R.W., Marios, J.J., Dlott, J.W., Bostock, R.M. and Morrison, J.C. (1991) Immunodetection and quantification of *Botrytis cinerea* on harvested wine grapes. *Phytopathology* 81, 404–411.

Salinas, J. and Schots, A. (1994) Monoclonal antibodies-based test for detection of conidia of *Botrytis cinerea* on cut flowers *Phytopathology*, 84, 351–356.

Schmechel, D., McCartney, H.A. and Halsy, K. (1994) The development of immunological techniques for the detection and evaluation of fungal disease inoculum in oilseed rape crops. In: Schots, A., Dewey, F.M. and Oliver, R. (eds), *Modern Assays for Plant Pathogenic Fungi: Identification, Detection and Quantification.* CAB International, Wallingford, UK, pp. 247–253.

Schmitthenner, A.F. and Miller, S. (1988) ELISA detection of *Phytophthora* from soil. *Phytopathology* 78, 1576 (Abs.).

Shane, W.W. (1991) Prospects for early detection of *Pythium* blight epidemics on turfgrass by antibody-aided monitoring. *Plant Disease* 75, 921–925.

Smith, C.M., Saunders, D.W., Allison, D.A., Johnson, L.F.B., Labit, B., Kensall, S.J. and Holloman, D.W. (1990) Immuno-diagnostic assay for cereal eyespot: novel technology for disease detection. *Proceedings of the Brighton Crop Protection Conference – Pests and Diseases*, pp. 763–770.

Stace-Smith, R., Bowler, G., MacKenzie, D.J. and Ellis, P. (1995) Monoclonal antibodies differentiate weakly virulent from highly virulent isolates of *Leptosphaeria maculans*, the organism causing blackleg of canola. *Canadian Journal of Plant Pathology* (in press).

Sundaram, S., Plasencia, J. and Bantarri, E.E. (1991) Enzyme-linked immunosorbent assay for detection of *Verticillium* spp. using antisera produced to *V. dahliae* from potato. *Phytopathology* 81, 1485–1489.

Thornton, C.R., Dewey, F.M. and Gilligan, C.A. (1993) Development of monoclonal antibody-based immunological assays for the detection of live propagules of *Rhizoctonia solani* in soil. *Plant Pathology* 42, 763–773.

Thornton, C.R., Dewey, F.M. and Gilligan, C.A. (1994) Development of a monoclonal antibody-based enzyme-linked immunosorbent assay for the detection of live propagules of *Trichoderma harzianum* in a peat/bran medium. *Soil Biology and Biochemistry*, 26, 909–920.

Timmer, L.W., Menge, J.A., Zitko, S.E., Pond, E., Miller, S.A. and Johnson, E.L.V. (1993) Comparison of ELISA techniques and standard isolation methods for *Phytophthora* detection in citrus orchards in Florida and California. *Plant Disease* 77, 791–796.

Van de Koppel, M.M. and Schots, A. (1994) A double (monoclonal) antibody sandwich ELISA for the detection of *Verticillium* species in roses. In: Schots, A., Dewey, F.M. and Oliver, R., (eds), *Modern Assays for Plant Pathogenic Fungi: Identification, Detection and Quantification*. CAB International, Wallingford, UK, pp. 99–104.

Velicheti, R.K. (1993) Immunodetection of *Phomopsis* species in asymptomatic soybean plants. *Plant Disease* 77, 70–73.

Werres, S. (1988) Enzyme-linked immunosorbent assay (ELISA) as a method for detection of *Phytophthora fragariae* Hickman in strawberry roots. *Nachrichtenblatt des Deutschen Pflanzenschutzdienstes* 40(10), 146–150.

Wong, W.C., White, M. and Wright, I.G. (1988) Production of monoclonal antibodies to *Fusarium oxysporum* f. sp. *cubense* race 4. *Letters in Applied Microbiology* 6, 39–42.

Xia, J.Q., Lee, F.N., Scott, H.A. and Raymond, L.R. (1992) Development of monoclonal antibodies specific for *Pylicularia grisea*, the rice blast pathogen. *Mycological Research* 96, 867–873.

Yuen, G.Y., Craig, M.L. and Avila, F. (1993) Detection of *Pythium ultimum* with a species-specific monoclonal antibody. *Plant Disease* 77(7), 692–698.

Antibody Approaches to Plant Viral Diagnostics

R.J. Sward and D.R. Eagling
Institute for Horticultural Development, Victorian Department of Agriculture, Private Bag 15, South Eastern Mail Centre, Victoria 3176, Australia.

8.1 Introduction

Plant viruses are known to cause serious losses in a wide range of horticultural and broad-acre field crops. In order to develop a suitable strategy for control of a plant virus disease detailed information on the ecology and epidemiology is vital. In the majority of cases the identity of the virus will provide the key for accessing existing data on known properties and relationships of the virus in question. However, it is often no longer adequate to determine the general name given to the virus because strains or isolate types of a single virus may exhibit major differences in certain of their properties. For example, the virus known as barley yellow dwarf virus (BYDV) has been shown to consist of a group of at least five variously-related virus types typified by specific named isolates each of which differs in its aphid–vector relationships and other properties (Conti *et al.*, 1990).

The earliest form of plant virus diagnosis was based on the appearance of specific symptoms in certain crop plants. However, many plant viruses do not produce specific symptoms in their hosts and it was therefore necessary to find suitable alternative methods to confirm the identity of the causal virus. Transmission to indicator host plants often provided valuable information and is still a widely used method, but it is relatively laborious, requires considerable glasshouse space and may take a number of weeks before a result is obtained.

Most plant viruses consist of an outer protein coat surrounding a nucleic acid. The coat protein, when purified and injected into a suitable laboratory animal, usually elicits a specific immune response which results in the formation of a circulating antibody in the animal's bloodstream. The antigenic properties of the virus coat protein and the antibodies produced in laboratory animals provide the basis of a whole range of serological tests that offer numerous advantages over traditional tests based on symptom recognition and the use of indicator plants.

A variety of serology-based diagnostic tests, in many cases developed initially as part of medical research programmes, have been adapted for use with plant viruses. Some of these tests are more useful than others, but all are benefiting from the increasing availability of high quality antisera to a wide range of viruses representing most of the known plant virus groups.

8.2 Immunoprecipitation

When an antigen and its homologous or closely related antibody are mixed under appropriate conditions, the resulting antigen–antibody complex forms a precipitate. Various tests that rely on the visualization of the precipitate have been used in plant virus diagnostics for more than 40 years. These tests are broadly known as immunoprecipitation tests and will only be briefly discussed here as they have been reviewed in detail by Ball (1974) and Van Regenmortel (1982).

8.2.1 Precipitin tests

These tests are carried out in a liquid phase, either in test tubes or in capillary tubes (Raizada *et al.*, 1989) and are dependent on relatively large amounts of both virus and antiserum.

8.2.2 Microprecipitin tests

These are performed in single droplets on a Petri dish and are relatively sensitive as well as being more economical in their use of both virus and antibody.

8.2.3 Immunodiffusion tests

A number of variations have been developed but all are performed in a gel, such as agar or agarose, rather than in a liquid phase. In the radical diffusion test devised by Oudin (1952), either the antibody or antigen is incorporated into the gel before it sets. A hole is cut in the gel and it is filled with the reciprocal reactant. A positive reaction is indicated by a ring of precipitation. A major disadvantage of this test is the large amount of each reactant used.

In the gel double-diffusion test both antigen and antibody are placed in wells cut in the gel. The reactants are allowed to diffuse through the gel and a line of precipitate is formed at the point where they meet. This technique, reviewed by Ouchterlony and Nilsson (1978), is still widely used by plant virologists to examine the identity and relationship of virus isolates and to provide information on the relative titre of an antiserum. Disadvantages with the test are reductions in sensitivity when viruses are present in crude extracts in low concentration (e.g. luteoviruses) and the need to dissociate the particles of filamentous viruses to allow diffusion through the gel. Gels are traditionally formed on glass surfaces of either microscope slides or Petri dishes, but in our laboratory we have often used the hydrophilic side of GelBond™ film (FMC Corporation, ME, USA). Gels may then be stained with an appropriate protein stain and dried to form a permanent record.

8.2.4 *Immunoelectrophoresis*

These techniques can be used to differentiate often complex mixtures of antigens on the basis of electrophoretic mobility and antigenic specificity and, hence, can be useful to characterize viruses or differentiate closely related virus strains. The mixture of antigens is first separated by electrophoresis and then antiserum is placed in a depression cut into the agar or agarose gel, parallel to the path of electrophoretic migration, following which immunodiffusion precipitin lines are allowed to develop.

8.2.5 *Agglutination*

The reacting antibody or the antigen itself can be coupled to a larger particle approximately the size of a cell. A positive antigen–antibody reaction then results in a clumping of the particles that is visible to the naked eye or under a stereo-microscope. This reaction is referred to as agglutination rather than precipitation.

When a specific antiserum is mixed with virus-infected crude plant sap, a positive reaction results in the chloroplasts and other cell debris clumping together. This is referred to as the chloroplast agglutination test. Alternative carrier particles may also be used for the agglutination test including erythrocytes (haemagglutination), bentonite, barium sulphate and the extremely useful polystyrene latex (latex flocculation test). The latex flocculation test is performed on the surface of a glass slide or Petri dish with preparations of latex particles that are now available from a range of commercial sources. This test is highly suited to the detection of plant viruses (Abu Salih *et al.*, 1968), and is reported to be up to 1000 times more sensitive than the microprecipitin or immunodiffusion tests and can be carried out with lower concentrations of reactants (Koenig *et al.*, 1979). More recently, Fukami *et al.* (1989) have described a gelatin particle agglutination test for the mass screening of Welsh onion for garlic latent virus, and Sander *et al.* (1989) have described a reverse passive haemagglutination test which involves a single step and provides results which can be read with the unaided eye after 90 min incubation at room temperature.

Another variant of the latex flocculation test is the protein-A-coated latex-linked antiserum (PALLAS) test. In this test, the antibodies are coupled to the latex by using an intermediate layer of protein-A from the bacterium, *Staphylococcus aureus* (Torrance, 1980). This allows the use of antisera with a low titre as the amount of antibody required is not critical as in the case of the conventional latex flocculation test. Ertunct (1989) found the PALLAS test to be more sensitive than the latex flocculation test when testing crude sap and purified preparations for cucumber mosaic virus.

A further variation of the agglutination test is the virobacterial agglutination test initially developed for the detection of plant viruses by Chirkov *et al.* (1984). In this test, *Staphylococcus aureus* bacteria particles conjugated with specific polyclonal antibodies are mixed with crude plant sap. A positive reaction is indicated by the clumping of bacterial particles within 30 s to 5 min after mixing. This test was recently used by Walkey *et al.* (1992) to identify a wide range of plant viruses and was found to be more sensitive and simpler to use than the latex particle agglutination test.

8.3 Immunoelectron Microscopy

The two techniques most commonly associated with immunoelectron microscopy are immunosorbent electron microscopy (ISEM) and decoration (Derrick, 1973; Milne and Luisoni, 1975; Milne, 1984). Both techniques continue to be popular in diagnostic services where they rapidly provide data on the amounts and kinds of virus particles present in even relatively crude preparations, and Milne (1986) lists a series of criteria for when either technique would be preferred over the simpler standard negative staining. Both techniques are also particularly useful in research, either in the initial phases of an investigation or in the development and/or validation of mass testing techniques such as ELISA. A new dimension has been added to immunoelectron microscopy (and other serological tests) with the development of a potyvirus group-specific monoclonal antibody (Jordan and Hammond, 1991) and Kiratiya-Angul and Gibbs (1992) have used this antibody in ISEM and ELISA to detect potyviruses in several wild plant species in southeastern Australia.

ISEM generally involves four steps: the immobilization of antibodies on to a grid support film; the removal of unadsorbed antibodies by washing; the 'trapping' of specific virus particles following exposure of the plant extract to the immobilized antibodies; and the visualization of the result by negative staining. In practice, conditions for making the support film, antiserum coating and virus trapping are legion, with workers tending to settle for a technique that functions reasonably well in their laboratory. For decoration, virus particles in a plant extract are immobilized on to a grid support film and unadsorbed material removed by washing. The virus particles are then exposed to antibodies and the result visualized by negative staining. ISEM and decoration are often combined into a single test to increase the number of virus particles that are initially trapped and then subsequently decorated. Labelled antibodies (gold, protein-A and others) are rarely used for diagnostic purposes.

8.4 Immunosorbent Assays

8.4.1 Enzyme-linked immunosorbent assays (ELISA)

In ELISA, either the antigen or antibody is attached to a solid surface and the sensitivity of detection of the ensuing antigen–antibody reaction is increased by attaching an enzyme, usually to the immunoglobulin G (IgG) fraction of the antibody. In a positive reaction, the enzyme is generally detected by observing a colour reaction that follows the addition of an appropriate enzyme substrate. The original form of the test was developed by Engvall and Perlmann (1971) for the quantitative assay of IgG.

The classic papers of Voller *et al.* (1976) and Clark and Adams (1977) directed the use of the ELISA technique into the new area of direct detection and assay of infectious agents such as viruses and thus heralded a revolution in plant virus diagnostics. Clark and Bar-Joseph (1984) provide a thorough summary of the principles of ELISA as well as variations in procedures and protocols that were developed for use in plant virology in the period to that date. Cooper and Edwards (1986) also provide a concise overview of the variations as well as some of the limitations to the use of ELISA.

The double-antibody-sandwich (DAS) ELISA of Clark and Adams (1977) remains the most commonly used method and is the logical starting point of those unfamiliar with the technique. In this method, the wells of a solid support, usually a polystyrene microtitre plate, are first coated with trapping antivirus IgG; non-adsorbed IgG is washed off. A test sample containing virus (the antigen) is then added to the wells and incubated; non-attached virus is washed off. The enzyme-labelled antivirus IgG conjugate is then added, followed by a colourless substrate which is hydrolysed to a coloured product upon reaction with the bound enzyme. The colour reaction is assessed visually or spectrophotometrically.

The DAS ELISA is highly strain specific (Koenig, 1978) and this may be a disadvantage for diagnostic work where a wide range of strains may need to be detected (Yanase *et al.*, 1986). In this case, another variation may be developed to suit a particular purpose, as for example an indirect form of ELISA incorporating an antiglobulin conjugate (Van Regenmortel and Burckard, 1980). The ELISA has now advanced to the stage where there are many variations and configurations available, and it has become almost impossible to list them all and discuss their relative merits. Cooper and

Edwards (1986) have diagrammatically represented eight selected variations in the design of sandwich enzyme immunoassays and these cover direct and indirect as well as those incorporating protein-A, the F(ab')$_2$ segments of IgG and the Clq component of complement. In the ensuing section, we will highlight some of the more important factors to be considered in terms of the use of ELISA for plant virus diagnostics as well as recent variations and refinements in each of the ELISA steps.

8.4.2 Important considerations for plant virus ELISAs

Sap extracts can be made from the majority of plant species using phosphate buffered saline (PBS) plus Tween 20 with polyvinylpyrrolidone (PVP) added to reduce the amount of non-specific reaction due to sap components such as tannins, although in certain host–virus combinations, other buffer additives are more suitable (Scott et al., 1989). The particular plant part or organ and how and when it is sampled must be the subject of careful choice. For example, when sampling fruit trees for the pollen-borne ilarviruses, it is important to be aware that the virus titre drops to relatively low levels within a short period after bud burst so testing must be completed within a narrow time period early in the growing season (Torrance and Dolby, 1984). Additionally, the virus distribution is reported to be uneven in plum trees infected with plum pox virus and composite samples should therefore be collected from various parts of the canopy (Adams, 1978). Particular plant parts may be incompatible with certain enzyme/substrate systems as shown by Jones and Mitchell (1987), where root extracts from physically injured plants tested using a horseradish peroxidase conjugate and tetramethylbenzidine (TMB) substrate were found to contain high levels of a factor able to oxidize TMB irrespective of the virus status of the plant.

Due to the difficulty of producing extremely pure preparations of many plant viruses, many polyclonal antisera to plant viruses are contaminated with low levels of antibodies to host plant proteins. A number of strategies are employed to reduce the non-specific background reaction to host components such as cross-absorbing the host proteins by preincubation of antisera with healthy leaf extract prior to fractionation (Rybicki and Von Wechmar, 1982), or the addition of healthy leaf extract to enzyme-conjugated gamma-globulin (Stein et al., 1987). The use of monoclonal antibodies (MAbs) alone or in combination with polyclonal antibodies (PAbs) has largely resolved this problem as the MAbs are specifically selected on the basis of their reaction against viral antigens only and exclude any reactions against plant components (Martin, 1987).

In terms of their specificity, MAbs offer another important feature when used for plant virus diagnostics. They can be selected on the basis of their high specificity to a single strain of the virus or alternatively on their ability to react to all viruses in a broad group or family. Jordan and Hammond

(1991) generated a panel of MAbs against a mixture of 12 potyvirus isolates and of the 30 MAbs selected for detailed analysis, one was found to recognize an epitope conserved on all aphid-transmissible potyviruses examined, while other MAbs reacted with either single isolates of bean yellow mosaic virus (BYMV), only with isolates within the BYMV subgroup, only with a range of BYMV strains, or with a BYMV isolate and at least one other potyvirus. The high specificity of many MAbs can be used to advantage for detection of particular virus strains and Rose *et al.* (1987) were able to detect the tobacco veinal necrosis strain of potato virus Y with MAbs but not with PAbs, while Sward and Lister (1988) required a MAb specific for the MAV isolates of barley yellow dwarf virus (BYDV) to differentiate between MAV- and PAV-like isolates because of cross-reactivity encountered with BYDV polyclonal antibodies. However, the high specificity of a single MAb can also be a disadvantage and Koenig and Torrance (1986) found that a mixture of MAbs was required to detect both intact and partially proteolysed particles of potato virus X and Gugerli and Fries (1983) used a cocktail of several MAbs to detect all strains of potato virus Y. In some cases it has been found that using a MAb for both coat and conjugate steps can give poor results and greater sensitivity is achieved when MAbs are used for one of the ELISA steps and PAbs for another (Somowiyarjo *et al.*, 1988; Barker, 1989). Spurious cross-reactions between plant viruses and MAbs may result when highly concentrated preparations of MAbs are used although these can be abolished by using defatted milk proteins as the blocking agent (Zimmermann and Van Regenmortel, 1989).

An important point to note is that for the largest plant virus group, the potyviruses, the findings of Shukla *et al.* (1992) suggest that serology, while useful for detection, is an imperfect criterion for identification and classification. Shukla *et al.* (1988) have shown that the virus-specific epitopes are usually located in the *N*-terminus region of the coat protein and that this region is removed during virus purification and storage. Polyclonal antisera subsequently produced to the partly degraded potyvirus particles exhibit variable cross-reactivity. Shukla *et al.* (1992) have proposed a number of solutions to these problems of potyvirus serology.

8.4.3 Optimization

Careful optimization of all steps of the ELISA can lead to a significant improvement in detection capability. Hewings and D'Arcy (1984) were able to improve the detection of beet western yellows virus from 37ngml^{-1} to 2ngml^{-1} by carefully varying the concentrations of coating and conjugated antibodies and the time and temperature of incubation for each step. One of the major limitations in optimizing an immunoassay to increase sensitivity is the ratio of IgG to enzyme obtained in preparing the conjugate IgG. Alkaline phosphatase is usually conjugated to the antibody in a one-step procedure

using glutaraldehyde (Avrameas, 1969). However, the efficiency of conjugation is low as the IgG conjugates retain 60–70% of the initial enzyme activity
but only 1–10% of the initial immunological activity (Engvall *et al.*, 1971).
Furthermore, conjugation of antibodies with an enzyme may cause spatial
impairment of the combining sites, thus reducing the binding ability of the
antibodies (Koenig, 1978). Some researchers (for example Wisdom, 1976)
recommend the removal of non-labelled IgG from the preparation using gel
filtration, although others (for example Clark, 1981) report no significant
improvement in either the efficiency of conjugation or virus detection using
this variation.

8.4.4 Enzyme–substrate combinations

There are a number of enzyme conjugates and substrates that have been used to
facilitate the assessment of results in ELISA tests. The most common substrate
is p-nitrophenyl phosphate which is used in combination with alkaline
phosphatase. Other reactants such as biotin–avidin (Diaco *et al.*, 1985), biotin–
streptavidin (Dietzgen and Herrington, 1991), 3,3',5,5'-tetramethylbenzidine
(Bos *et al.*, 1981), nitrophenyl galactose (Yolken, 1982), glucose and
5-aminosalicylic acid (Johnson *et al.*, 1980), urea (Chandler *et al.*, 1982),
fluorogenic compounds such as 4-methylumbelliferyl phosphate (Dolores-
Talens *et al.*, 1989) and Bromothymol blue and a starch–iodine complex in
conjunction with penicillin in penicillinase-based ELISA (Sudarshana and
Reddy, 1989) have been used. Penicillin is reported (Sudarshana and Reddy,
1989) to be available at substantially lower costs than p-nitrophenyl phosphate
and while urea substrate is also inexpensive and can be rapidly detected by eye
with a pH indicator (Cooper and Edwards, 1986), considerable care must be
taken in the washing steps prior to addition of the substrate to guard against a
false pH change (Chandler *et al.*, 1982).

   The use of spectrophotometry to detect alkaline phosphatase and
peroxidase substrates has been well documented to be less sensitive than
fluorimetry (for example, Ishikawa and Kato, 1978) although the detection
of fluorogenic substrates requires more sophisticated equipment (Tijssen,
1985). The use of fluorogenic, chemiluminescent or radioactive labels has
significantly increased the sensitivity of immunoassay systems, and detection
of femtomolar (10^{-15}) levels of peroxidase has been achieved (Puget *et al.*,
1977) although Tijssen (1985) has cautioned that the sensitivity of immunoassay systems using these labels and substrates appears to be limited by the
ability of the antigen to bind to the antibody on the solid phase. Another
system involves the bound enzyme-labelled antibody catalysing the dephos-
phorylation of $NADP^+$ to NAD^+, which is then used in a secondary
enzyme-mediated cyclic reaction to produce a red-coloured formazan dye
(Self, 1985). This enzyme-amplification technique has been used to detect
plant viruses even in individual insect vectors (Torrance, 1987).

8.4.5 Statistical considerations

The interpretation of ELISA data and methods used to establish the positive–negative threshold has been examined by Sutula *et al.* (1986) who were concerned that the careful attention given to the setting of test thresholds in the clinical (human) laboratory is not as well recognized in the plant sciences. They consider that it is important to replicate samples, always to include a positive control, to use a number of different individuals for the negative standards to ensure adequate coverage of the possible range of healthy (background) values and to match control samples and test samples with respect to host type, tissue type, age and position on the plant. The positive–negative threshold should be chosen on the basis of an acceptable reference, not arbitrarily, and when reporting ELISA data the threshold used should be stated.

When very large numbers of plants are required to be tested for virus, it may be possible to reduce the workload by batch testing, in which the plant samples are pooled before testing by ELISA. If high infection levels are expected, test batches comprising relatively small numbers of samples should be used (Gibbs and Gower, 1960). If low levels of virus are expected, large test batches can be used provided the ELISA system being used is sufficiently sensitive to detect virus at a high dilution (Moran *et al.*, 1983). If the incidence of virus is expected to vary dramatically between crops to be sampled, it may be preferable to adopt a scheme of sequential batch testing, in which progressively smaller batches are tested only if necessary (Rodoni *et al.*, 1995).

8.4.6 Solid phases

The polystyrene, 96 well microwell plate is still the most commonly used solid support and is particularly useful for large-scale testing where automation is used. A range of microwell plates is commercially available with different properties that may affect their performance in a particular set of tests and Craig *et al.* (1989) conclude that the solid phase should be considered as part of the standardization of ELISAs. More recently, the traditional polystyrene microwell plate and the even more recent filter membrane are being replaced with new technologies such as magnetic microspheres and microparticles. The magnetic microspheres enzyme immunoassay comprises four steps: the immobilization of antibody on to magnetizable beads; the incorporation of the beads into a mixture of plant sap extract and antibody–enzyme conjugate; the removal of the beads from the mixture; and the visualization of the result with an appropriate substrate. The microparticle immunoassay system has a similar series of steps to the magnetic microsphere immunoassay, with carboxylated latex particles of approximately 0.2 µm replacing the magnetic microspheres as the solid phase. Both systems are reported (Bantarri *et al.*, 1991; Robbins *et al.*, 1991) to deliver sensitivity comparable to either a conventional ELISA or dot-ELISA while reducing assay time to 45 min. The magnetic microsphere immunoassay is reported (Bantarri *et al.*, 1991) to be especially useful in assaying samples that contain high levels of impurities or interfering substances, while the microparticle immunoassay has been used primarily in the medical field for the detection of hepatitis A (Robbins *et al.*, 1991) and B virus (Robbins *et al.*, 1992).

8.5 Immunoblotting

Immunoblotting, in its broadest sense, was first applied to plant virus diagnostics in 1982 (O'Donnell *et al.*). Initially the technique involved four steps: the separation of viral proteins in a sample of plant sap by gel electrophoresis; the transfer of the proteins to a protein binding membrane; the blocking of free-protein binding sites on the membrane; and the visualization of the proteins either directly with a labelled virus-specific antibody or indirectly with labelled globulin-specific antibodies or protein-A. In plant virus diagnostics, this form of immunoblotting, often referred to as an electro-blot immunoassay, has a wide variety of documented applications (Koenig and Burgmeister, 1986). More recently, a simplified version of the technique, referred to as dot-blot immunoassay, has been developed (Gumpf *et al.*, 1984). In this version, the antigens are applied directly to the membrane without any prior electrophoretic separation. When used in conjunction with enzyme-labelled antibodies the test is termed 'dot-ELISA' (Bantarri and Goodwin, 1985). The dot-blot and dot-ELISA techniques have lost the advantage of the high specificity of the immunoblotting techniques that involve electrophoretic separation of the virus coat proteins from the non-specific host proteins. Even so, dot immuno binding on nitrocellulose as used by Lange and Heide (1986) was found to be highly suitable for routine seed health testing even in poorly equipped laboratories and it was found to be a simple and relatively inexpensive procedure, gave a clear-cut answer and did not require any special equipment. A dot-ELISA used for the detection of a number of potato viruses was equally sensitive on plain paper and nitrocellulose, although both of these methods were 2–8 times less sensitive than double-antibody-sandwich (DAS) ELISA (Heide and Lange, 1988). Other detection systems are also possible (Hsu, 1984), and Reichenbacher *et al.* (1990) have developed a relatively simple ELISA using dry reagent carriers immobilized on PVC strips, while Tsuda *et al.* (1992) have developed a rapid immunofilter paper assay that utilizes antibody-coated latex beads immobilized on a filter paper strip.

8.6 Future Prospects

Enzyme immunoassays remain the most widely accepted of the methods for the detection of plant viruses. The major drawbacks of the conventional ELISA are its relatively high cost, the time taken to complete tests and the use of relatively large amounts of reagents. In addition, there is a requirement of purified antigens and antibodies (polyclonal or monoclonal) while the varying binding characteristics of the solid phases for different proteins influence the reproducibility and standardization of the ELISA (Burt *et al.*, 1979). In the medical field, new advances such as the microparticle immunoassay have significantly reduced the time taken to complete tests and offered opportunities for automation (Robbins *et al.*, 1992). However, this trend is contrasted by the move towards simplifying the ELISA to reduce costs and time, especially in developing countries. Of relevance are advances such as dipstick ELISA (Sumathy *et al.*, 1992) which are expanding the use of ELISA as a diagnostic tool in certain situations.

Another advance in the medical field is the development of synthetic peptides which mimic intact proteins (Leinikki *et al.*, 1993). These peptides are based on published sequences and have a number of advantages including their increased specificity and sensitivity, their stability and their low assay variation and are therefore valuable tools for the production of highly defined antibodies for use in viral diagnosis. In cases such as the potyviruses, where the protein coat amino acid sequences of several key isolates are available, the opportunity exists to react monoclonal antibodies with synthetic peptides and thus identify unique linear epitopes along the linearized protein coat molecule (Gugerli, 1990).

8.7 Antisera Available Commercially

There is a wide range of plant virus antisera now available from a number of commercial sources throughout the world (see Table 8.1). These antisera comprise both polyclonal and monoclonal antibodies and represent most of the major plant virus groups as cited in *Descriptions of Plant Viruses* (Murant and Harrison, 1989). It is interesting to note that certain companies appear to have specialized in developing antisera to a large number of viruses within certain groups. For example, the American Type Culture Collection (ATCC) has antisera to nine different sobemoviruses and Deutsche Sammlung von Mikroorganismen und Zellkulturen (DSM), Germany, has antisera to nine different carlaviruses. Other companies, such as Boehringer Mannheim, have concentrated on developing antisera to a particular crop. Some virus groups, notably the parsnip yellow fleck group, the plant reoviruses subgroup II (Fijiviruses) and the tenuiviruses, do not yet have antisera available from the commercial sources from which we have drawn our information. While

Table 8.1. Commercially available antisera for plant viruses.

Virus	Agdia[1]	Sanofi[2]	Loewe[3]	ATCC[4]	DSM[5]
Alfalfa mosaic group					
Alfalfa mosaic	●	●	●	●○	●
Barley yellow mosaic group					
Barley yellow mosaic		●	●		●
Bromoviruses					
Broad bean mottle				●○	
Brome mosaic			●		●
Cowpea chlorotic mottle				●	●
Capilloviruses					
Apple stem grooving			●	●	
Potato T					●
Carlaviruses					
Carnation latent	●		●	●	●
Chrysanthemum B	●	●	●	●	●
Cowpea mild mottle					●
Helenium S					●
Kalanchoe I	●				
Lily symptomless	●			●	
Narcissus latent				●	
Passiflora latent					●
Pea streak					●
Poplar mosaic					●
Potato M	●	●	●		●
Potato S	●				●
Carmoviruses					
Carnation mottle	●	●	●	●	●
Cucumber leaf spot				●	
Galinsoga mosaic				●	
Glycine mottle				●	
Hibiscus chlorotic ringspot	●			●	
Melon necrotic spot		●			●
Pelargonium flower break	●	●	●		●
Saguaro cactus				●	
Turnip crinkle					●
Caulimoviruses					
Carnation etched ring		●		●○	●
Cauliflower mosaic	●		●	●	
Dahlia mosaic	●				
Closteroviruses I					
Carnation necrotic fleck	●	●		●○	
Citrus tristeza	●	●			●
Closteroviruses II					
Apple chlorotic leafspot		●	●	●○	
Grapevine leafroll-associated	●	●			
Grapevine A	●			●	

	1	2	3	4	5
Closteroviruses III					
Lilac chlorotic leafspot				●	●
Cocksfoot mild mosaic group					
Cocksfoot mild mosaic					●
Comoviruses					
Andean potato mottle					●
Bean pod mottle				●	
Broad bean stain				●	
Cowpea mosaic				●	●
Cowpea severe mosaic				●○	●
Echtes Ackerbohnenmosaik					●
Glycine mosaic				●	
Radish mosaic			●		●
Red clover mottle			●		
Squash mosaic	●	●	●	●	
Cryptoviruses					
Carnation cryptic				●	
Red clover cryptic (?)					●
Ryegrass cryptic (?)					●
White clover cryptic 1				●	●
White clover cryptic 2				●	
Cucumoviruses					
Cucumber mosaic	●	●○	●	●	●
Peanut stunt				●	●
Tomato aspermy	●	●	●	●	●
Dianthoviruses					
Carnation ringspot	●		●	●	●
Red clover necrotic mosaic				●	
Fabaviruses					
Broad bean wilt		●	●		●
Furoviruses					
Beet necrotic yellow vein (?)	●	●	●		
Wheat soil-borne mosaic		●			
Geminiviruses					
Bean golden mosaic					●
Beet curly top				●	
Maize streak				●	
Wheat dwarf		●			
Hordeiviruses					
Barley stripe mosaic	●			●	
Ilarviruses					
American plum line pattern				●	
Apple mosaic	●	●	●	●○	●
Asparagus II	●			●	
Citrus leaf-rugose				●	●
Prune dwarf	●	●	●	●○	●
Prunus necrotic ringspot	●	●	●	●○	●
Raspberry bushy dwarf (?)	●		●	●	●
Tobacco streak	●	●		●○	
Tulare apple mosaic				●	●

Table 8.1 Continued.

Virus	Agdia[1]	Sanofi[2]	Loewe[3]	ATCC[4]	DSM[5]
Luteoviruses					
Barley yellow dwarf	●	●	●	●○	
Bean leafroll				●	
Beet western yellows	●	●	●	●○	●
Potato leafroll	●	●	●	●○	●
Soyabean dwarf				●○	
Tobacco necrotic dwarf				●	
Maize chlorotic dwarf group					
Maize chlorotic dwarf				●	
Marafiviruses					
Maize rayado fino				●	
Necroviruses					
Carnation yellow stripe (?)				●	
Tobacco necrosis			●	●	●
Nepoviruses					
Arabis mosaic	●	●	●	●	●
Aracacha B (?)					●
Artichoke Italian latent		●		●	
Blueberry leaf mottle	●			●	
Cherry leafroll	●	●	●	●	●
Cherry rasp leaf				●	
Dogwood mosaic				●	
Grapevine fanleaf	●	●	●	●	
Myrobalan latent ringspot				●	
Peach rosette mosaic	●			●	
Potato black ringspot					●
Raspberry ringspot	●	●	●		●
Strawberry latent ringspot (?)	●	●	●		●
Tobacco ringspot	●			●	●
Tomato black ring		●	●		●
Tomato ringspot	●	●	●	●	●
Parsnip yellow fleck group					
Pea enation mosaic group					
Pea enation mosaic			●		●
Plant preoviruses I (Phytoreoviruses)					
Wound tumor				●	
Plant reoviruses II (Fijiviruses)					
Plant reoviruses III					
Rice ragged stunt				●	●
Potexviruses					
Argentine plantago				●	
Cactus X				●	●
Cassava common mosaic				●	
Clover yellow mosaic				●	

	1	2	3	4	5
Cymbidium mosaic	●	●		●	●
Discorea latent				●	
Hydrangea ringspot			●		●
Narcissus mosaic				●	
Nerine X				●	
Papaya mosaic	●				
Potato X	●	●	●	●	●
Viola mottle					●
White clover mosaic				●	
Potyviruses					
Alstroemeria mosaic					●
Asparagus I			●	●	
Barley mild mosaic				●	
Barley yellow mosaic				●	
Bean common mosaic			●		
Bean yellow mosaic		●	●	●○	
Beet mosaic			●		●
Bidens mottle				●	
Blackeye cowpea mosaic				●	
Carnation vein mottle			●		●
Carrot thin leaf				●	
Celery mosaic			●	●	●
Clover yellow vein				●	
Eggplant green mosaic				●	
Freesia mosaic			●		●
Iris severe				●	
Leek yellow stripe		●			
Lettuce mosaic	●	●	●	●○	●
Maize dwarf mosaic		●	●		
Onion yellow dwarf		●		●	
Papaya ringspot / (watermelon mosaic I)	●	●	●	●	●
Pea seedborne mosaic		●	●	●	●
Peanut green mosaic				●	
Peanut mottle				●○	●
Peanut stripe				●	●
Pepper veinal mottle					●
Plum pox	●	●	●	●	●
Potato A	●	●	●	●	●
Potato V	●				●
Potato Y		●	●	●	●
Sorghum mosaic	●			●	
Soyabean mosaic		●	●	●	
Sugarcane mosaic	●		●	●	●
Tobacco etch		●		●	●
Tulip breaking	●			●○	●
Turnip mosaic	●		●	●	●
Watermelon mosaic II		●		●	●
Washington tulip				●	
Wheat spindle streak mosaic				●	
Wheat streak mosaic				●	●
Yam mosaic				●	

Table 8.1 Continued.

Virus	Agdia[1]	Sanofi[2]	Loewe[3]	ATCC[4]	DSM[5]
Zucchini yellow fleck			•		
Zucchini yellow mosaic	•	•	•	•	
Rhabdoviruses					
Cereal chlorotic mottle				•	
Eggplant mottled dwarf					•
Potato yellow dwarf				•	•
Sonchus yellow net					•
Sobemoviruses					
Blueberry shoestring (?)	•			•	
Ginger chlorotic fleck				•	
Lucerne transient streak				•	
Panicum mosaic				•	
Rice yellow mottle				•	•
Solanum nodiflorum mottle				•	
Southern bean mosaic				•	•
Sowbane mosaic				•	
Velvet tobacco mosaic				•	
Tenuiviruses					
Tobamoviruses					
Cucumber green mottle mosaic				•	•
Frangipani mosaic				•	
Kyuri green mottle mosaic				•	
Odontoglossum ringspot	•	•	•	•	•
Pepper mild mottle	•	•		•	•
Ribgrass mosaic				•	
Sammons' opuntia				•	
Sun-hemp mosaic				•	•
Tobacco mild green mosaic					•
Tobacco mosaic	•	•	•	•	•
Tomato mosaic	•		•	•	•
Tobraviruses					
Tobacco rattle	•		•	•	•
Tombusviruses					
Carnation Italian ringspot					•
Cucumber necrosis (?)				•	•
Cymbidium ringspot (?)	•			•	•
Pelargonium leaf curl			•		•
Petunia asteroid mosaic			•		•
Tomato bushy stunt				•	•
Tospoviruses					
Tomato spotted wilt	•	•	•	•○	•
Tymoviruses					
Andean potato latent				•	
Belladonna mottle				•	•
Desmodium yellow mottle				•	
Eggplant mosaic					•
Kennedya yellow mosaic				•	

Okra mosaic		●	●
Plantago mottle		●	
Poinsettia mosaic (?)	●	●	●
Scrophularia mottle		●	
Turnip yellow mosaic	●	●	●

●, polyclonal antisera; ○, monoclonal antisera.

Agdia also distributes a number of *Bioreba* (Baumgartenstrasse 5 D-7801 Ebringen, Germany) polyclonal antisera to plant viruses. These are mainly to viruses in potatoes (Potato A potyvirus, Potato M carlavirus, Potato S carlavirus, Potato V potyvirus, Potato X potexvirus, Potato Y potyvirus and Potato leafroll luteovirus) as well as to Arabis mosaic nepovirus, Barley yellow dwarf luteovirus, Beet necrotic yellow vein furovirus, Citrus tristeza closterovirus, Grapevine A closterovirus, Grapevine fanleaf nepovirus, Grapevine leafroll associated closterovirus and Raspberry ringspot nepovirus.

Boehringer Mannheim GmbH Biochemica, Sandhofer Str, 116, Postfach 310120, 6800 Mannheim 31, Germany, also supply a number of polyclonal antisera, mainly to viruses in potatoes (Potato A potyvirus, Potato M carlavirus, Potato S carlavirus, Potato V potyvirus, Potato X potexvirus, Potato Y potyvirus and Potato leafroll luteovirus) but also to Apple mosaic ilarvirus, Beet necrotic yellow vein furovirus, Plum pox potyvirus and Prune dwarf ilarvirus.

[1] *Agdia, Inc.* 30380 County Rd. 6, Elkhart, IN 46514, USA. In addition to their potyvirus specific antisera, Agdia also supply a general potyvirus group antisera that reacts to Alstroemeria mosaic, Amaranthus leaf mottle, Arabis mosaic, Asparagus I, Bean common mosaic, Bean yellow mosaic, Bearded iris mosaic, Beet mosaic, Bidens mottle, Blackeye cowpea mosaic, Cardamon mosaic, Carnation vein mottle, Carrot thin leaf, Celery mosaic, Clover yellow vein, Cocksfoot streak, Colombian datura, Commelina mosaic, Cowpea aphid-borne mosaic, Daphne Y, Datura shoestring, Garlic mosaic, Gloriosa stripe mosaic, Groundnut eyespot, Guinea grass mosaic, Helenium Y, Henbane mosaic, Hippeastrum mosaic, Hyacinth mosaic, Iris fulva mosaic, Iris mild mosaic, Iris severe mosaic, Leek yellow stripe, Maize dwarf mosaic, Malva vein clearing, Narcissus degeneration, Narcissus yellow stripe, Onion yellow dwarf, Ornithogalum mosaic, Parsnip mosaic, Passionfruit woodiness, Pea mosaic, Pea seedborne mosaic, Peanut mottle, Peanut stripe, Pepper mottle, Pepper severe mosaic, Pepper veinal mottle, Statice Y, Sugarcane mosaic, Sweet potato feathery mottle, Sweet potato latent, Tamarillo mosaic, Tobacco vein mottling, Tulip breaking, Tulip chlorotic blotch, Vallota mosaic, White lupin mosaic, Wisteria vein mosaic, Yam mosaic and Zucchini yellow fleck viruses.

[2] *Sanofi Phyto-diagnostics* Z.I. de la Ballastiere, BP 126, F-33501 Libourne cedex, France.

[3] *Loewe Biochemica GmbH* Nordring 38, Postfach 9, W-8156 Otterfing (via Munich), Germany, also supply antisera to Beet yellows mosaic, Cucumber green mottle and Grapevine Algerian latent viruses which are as yet ungrouped.

[4] *American Type Culture Collection* (ATCC), 12301 Parklawn Drive, Rockville, Maryland 20852-1776, USA, also supply antisera to Cowpea mottle, Cucumber fruit streak virus, Maize chlorotic mottle, Maize white line mosaic, Malvastrum mottle, Ourmia melon and Sorghum yellow banding viruses which are as yet ungrouped as well as five potyvirus group cross-reactive monoclonal antisera.

[5] *Deutsche Sammlung von Mikroorganismen und Zelkulturn GmbH.* (DSM), Arbeitsgruppe Pflanzenviren, Messeweg 11/12, D-3300 Braunschweig, Germany, also supply antisera to Barley mild mosaic, Barley stripe, Bell pepper mottle, Chilli veinal mottle, Impatiens necrotic spot, Maracuja mosaic, Neckar river, Pelargonium line pattern, Pelargonium ringspot, Pelargonium zonate spot, Sieg river, Sugarbeet yellows, Tobacco vein banding and the satellite of tobacco necrosis virus which are as yet ungrouped.

Table 8.1 covers the major suppliers of plant virus antisera known to us, we do not claim that it is totally exhaustive as antisera may sometimes be purchased from government, university or other research laboratories.

8.8 Concluding Remarks

There are now a wide range of tests available for use in plant virus diagnostics and some of these are highly advanced and sophisticated. Some of the most useful and versatile tests are based on the antibody approach to virus diagnostics and utilize major advances that have occurred in the area of serology over the last two decades. Increased commercial availability of high quality polyclonal antisera, rapid advances in monoclonal antibody technology, development of synthetic peptides, production of new types of solid phases and numerous other advances provide powerful aids for a variety of diagnostic tests. However, it is extremely important to be aware of the limitations inherent in every test. In situations where it is critical that the probability of virus detection is maximized, it is often prudent at least in the initial phase of test validation to conduct parallel tests employing quite different strategies for diagnosis of the suspect virus.

References

AAB (Association of Applied Biologists) (1989) *Descriptions of Plant Viruses.* Murant, A.F. and Harrison, B.D. (eds). Institute of Horticultural Research, Wellesbourne, UK.

Abu-Salih, H.S., Murant, A.F. and Daft, M.J. (1968) The use of antibody-sensitized latex particles to detect plant viruses. *Journal of General Virology* 3, 299–302.

Adams, A.N. (1978) The detection of plum pox virus in *Prunus* species by enzyme-linked immunosorbent assay (ELISA). *Annals of Applied Biology* 90, 215–221.

Avrameas, S. (1969) Coupling of enzymes to proteins with glutaraldehyde. Use of the conjugates for the detection of antigens and antibodies. *Immunochemistry* 6, 43–52.

Ball, E.M. (1974) Serological tests for the identification of plant viruses. *American Phytopathological Society Monograph*, 31pp.

Bantarri, E.E. and Goodwin, P.H. (1985) Detection of potato viruses S, X and Y by enzyme-linked immunosorbent assay on nitrocellulose membranes (dot-ELISA). *Plant Disease* 69, 202–205.

Bantarri, E.E., Clapper, D.L., Hu, S.P., Daws, K.M. and Khurana, S.M.P. (1991) Rapid magnetic microsphere enzyme immunoassay for potato virus X and potato leafroll virus. *Phytopathology* 81, 1039–1042.

Barker, I. (1989) Development of a routine immunoassay for the detection of beet necrotic yellow vein virus using monoclonal antibodies. *Committee paper–Sugar Beet Research and Education Committee* No. 2371, 27–35.

Bos, E.S., Van der Doelen, A.A., Van Rooy, N. and Schuurs, A.H.W.M. (1981) 3,3',
5,5'-tetramethylbenzidine as an Ames test negative chromogen for horseradish
peroxidase in enzyme-immunoassay. *Journal of Immunoassay* 2, 187–204.

Burt, S.M., Carter, T.J.N. and Kricka, L.J. (1979) Thermal characteristics of microtitre
plates used in immunological assays. *Journal of Immunological Methods* 31,
231–236.

Chandler, H.M., Cox, J.C., Healey, K., MacGregor, A., Premier, R.R. and Hurrell,
J.G.R. (1982) An investigation of the use of urease–antibody conjugates in
enzyme immunoassays. *Journal of Immunological Methods* 53, 187–194.

Chirkov, S.N., Olovnikov, A.M., Surguchyova, N.A. and Atabekov, J.G. (1984)
Immunodiagnosis of plant viruses by a virobacterial agglutination test. *Annals of
Applied Biology* 104, 477–483.

Clark, M.F. (1981) Immunosorbent assays in plant pathology. *Annual Review of
Phytopathology* 19, 83–106.

Clark, M.F. and Adams, A.M. (1977) Characteristics of the microplate method of
enzyme-linked immunosorbent assay for the detection of plant viruses. *Journal
of General Virology* 34, 475–483.

Clark, M.F. and Bar-Joseph, M. (1984) Enzyme immunosorbent assays in plant
virology. In: Maramorosch, K. and Koprowski, H. (eds), *Methods in Virology*.
Academic Press, New York, pp. 51–85.

Conti, M., D'Arcy, C.J. and Jedlinski, H. (1990) The 'Yellow Plague' of cereals, barley
yellow dwarf virus. In: Burnett, P.A. (ed.), *World Perspectives on Barley Yellow
Dwarf*. CIMMYT, Mexico.

Cooper, J.I. and Edwards, M.L. (1986) Variations and limitations of enzyme-
amplified immunoassays. In: Jones, R.A.C. and Torrance, L. (eds), *Developments
and Applications in Virus Testing*. Association of Applied Biologists, Wellesbourne, UK, pp. 139–154.

Craig, J.C., Parkinson, D. and Knowles, N. (1989) The suitability of different
microtitre plates for detection of antibody to virus antigens by indirect ELISA.
Journal of Biological Standardization 17, 125–135.

Derrick, K.S. (1973) Quantitative assay for plant viruses using serologically specific
electron microscopy. *Virology* 56, 652–653.

Diaco, R., Hill, J.H., Hill, E.K., Tachibana, H. and Durand, D.P. (1985) Monoclonal
antibody-based biotin-avidin ELISA for the detection of soybean mosaic virus
in soybean seeds. *Journal of General Virology* 66, 2089–2094.

Dietzgen, R.G. and Herrington, M.E. (1991) A sensitive semi-quantitative biotin–
streptavidin ELISA for the detection of potyviruses infecting cucurbits. *Australian Journal of Agricultural Research* 42, 417–427.

Dolores-Talens, A.C., Hill, J.H. and Durand, D.P. (1989) Application of enzyme-
linked fluorescent assay (ELFA) to detection of lettuce mosaic virus in lettuce
seeds. *Journal of Phytopathology* 124, 149–154.

Engvall, E. and Perlmann, P. (1971) Enzyme-linked immunosorbent assay (ELISA):
quantitative assay of immunoglobulin G. *Immunochemistry* 8, 871–874.

Engvall, E., Jonson, K. and Perlmann, P. (1971) Enzyme-linked immunosorbent assay
II. Quantitative assay of protein antigen, immunoglobin G, by means of
enzyme-labelled antigen and antibody coated tubes. *Biochemica et Biophysica
Acta* 251, 427–434.

Ertunct, F. (1989) Detection of cucumber mosaic virus strains by latex flocculation and protein A-coated latex-linked antisera tests. *Journal of Turkish Phytopathology* 18, 87–92.

Fukami, M., Natsuaki, K.T., Motoyoshi, F. and Tomaru, K. (1989) Simple and rapid detection of garlic latent virus from Welsh onions by gelatin particle agglutination test. *Annals of the Phytopathological Society of Japan* 55, 671–675.

Gibbs, A.J. and Gower, J.C. (1960) The use of a multiple transfer method in plant virus transmission studies – some statistical points arising in the analysis of results. *Annals of Applied Biology* 48, 75–83.

Gugerli, P. (1990) Advanced immunological techniques for virus detection. In: *Control of Virus and Virus-like Diseases of Potato and Sweet Potato*. Report of the III Planning Conference. International Potato Center, Lima, Peru. 20–22 Nov. 1989, pp. 29–33.

Gugerli, P. and Fries, P. (1983) Characterization of monoclonal antibodies to potato virus Y and their use for virus detection. *Journal of General Virology* 64, 2471–2477.

Gumpf, D.J., Kositratana, W. and Zheng, G.Y. (1984) Dot-immunobinding assay for virus detection. *Phytopathology* 74, 847 (Abstract only).

Heide, M. and Lange, L. (1988) Detection of potato leafroll virus and potato viruses M, S, X and Y by dot immunobinding on plain paper. *Potato Research* 31, 367–373.

Hewings, A.D. and D'Arcy, C.J. (1984) Maximizing the detection capability of a beet western yellows virus ELISA system. *Journal of Virological Methods* 9, 131–142.

Hsu, Y.H. (1984) Immunogold for detection of antigen on nitrocellulose paper. *Analytical Biochemistry* 142, 221–225.

Ishikawa, E. and Kato, K. (1978) Ultrasensitive enzyme immunoassay. *Scandinavian Journal of Immunology* 8, Supplement 7, 43–55.

Johnson, R.B., Libby, R.M. and Nakamura, R.M. (1980) Comparison of glucose oxidase and peroxidase as labels for antibody in enzyme-linked immunosorbent assay. *Journal of Immunoassay* 1, 27–37.

Jones, A.T. and Mitchell, M.J. (1987) Oxidising activity in root extracts from plants inoculated with virus or buffer that interferes with ELISA when using the substrate 3,3',5,5'-tetramethylbenzidine. *Annals of Applied Biology* 111, 359–364.

Jordan, R. and Hammond, J. (1991) Comparison and differentiation of potyvirus isolates and identification of strain-, virus-, subgroup-specific and potyvirus group-common epitopes using monoclonal antibodies. *Journal of General Virology* 72, 25–36.

Kiratiya-Angul, S. and Gibbs, A. (1992) Group-specific antiserum detects potyviruses in several wild plant species in south-eastern Australia. *Australasian Plant Pathology* 21, 118–119.

Koenig, R. (1978) ELISA in the study of homologous and heterologous reactions of plant viruses. *Journal of Immunological Methods* 60, 243–254.

Koenig, R. and Burgmeister, W. (1986) Applications of immuno-blotting in plant virus diagnosis. In: Jones, R.A.C. and Torrance, L. (eds), *Developments and Applications in Virus Testing*. Association of Applied Biologists, pp. 121–137.

Koenig, R. and Torrance, L. (1986) Differences in the detectability of intact and partially proteolyzed potato virus X in different serological tests by means of monoclonal antibodies. *Nachrichtenblatt des Deutschen Pflanzenschutzdienstes* 38, 136–137.

Koenig, R., Fribourg, C.E. and Jones, R.A.C. (1979) Symptomatological, serological and electrophoretic diversity of isolates of Andean potato latent virus from different regions of the Andes. *Phyton* 69, 748–752.

Lange, L. and Heide, M. (1986) Dot immuno binding (DIB) for detection of virus in seed. *Canadian Journal of Plant Pathology* 8, 373–379.

Leinikki, P., Lehtinen, M., Hyoty, H., Parkkonen, P., Kantanen, M.-L. and Hakulinen, J. (1993) Synthetic peptides as diagnostic tools in virology. In: *Advances in Virus Research*, vol. 42. Academic Press, London, pp. 149–186.

Martin, R.B. (1987) The use of monoclonal antibodies for virus detection. *Canadian Journal of Plant Pathology* 9, 177–181.

Milne, R.G. (1984) Electron microscopy for the identification of plant viruses in *vitro* preparations. *Methods in Virology* 7, 87–120.

Milne, R.G. (1986) New developments in electron microscope serology and their possible applications. In: Jones, R.A.C. and Torrance, L. (eds), *Developments and Applications in Virus Testing*. Association of Applied Biologists, Wellesbourne, UK, pp. 179–191.

Milne, R.G. and Luisoni, E. (1975) Rapid high resolution immune electron microscopy of plant viruses. *Virology* 68, 270–274.

Moran, J.R., Garrett, R.G. and Fairweather, J.V. (1983) Strategy for detecting low levels of potato viruses X and S in crops and its application to the Victorian Certified Seed Potato Scheme. *Plant Disease* 67, 1325–1327.

Murant, A.F. and Harrison, B.D. (eds) (1989) *Descriptions of Plant Viruses*. Association of Applied Biologists, Wellesbourne, UK.

O'Donnell, I.J., Shukla, D.D. and Gough, K.H. (1982) Electro-blot radioimmunoassay of virus-infected plant sap – a powerful new technique for detecting plant viruses. *Journal of Virological Methods* 4, 19–26.

Ouchterlony, O. (1968) *Handbook of Immunodiffusion and Immunoelectrophoresis*. Ann Arbor Scientific Publications. Ann Arbor, MI, USA.

Ouchterlony, O. and Nilsson, L.A. (1978) Immunodiffusion and immunoelectrophoresis. In: Weir, D.M. (ed.), *Handbook of Experimental Immunology*, 3rd edn. Blackwell, Oxford.

Oudin, J. (1952) Specific precipitation in gels and its application to immunochemical analysis. *Methods in Medical Research* 5, 335–378.

Puget, K., Michelson, A.M. and Avrameas, S. (1977) Light emission techniques for the microestimation of femtogram levels of peroxidase. Application to peroxidase (and other enzymes) – coupled antibody – cell antigen interactions. *Analytical Biochemistry* 79, 447–456.

Raizada, R.K., Srivastava, K.M., Govind-Chandra, Singh, B.P. and Chandra, G. (1989) Comparative evaluation of serodiagnostic methods for detection of chrysanthemum virus B in chrysanthemum. *Indian Journal of Experimental Biology* 27, 1094–1096.

Reichenbacher, D., Reiss, E., Hunger, H.D. and Richter, J. (1990) Detection of plant viruses by ELISA using different reagent strips. *Journal of Phytopathology* 128, 333–339.

Robbins, D.J., Krater, J., Kiang, W., Alcalde, X., Helgesen, S., Carlos, J. and Mimms, L.T. (1991) Detection of total antibody against hepatitis A virus by an automated microparticle enzyme immunoassay. *Journal of Virological Methods* 32, 255–263.

Robbins, D., Wright, T., Coleman, C., Umhoefer, L., Moore, B., Spronk, A., Douville, C., Kuramoto, I.K., Rynning, M., Gracey, D., Salbilla, V., Nehmadi, F. and Mimms, L.T. (1992) Serological detection of HBeAg and anti-HBe using automated microparticle enzyme immunoassays. *Journal of Virological Methods* 38, 267–281.

Rodoni, B.C., Hepworth, G., Richardson, C. and Moran, J.R. (1995) The use of sequential batch testing procedure and ELISA to determine the incidence of five viruses in Victorian cut-flower Sim carnations. *Australian Journal of Agricultural Research* 45, 1–8.

Rose, D.G., McCarra, S. and Mitchell, D.H. (1987) Diagnosis of potato virus Y^N: a comparison between polyclonal and monoclonal antibodies and a biological assay. *Plant Pathology* 36, 95–99.

Rybicki, E.P. and Von Wechmar, M.B. (1982) Enzyme-assisted immune detection of plant virus proteins electroblotted onto nitrocellulose paper. *Journal of Virological Methods* 5, 267–278.

Sander, E., Dietzgen, R.G., Cranage, M.P. and Coombs, R.R.A. (1989) Rapid and simple detection of plant viruses by reverse passive haemagglutination. I. Comparison of ELISA (enzyme-linked immunosorbent assay) and RPH (reverse passive haemagglutination) for plant virus diagnosis. *Zeitschrift für Pflanzenkrankheitenund Pflanzenschutz*. 96, 113–123.

Scott, S.W., Burrows, P.M. and Barnett, O.W. (1989) Effects of plant sap on antigen concentrations calibrated by ELISA. *Phytopathology* 79, 1175.

Self, C.H. (1985) Enzyme amplification: a general method applied to provide an immunoassisted assay for placental alkaline phosphatase. *Journal of Immunological Methods* 76, 389–393.

Shukla, D.D., Strike, P.M., Tracy, S.L., Gough, K.H. and Ward, C.W. (1988) The N- and C-termini of the coat proteins of potyviruses are surface located and the N-terminus contains the major virus-specific epitopes. *Journal of General Virology* 69, 1497–1508.

Shukla, D.D., Lauricella, R. and Ward, C.W. (1992) Serology of potyviruses: current problems and some solutions. In: Barnett, O.W. (ed.), *Potyvirus Taxonomy. Archives of Virology*. Supplement 5. Springer-Verlag, New York, pp. 57–69.

Somowiyarjo, S., Sako, N. and Nonaka, F. (1988) The use of monoclonal antibody for detecting zucchini yellow mosaic virus. *Annals of the Phytopathological Society of Japan*, 54, 436–443.

Stein, A., Levy, S. and Loebenstein, G. (1987) Detection of prunus necrotic ringspot virus in several rosaceous hosts by enzyme-linked immunosorbent assay. *Plant Pathology* 36, 1–4.

Sudarshana, M.R. and Reddy, D.V.R. (1989) Penicillinase-based enzyme-linked immunosorbent assay for the detection of plant viruses. *Journal of Virological Methods* 26, 45–52.

Sumathy, S., Thyagarajan, S.P., Latif, R., Madanagopalan, N., Raguram, K., Rajasambandam, P. and Gowans, E. (1992) A dipstick immunobinding enzyme-linked immunosorbent assay for serodiagnosis of hepatitis B and delta virus infections. *Journal of Virological Methods* 38, 145–152.

Sutula, C.L., Gillett, J.M., Morrissey, S.M. and Ramsdell, D.C. (1986) Interpreting ELISA data and establishing the positive–negative threshold. *Plant Disease* 70, 722–726.

Sward, R.J. and Lister, R.M. (1988) The identity of barley yellow dwarf virus isolates in cereals and grasses from mainland Australia. *Australian Journal of Agricultural Research* 39, 375–384.

Tijssen, P. (1985) Practice and theory of enzyme immunoassays. In: Burdon, R.H. and van Knippenberg, P.H. (eds), *Laboratory Techniques in Biochemistry and Molecular Biology*, vol 15. Elsevier, Amsterdam 549 pp.

Torrance, L. (1980) Use of protein A to improve sensitisation of latex particles with antibodies to plant viruses. *Annals of Applied Biology* 96, 45–50.

Torrance, L. (1987) Use of enzyme amplification in an ELISA to increase sensitivity of detection of barley yellow dwarf virus in oats and in individual vector aphids. *Journal of Virological Methods* 15, 131–138.

Torrance, L. and Dolby, C.A. (1984) Sampling conditions for reliable routine detection by enzyme-like immunosorbent assay of three ilarviruses in fruit trees. *Annals of Applied Biology* 104, 267–276.

Tsuda, S., Kameya-Iwaki, M., Hanada, K., Kouda, Y., Hikata, M. and Tomaru, K. (1992) A novel detection and identification technique for plant viruses: rapid immunofilter paper assay (RIPA). *Plant Disease* 76, 466–469.

Van Regenmortel, M.H.V. (1982) *Serology and Immunochemistry of Plant Viruses.* Academic Press, London.

Van Regenmortel, M.H.V. and Burckard, J. (1980) Detection of a wide spectrum of tobacco mosaic virus stains by indirect enzyme immunosorbent assays (ELISA). *Virology* 106, 327–334.

Voller, A., Bartlett, A., Bidwell, D.E., Clark, M.F. and Adams, A.N. (1976) The detection of viruses by enzyme-linked immunosorbent assay (ELISA). *Journal of General Virology* 33, 165–167.

Walkey, D.G.A., Lyons, N.F. and Taylor, J.D. (1992) An evaluation of a virobacterial agglutination test for the detection of plant viruses. *Plant Pathology* 41, 462–471.

Wisdom, G.B. (1976) Enzyme immunoassay. *Clinical Chemistry* 22, 1243–1255.

Yanase, H., Nakatani, F., Munakata, T. and Machida, K. (1986) Detection of different strains of apple chlorotic leafspot virus and apple stem grooving virus by direct double antibody sandwich method and modified indirect ELISA (F(ab')2 procedure). *Bulletin of the Fruit Tree Research Station, Japan* 13, 69–81.

Yolken, R.H. (1982) Enzyme immunoassays for the detection of infectious antigens in body fluids: current limitations and future prospects. *Reviews of Infectious Disease* 4, 35–68.

Zimmermann, D. and Van Regenmortel, M.H.V. (1989) Spurious cross-reactions between plant viruses and monoclonal antibodies can be overcome by saturating ELISA plates with milk proteins. *Archives of Virology* 106, 15–22.

Nucleic-acid-based Approaches to Plant Virus and Viroid Diagnostics

P.M. Waterhouse and P.W.G. Chu

CSIRO, Division of Plant Industry, PO Box 1600, Canberra, ACT 2601, Australia.

9.1 Introduction

Plant viruses and viroids cause a large number of important natural diseases among major crops (Agrios, 1988; Commonwealth Mycological Institute/ Association of Applied Biologists *Descriptions of Plant Viruses* series, 1970–1988). These include virus diseases of cereals such as barley yellow dwarf and rice tungro, numerous diseases in temperate fruits (Foster, 1988), small fruits (Converse, 1987) and legumes (Edwardson and Christie, 1986), and tomato spotted wilt and potato leafroll in potatoes (de Bokx, 1981). Coconut cadang cadang viroid (CCCV) is estimated to have killed more than 20 million palm trees between 1926 and 1978. Similarly, considerable losses have been recorded for citrus and potatoes infected by citrus exocortis viroid (CEV) and potato spindle tuber viroid (PSTV) respectively.

Losses in crops due to virus or viroid infection can be minimized by taking measures to prevent the spread of these pathogens within the crop. This requires an early and accurate diagnosis. Traditional diagnosis of plant viruses required bioassays on indicator plants, observing symptoms, and determining host range, virus particle morphology and vector relations. Over the last decade, progress in molecular biology, biochemistry and immunology has promoted the development of many new more accurate, rapid and less labour-intensive methods of virus and viroid detection. These assays have not only improved the speed and accuracy of diagnosis of viruses and viroids in the field but also have facilitated the control of many of these pathogens in important crops by enabling rapid monitoring of pathogen-free schemes.

The enzyme-linked immunosorbent assay (ELISA) was introduced into plant virology almost 20 years ago (Clark and Adams, 1977) and with relatively little alteration has almost completely replaced any other technique for routine diagnosis of all major viruses (see Chapter 8). However, there are circumstances when nucleic-acid-based detection of viruses or viroids is either a preferred or an only option. In this chapter we concentrate on such assays, their applications, advantages and pitfalls.

9.2 Targets for Detection

Viruses are extremely small acellular infectious agents visible only with the electron microscope. They differ from other cellular organisms in that they lack the machinery for self-replication (Matthews, 1981). Plant viruses are highly diversified causing many diseases of economic significance (see Chapter 8). There are now 26 named groups and 15 or more ungrouped plant viruses among 450 or so plant viruses known (Francki *et al.*, 1985; Buchen-Osmond *et al.*, 1988). These viruses are classified according to the nucleic acid compositions of their genomes and particle morphology (Matthews, 1991). The sizes of these particles range from 17 nm in diameter for satellite tobacco necrosis virus to 85 nm for tomato spotted wilt virus. The particles may be elongated rods (e.g. tobacco mosaic virus; TMV) or filamentous (e.g. potato virus X) or spherical (e.g. tobacco ringspot virus) (Francki, 1985) or paired (e.g. geminiviruses) or bacilliform (e.g. rhabdoviruses) and with or without an outer membrane envelope. They may contain genomes of single-stranded RNA (e.g. TMV) or dsRNA (reoviruses) or single-stranded DNA (geminiviruses) or dsDNA (cauliflower mosaic virus).

Viroids are probably the most intensively studied group of pathogens since their discovery in 1971 by Diener (Diener, 1971). They are the smallest infectious agents known, being composed of only nucleic acids. Today there are 17 distinct viroids causing 16 naturally occurring diseases in 12 species of plants. Purified viroid preparations, when spread under denaturing conditions and viewed under the electron microscope, are found to contain circular molecules of single-stranded RNA (Sanger *et al.*, 1976) of sizes ranging from 246 to 375 bases (Keese and Symons, 1987). In their native state, viroids probably exist as rod-like structures (Sogo *et al.*, 1973; Reisner, 1987). Viroids are assumed to replicate via a RNA–RNA pathway utilizing existing enzymes from the host cells since they themselves do not code for any proteins (Sanger, 1987).

The nucleotide sequence of the genomes of many plant viruses and viroids (including different strains of the same virus or viroid) have now been completely determined (Buchen-Osmond *et al.*, 1988; Koltunow and Reza-ian, 1989). Many of these sequences have been lodged in the GenBank/EMBL databases.

The genomes of plant viruses are packaged in coat proteins to form virus particles. These particles are often good antigens and lend themselves to serological detection. However, viroids have no such protein packaging nor do their genomes encode any proteins. Thus nucleic-acid-based detection is the only viable option.

9.3 Detection Methods

The detection of viruses and viroids based on their nucleic acids can be divided into specific and non-specific methods. Specific methods are those used to detect a particular virus or viroid in circumstances where the pathogen is likely to occur or after preliminary evidence suggests the presence of that pathogen. Such specific detection methods are either based on using labelled nucleic acid probes that will hybridize to the genomic material of the pathogen or on the specific amplification of the genomic material using the polymerase chain reaction (PCR). For either method nucleotide sequence information about the pathogen genome is essential. Non-specific methods include those used to detect unknown viruses or viroids. This type of method is based on the extraction from plants of nucleic acid fractions that are diagnostic of infection.

9.3.1 Specific detection methods using nucleic acid hybridization methods

The nucleic acid hybridization (NAH) method has been used widely for the detection of viruses and viroids (Bar-Joseph and Rosner, 1984; Hull, 1984; Lakshman *et al.*, 1986; Waterhouse *et al.*, 1986; Bernardy *et al.*, 1987; Haber *et al.*, 1987; Mills, 1987; Varveri *et al.*, 1987). NAH was first used to analyse viral RNA in solution by hybridization with radioactive probes. However, non-radioactive probing of virus or viroid samples, immobilized on a solid membrane support, is becoming the most common method.

The principle

Viral genomes are made of nucleic acids. Most plant viruses have single-stranded (ss) RNA genomes but there are groups of viruses with genomes comprised of double-stranded (ds) RNA, ssDNA or dsDNA. The nucleic acids are made up of nucleotides arranged in specific sequences joined by covalent phosphodiester bonds. The specific sequences of the nucleotides in the nucleic acids of the virus determine the structure and functions of its genes. Similarly the nucleotide sequence of a viroid in some way determines its structure and function. Thus a specific virus or viroid will have a genome composed of unique nucleotide sequences and viruses or viroids which are closely related share a greater nucleotide sequence similarity than those which are distantly related. The principle of nucleic acid hybridization (NAH) in diagnosis is based on this assumption and observation. NAH is based on the formation of double-stranded nucleic acids by specific hybridization between the target single-stranded nucleic acid sequence in the genome (either DNA or RNA) and the complementary single-stranded nucleic acid probe (either DNA or RNA). The kinetics of hybridization and the factors affecting the stability of the duplex are reviewed elsewhere (Britten and Davidson, 1985). If one of the strands (probe) in the duplex is labelled with a detectable marker, e.g. a radioactive moiety, then the formation of the duplex can be assayed after removal of unhybridized sequences.

The hybridization may be performed either in solution or on solid filter supports. When the hybridization is done in solution, the hybrids formed are analysed by gel electrophoresis or by liquid scintillation counting. With filter support hybridization the target molecules are immobilized on the filter and the labelled probe allowed to bind to them. The bound probe is then detected by the appropriate method. The filter hybridization method is used most commonly and there are many types of probes and methods of labelling them. Probes include cloned and uncloned nucleic acids, oligonucleotides, *in vitro* RNA transcripts, radioactive and non-radioactive probes. Their comparative advantages and disadvantages have been reviewed extensively (Al-Hakim and Hull, 1986; Zwadyk *et al.* 1986; Mifflin *et al.*, 1987; Tabares, 1987). The choice depends on the hybridization strategy, availability of material and equipment for preparation of probe, the concentration of target sequences to be detected, specificity and sensitivity required, and scale of operation. In general, however, ease of handling is usually the most important factor.

Radioactive probes

Use of radioactive isotopes is the traditional labelling method of nucleic acids and the isotopes ^3H, ^{35}S, ^{32}P and ^{125}I are generally used. The ^{32}P label is the isotope of choice since its high energy results in shorter detection time. Generally detection is by autoradiographic exposure.

There are a variety of ways of incorporating radioactive labels into a probe. Nick translation (Rigby *et al.*, 1977) is the traditional labelling method for double-stranded DNA (cloned or uncloned) and many commercial kits are available which can produce specific activity of 5×10^8 dpmμg^{-1}DNA when using ^{32}P-labelled deoxynucleotide triphosphates (dNTPs) (Arrand, 1985). More recently, commercial kits based on multipriming of denatured/ single-stranded DNA with oligonucleotide primers (random primers or hexamer primers) using Klenow fragment can produce specific activity of 5×10^9/dpmμg^{-1}|DNA probe (Feinberg and Vogelstein, 1983, 1984; Hodgson and Fisk, 1987).

Linear single-stranded nucleic acids (oligonucleotides, single-stranded RNA, single-stranded DNA) and double-stranded RNA can be end labelled using the T4 polynucleotide kinase which transfers the gamma-phosphate of ^{32}P ATP to a free 5' OH group in either DNA or RNA (Richardson, 1965; Maxam and Gilbert, 1980). Both a forward and an exchange reaction are possible. Specific activity of 5×10^5 cpm pmol^{-1} and 8×10^5 cpm pmol^{-1} can be achieved with the exchange and forward reaction respectively. Thus, labelling is not very efficient for large single-stranded DNA or double-stranded DNA molecules which can be labelled by other means (see above). The advantages of end labelling are that the probe is strand specific for single-stranded templates and small pieces of nucleic acids (oligonucleotides) can be labelled.

For the synthesis of high specific activity probes for RNA, the enzyme reverse transcriptase is used to synthesize cDNA using appropriate primers (Maniatis *et al.*, 1982), followed by labelled DNA synthesis as described above.

Single-stranded RNA probes can be easily prepared from transcription vector plasmids. These plasmids contain either or both of the T7 or SP6 polymerase promoter sequences. A few bases away from the promoter is a multiple cloning site into which a selected fragment of the DNA required as a NAH probe can be subcloned using standard cloning methods. Multiple copies of single-stranded, radioactively labelled, RNA representing the insert will be transcribed by the addition of SP6 or T7 RNA polymerase and ^{32}PrNTP. Transcription vectors are commercially available, e.g. riboprobe vectors from Promega Biotech.

Non-radioactive probes

Non-radioactive probes for NAH are a relatively new development and they are beginning to replace radiolabelled probes. Such probes include various biotin labelling systems (Al-Hakim and Hull, 1986; Dahlen, 1987; Gebeyehu *et al.*, 1987; McInnes *et al.*, 1987), fluorescein labels (Zuckermann *et al.*, 1987), enzyme labels (Li *et al.*, 1987; Mclaughlin *et al.*, 1987; Seriwatana *et al.*, 1987) or more recently labelling with steroid antigens (Boehringer-Mannheim, 1988; Schafer *et al.*, 1988). A typical result of an assay using a photobiotin-labelled probe is shown in Fig. 9.1.

The basic methods are:

1. *Biotinylation of nucleotides and streptavidin detection.* One method of making biotinylated probes involves using the same techniques as described for making radioactive probes (such as nick translation, oligolabelling or end labelling) except the radiolabelled nucleotides are replaced with biotin-11-UTP (Langer *et al.*, 1981; Leary *et al.*, 1983). Another method is the use of PCR (see section 9.3.3) to generate large amounts of DNA from a specific region of the virus or viroid genome which are biotinylated by including either biotin-11-UTP or biotin-14-dATP in the amplification reaction

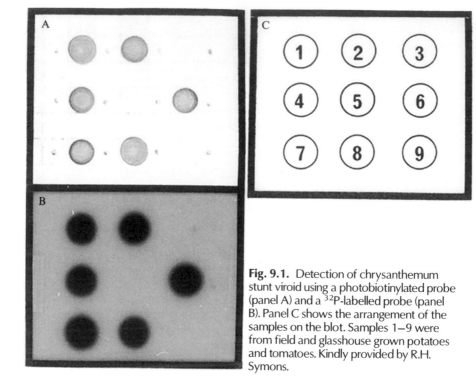

Fig. 9.1. Detection of chrysanthemum stunt viroid using a photobiotinylated probe (panel A) and a ^{32}P-labelled probe (panel B). Panel C shows the arrangement of the samples on the blot. Samples 1–9 were from field and glasshouse grown potatoes and tomatoes. Kindly provided by R.H. Symons.

(Kanematsu *et al.*, 1991). Further methods include the use of a photo-reactive biotin derivative which binds to single-stranded or double-stranded DNA or RNA when exposed to visible light (Forster *et al.*, 1985; McInnes *et al.*, 1987), or chemically linking biotin derivatives to oligonucleotides or DNA (Al-Hakim and Hull, 1986). The probe is detected by incubation with avidin or streptavidin which has been chemically coupled to an enzyme (e.g. phosphatase, luciferase, β-galactosidase, peroxidase, etc.) which catalyses a reaction that can be detected colorimetrically, fluorometrically or photometrically. Streptavidin conjugates appear to be more stable than avidin conjugates (Shi *et al.*, 1988). As little as 1–5 pg of target sequence can be detected.

2. *Covalent linking of a reporter enzyme*, such as alkaline phosphatase, to single-stranded nucleic acids produces a probe which can be used directly as a hybridization probe (Li *et al.*, 1987; McLaughlin *et al.*, 1987; Seriwatana *et al.*, 1987). No enzymatic synthesis of the nucleic acids is required and the final stage of the procedure is to develop a colorimetric or other appropriate reaction product. Approximately 1–5 pg of target sequence can be detected (Renz and Kurz, 1984) provided the enzyme retains its activity after being subjected to the hybridization conditions.

3. *The incorporation of antigen-conjugated nucleotide* (e.g. digoxigenin-dUTP) into the nucleic acid probes allows detection by enzyme-immunological methods (Boehringer-Mannheim, 1988; Schafer *et al.*, 1988). Another immunological approach to NAH is to use antibodies to chemically modified DNA (Gratzner, 1982; Tchen *et al.*, 1984). When such a probe is hybridized to a target DNA, the hybrids can be detected serologically. A probe based on this technique is commercially available (Chemiprobe, Orgenics Ltd, Yavne, Israel) and is stable with sensitivity at the subpicogram level.

9.3.2 *Advantages and disadvantages of non-radioactive labelling procedures*

The advantages of non-radioactive labelling and detection methods include:

1. The assays are easily performed without the need for the protective precautions required with the use of radioisotopes.
2. The probes are stable and can be stored for over 12 months, without the restriction of half-life determined by a radioactive isotope.
3. The sensitivity obtained is equal or better than ^{32}P-labelled probe (McInnes *et al.*, 1987; Boehringer-Mannheim, 1988).
4. Detection can be rapid since no long autoradiographic development steps are involved.
5. The reagents are inexpensive and commercially available in a kit form.
6. Like the enzyme and chemical immunological assays, there is a whole spectrum of methods available for detecting the non-radioactive markers on the nucleic acid probes. Recent innovations include:
(a) the use of a novel biotinylated nucleotide analogue that can be detected directly by colorimetric means (Gebeyehu *et al.*, 1987);
(b) use of Eu-labelled streptavidin to detect biotin by time-resolved fluorometry (Dahlen, 1987) and fluorescent staining (Pinkel *et al.*, 1986);
(c) amplification of signal, e.g. by using biotinylated polymers of alkaline phosphatase (Hsu *et al.*, 1981).

A recent study compared six different non-radioactive NAH probes for the detection of potato spindle tuber viroid (Kanematsu *et al.*, 1991). The probes were either digoxigenin labelled or biotinylated by photo-reaction, nick translation, random priming, PCR using biotin-11-UTP or PCR using biotin-14-dATP. All these probes were of at least equal sensitivity to those of radioactive probes (*c.* 20pg of viroid RNA) and the probes made using PCR were found to be the most suitable for practical diagnosis. These gave the highest sensitivity (0.2–2pg) and were rapidly prepared in large quantities.

The main disadvantages of non-radioactive probes are:

1. The streptavidin detection system can give a high background against plant components in crude extracts.

2. When a colorimetric product is produced on filter, it is more difficult to record and quantify the results and the signal may not be removable for reprobing of the filter.

9.3.3 Specific detection using polymerase chain reaction amplification

The principle

The PCR (Saiki *et al.*, 1985, 1988; Scharf *et al.*, 1986) is a recently developed method that amplifies specific regions of DNA. Typically, the target dsDNA is heat denatured, a pair of synthetic oligonucleotide primers are then hybridized to both strands of the target DNA, one to the 5' end of the sense strand and one to the 5' end of the antisense strand by an annealing step. The enzyme then synthesizes new DNA on the templates to produce theoretically twice the number of target DNA molecules. The sample is then heated to dissociate the DNA, followed by annealing of the excess primers. If this cycle of dissociation, annealing and extension is repeated many times there is an exponential increase in the number of target DNA molecules produced. The method is simple and versatile and reproduces DNA sequences between the primers with high fidelity and with an efficiency of up to 85% per cycle (Weier and Gray, 1988). The PCR product can be used not only as a detection method based on the production of a diagnostic sized band but also to produce DNA for a NAH target, for direct sequencing (Simpson *et al.*, 1988), to determine strain variations, to facilitate cloning (Simpson *et al.*, 1988) or for use as a specific probe. There are several enzymes used for PCR. They include Klenow enzyme (Saiki *et al.*, 1985; Keohavong *et al.*, 1988), *Taq* DNA polymerase (Scharf *et al.*, 1986) and T4 DNA polymerase (Keohavong *et al.*, 1988). The Klenow enzyme was originally used when *Taq* DNA polymerase was not available. It only allows extension of 250 bp (Saiki *et al.*, 1988) and can produce non-specific extensions. *Taq* polymerase has the advantage of being stable to heating at 95°C so that it will survive the denaturation step thus obviating the need for addition for fresh enzyme during each cycle. Its optimal operating temperature is between 60 and 85°C (Chien *et al.*, 1976), so that more stringent selection of target DNA can be obtained while achieving maximum yield and length of products (Saiki *et al.*, 1988). The enzyme is commercially available in ultrapure form (Perkin Elmer Cetus) and can yield amplification of DNA sequences up to 2 kbp by 10^5–10^7 fold (Saiki *et al.*, 1985, 1988; Rochlitz *et al.*, 1988).

PCR has been used for detection of single-stranded DNA viruses such as the geminiviruses, tomato leaf curl virus (Navot *et al.*, 1992) and maize streak virus (Rybicki and Hughes, 1990), and the as yet unclassified virus subterranean clover stunt virus (Bariana *et al.*, 1994). However, *Taq*

Table 9.1. Typical protocol for extraction and RT-PCR detection of virus or viroid-infected plants.

---

1. *Sample extraction*
    Grind plant material in extraction buffer
    Remove proteins by phenol/chloroform extraction
    Precipitate nucleic acid with ethanol
    Resuspend nucleic acid in reverse transcription buffer

2. *Reverse transcription*
    Add primers, nucleotides and reverse transcriptase and incubate for 1 h at 42°C

3. *PCR*
    Add thermostable Taq polymerase and buffer
    Place in PCR machine

    The machine will heat and cool the samples as follows:
       (a) Denaturation: 3 min at 94°C
       (b) Denaturation: 1 min at 94°C
       (c) Annealing: 1 min at 37°C
       (d) Extension: 1 min at 72°C

    Repeat cycles (b), (c) and (d) 35 times
       (e) Extension: 5 min at 72°C

4. *Analysis*
    Run whole or a portion of reaction sample on 1% agarose gel with size markers and stain with ethidium bromide
    View on UV light box

---

polymerase will only extend off a DNA template and as the genomes of most plant viruses and all viroids are composed of RNA, the target sequence must first be converted into DNA with a reverse transcription step. This method is called reverse transcription-polymerase chain reaction (RT-PCR) and a typical protocol for the detection of an RNA plant virus or a viroid is shown in Table 9.1. The most commonly used RT-PCR assay for the detection of virus or viroid-infected material is to reverse transcribe a portion of the pathogen's genome into DNA, amplify the DNA by PCR, and then analyse the product on an agarose gel (e.g. Vunsh *et al.*, 1990). There are three important factors when designing a PCR assay for a particular pathogen: the choice of primers, the method of extraction of the samples and the method of visualizing the results.

Primer design

In assays of field material the usual goal is to detect all strains of the pathogen sought. Therefore primers for RT-PCR assays should be designed to amplify efficiently all strains of the pathogen but not give spurious amplification of other nucleic acids present in the plant. It has been a popularly held belief that such strains would have such sequence degeneracy in their genomes that RT-PCR assays would not be suited to this application. However, in cases

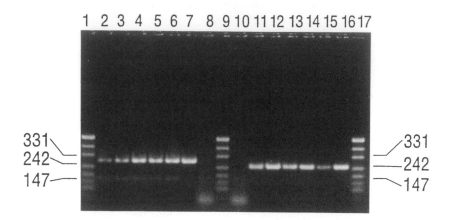

Fig. 9.2. PCR products, amplified from total nucleic acids purified from plants infected with either alfalfa mosaic virus (six isolates; lanes 2–7) or bean yellow mosaic virus (six isolates; lanes 11–16) electrophoresed in 1.5% Agarose/TBE and stained with ethidium bromide. Reaction mixture contained primers specific to each virus. Lanes 8 and 10 are PCR products from nucleic acids from uninoculated plants amplified using AMV and BYMV primers respectively. Lanes 1, 9 and 17 contain size markers of *Hpa*II digested pUC19 DNA (Bresatec). Reproduced with permission from Bariana *et al.* (1994).

studied so far, there are conserved regions of sequence within a virus or within a virus group that can be used as primer targets using a population of primers with base degeneracy in specific positions. Such degenerate primers have been developed for both potyviruses (Langeveld *et al.*, 1991) and luteoviruses (Robertson *et al.*, 1991) which give broad detection of members of the respective groups. It has also been found that non-degenerate primers designed to conserve regions are able to detect a number of different strains (Wetzel *et al.*, 1991; Bariana *et al.*, 1994; see Fig. 9.2). Unfortunately, in most cases the design of such primers is severely limited by the sequence information available on strains of the pathogen in question. PCR amplification becomes less efficient with increasing length of the target region. Therefore it is better to design the primers to amplify regions of less than 1000 bases.

Methods of extraction

RT-PCR amplification reactions are more susceptible to failure due to inhibitors in plant extracts than either ELISA or NAH. This necessitates more thorough preparation of the sample extracts for RT-PCR than for the other two methods and a phenol/choroform extraction step is almost essential. However, the thoroughness of the extraction depends on the target pathogen and the host. The detection of plum pox virus (PPV) from infected apricots, for example, did not require phenol/chloroform extraction, had a very simple one tube extraction and the RT-PCR assay was capable of detecting as little as 200 molecules of PPV RNA per sample (Wetzel *et al.*, 1991). Yet detection of the same virus from tree bark required extensive preparation followed by a QIAGEN anion exchange column (Korschineck *et al.*, 1991).

Visualizing results

The most common method of visualizing the result of RT-PCR is to run the product on an agarose gel and stain with ethidium bromide to see the PCR and size marker bands (Fig. 9.2). The gels can be stained with silver stain (Wetzel *et al.*, 1991) or probed by NAH on Southern blots of the gels (Vunsch *et al.*, 1990). In the latter case this PCR/NAH assay was 3–4 orders of magnitude more sensitive than conventional NAH or ELISA. Another sensitive method of visualization is the PCR membrane spot assay. This assay incorporates biotin-UTP into the RT-PCR product which is spotted on to nitrocellulose. The filter is then incubated with streptavadin-alkaline phosphatase, washed and incubated with a substrate that can be hydrolysed to give an insoluble blue precipitate. This assay has been used for the detection of PPV and was in excess of 16-fold more sensitive than conventional NAH or ELISA.

Variations of PCR

A minor variation of RT-PCR is to digest the product with restriction endonucleases prior to running on agarose gels. If the products are produced from the same strain of virus they will give identical band patterns on the gel. However, if there is sequence variation within the product this can give rise to different restriction digest patterns. This has been used for studying different strains of PPV (Wetzel *et al.*, 1991). A more radical variation is first to trap virus particles either in antibody-coated ELISA plate wells or plastic tubes, wash away the unbound material, and then perform RT-PCR on the recovered samples. This method is a hybrid between ELISA and PCR. It gives the sensitivity of PCR and eliminates the plant inhibitors by the antibody-trapping step. When used to detect PPV, this method had a 250-fold, 625-fold and a 5000-fold advantage over direct PCR, NAH and ELISA, respectively (Wetzel *et al.*, 1992).

A PCR system with other ELISA-like features has been developed for human immunodeficiency virus (Kemp *et al.*, 1989) which could find application in plant pathology. This system makes use of a biotinylated primer and a DNA binding protein (GCN4) for which the sequence to which it binds is known. The HIV segment is amplified with an upstream

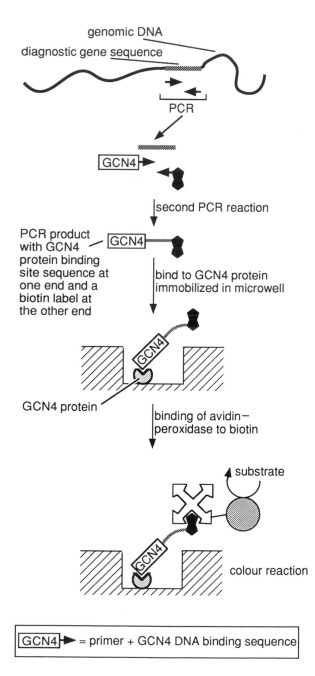

Fig. 9.3 Principle of GCN4 PCR ELISA. In step 1 the DNA is amplified using oligonucleotide primers. The product is reamplified with a biotinylated primer and a primer including the GCN4 binding sequence. If amplification has occurred the product will be bound to the plastic wells precoated with GCN4 binding protein as shown in step 3. The bound product is detected by avidin–peroxidase conjugate.

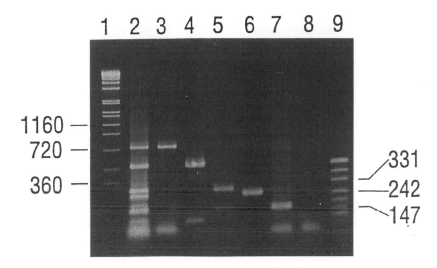

Fig. 9.4. PCR products, amplified from total nucleic acids purified from plants infected with clover yellow vein virus (lane 3), cucumber mosaic virus (lane 4), alfalfa mosaic virus (lane 5), bean yellow mosaic virus (lane 6), subterranean clover mottle virus (lane 7) or a pooled sample of all five extracts (lane 2) electrophoresed in 1.5% Agarose/TBE and stained with ethidium bromide. Reaction mixture contained a cocktail of primers each specific to one of the five viruses. Lane 8 contains PCR products from nucleic acids from uninoculated plants amplified with the same primer cocktail. Lane 1 contains size markers of *Eco*RI digested SPP1 DNA (Bresatec); lane 9 contains size markers of *HPa*II digested pUC19 DNA (Bresatec). (Reproduced with permission, from Bariana *et al.*, 1994.)

HIV-specific primer that has been previously biotinylated and a downstream primer that includes at its 5' end the sequence to which GCN4 protein will bind. The PCR product is then incubated in the well of a microwell plate which has been previously coated with GCN4 fusion protein. The well is then processed with avidin peroxidase conjugate and peroxidase substrate as in conventional ELISA. If the virus is present dsDNA molecules will be made that are trapped on to the well via the GCN4 sequence at one end and are recognized by the avidin–peroxidase conjugate via the biotin at the other end (Fig. 9.3). This system thus has the sensitivity of PCR while giving a colour reaction in an ELISA format.

Another variation of RT-PCR is a multiplex approach. This system is an attempt to produce a single assay that will specifically detect all the viruses known to infect a crop. The RT-PCR assay uses a cocktail of primers (two for each virus) that will amplify all the viruses known to infect the crop but gives a different sized amplification product for each of the different viruses. Thus the number and size of the bands produced are diagnostic of the number and identity of the viruses present in the sample. This system

(Bariana *et al.*, 1994) has been developed for viruses infecting subterranean clover and medics and detects all five seedborne viruses of subterranean clover simultaneously in a one-tube section (Fig. 9.4).

9.3.4 Non-specific detection methods

These methods are based on the nucleic acids of a virus or viroid having different properties to those present in uninfected plants. Viroids, for example, have small circular RNA genomes with considerable amounts of secondary structure. Similar RNA is not found in uninfected plants and it is therefore possible to detect the presence of the viroid by gel electrophoresis based on simplified versions of viroid purification schemes (Singh, 1982). The improved methods include a two-dimensional gel electrophoresis method (Schumacher *et al.*, 1983) which has a sensitivity of 60ng viroidg^{-1} tissue and a return gel electrophoresis method with a sensitivity of 0.8ng viroidg^{-1} tissue (Schumacher *et al.*, 1986).

RNA plant viruses either have double-stranded RNA as their genome (e.g. phytoreoviruses) or use double-stranded RNA as an intermediate for replication (single-stranded RNA viruses) and uninfected plants contain very little double-stranded RNA. Thus extraction of double-stranded RNA can be used as a diagnostic assay of these viruses. The technique is mainly used for substantiating that a disease phenotype is caused by a virus or for detecting poorly characterized viruses for which there is no alternative assay. It may also be used to detect viral diseases before symptoms appear (Lejour and Kummert, 1986). The yield of double-stranded RNA varies with different viruses, e.g. potyviruses had the lowest yield of double-stranded RNA among viruses with elongated monopartite genomes (Valverde *et al.*, 1986).

Plant extracts can be passed through a Sephadex column in conditions that promote the binding of double-stranded RNA but not single-stranded RNA or double-stranded DNA (Valverde *et al.*, 1986). The double-stranded RNA is then eluted and run on an acrylamide gel and treated with a sensitive silver stain. The double-stranded RNA extract contains not only the replicative form of the full length genome but also double-stranded forms of the viral subgenomic messenger RNAs. This gives a double-stranded RNA profile that is usually diagnostic of the virus group. It is also advocated that strains of viruses can be differentiated from this pattern (Valverde *et al.*, 1986). This type of analysis may also reveal the presence of infection complexes such as satellite RNA, multiple infections and cryptic virus (Lejour and Kummert, 1986). The problem with this method is that double-stranded RNA can be found in apparently virus-free plants (Kurppa and Martin, 1986; Valverde *et al.*, 1986; Spiegel, 1987). However, these non-viral double-stranded RNAs can be distinguished from viral double-strained RNA if adequate controls are used.

With many uncharacterized viruses which are difficult to purify or transmit and where no antiserum or nucleic acid probe is available, double-stranded RNA profiles, transmission and symptom expression are used. Many of these viruses have not only been detected by double-stranded RNA analysis recently but cloned diagnostic probes have also been prepared from the isolated double-stranded RNA, e.g. strawberry mild yellow (Spiegel, 1987), raspberry bushy dwarf (Kurppa and Martin, 1986), grapevine leafroll (Mossop *et al.*, 1985) and peanut rosette (Breyel *et al.*, 1988). The amount of double-stranded RNA associated with plants infected with peanut rossette is so high (5–10%) that it can be detected without the use of enzyme, or extensive purification through ion-exchange chromatography.

9.4 Conclusions

The advent of ELISA for plant viruses has revolutionized the detection and identification of these pathogens in most countries around the world. However, there are cases where ELISA cannot be used such as in the detection of viroids, umbraviruses, and deviant or heterologously encapsidated viruses which either have no coat protein or have acquired the coat protein of a different virus. In such cases, detection of the nucleic acid genome is an attractive option. Non-radioactive probes for NAH have become widely accepted and will probably soon replace radioactively labelled probes. Certainly this will be the case for routine field diagnosis as they are sensitive, robust and easy to use. RT-PCR assays are extremely sensitive being able to detect as few as 200 molecules in infected plant samples (Wetzel *et al.*, 1991). This coupled with the ability to produce multiplex assays that can screen samples simultaneously for numerous viruses and the potential for rapid colorimetric display of results suggest that such assays may in the future become as routinely used as ELISA. However, at present the RT-PCR assay requires more labour to prepare the samples for assay and is more sensitive to inhibitors in the extracts. Unless these are overcome, the application of nucleic-acid-based probes is likely to be confined to situations where ELISA is not possible or where sensitivity is of paramount importance.

References

Agrios, G.N. (1988) *Plant Pathology*, 3rd edn. Academic Press, San Diego, CA, USA.
Al-Hakim, A.H. and Hull, R. (1986) Studies towards the development of chemically synthesized non-radioactive biotinylated nucleic acid hybridization probes. *Nucleic Acids Research* 14, 9965–9976.
Arrand, J.E. (1985) Preparation of nucleic acid probes. In: Hames, B.D. and Higgins, S.J. (eds) *Nucleic Acid Hybridization – A Practical Approach*. IRL Press, Oxford, UK, pp. 17–45.

Bariana, H.S., Shannon, A.L., Chu, P.W.G. and Waterhouse, P.M. (1994) Detection of five seedborne legume viruses in one sensitive multiplex polymerase chain reaction test. *Phytopathology* 84,1201–1205.

Bar-Joseph, M. and Rosner, A. (1984) Diagnosis of plant viruses by the molecular hybridization method. *Phytoparasitica* 12, 214.

Bernardy, M.G., Jacoli, G.G. and Ragetli, H.W.J. (1987) Rapid detection of potato spindle tuber viroid (PSTV) by dot blot hybridization. *Journal of Phytopathology* 118, 171–180.

Boehringher-Mannheim, (1988) Instructions for DNA labelling and detection kit: non-radioactive. Random primed DNA labelling with digoxigenin-dUTP, detection of hybrids by enzyme immunoassay. Biochemica, Germany.

Breyel, E., Casper, R., Ansa, O.A., Kuhn, C.W., Misari, S.M. and Demski, J.W. (1988) A simple procedure to detect a double stranded RNA associated with groundnut rosette. *Journal of Phytopathology* 121, 118–124.

Britten, R.J. and Davidson, E.H. (1985) Hybridization strategy. In: Hames, B.D. and Higgins, S.J. (eds), *Nucleic Acid Hybridization – A Practical Approach*. IRL Press, Oxford, UK, pp. 3–15.

Buchen-Osmond, C., Crabtree, K., Gibbs, A. and McLean, G. (1988) *Viruses of Plants in Australia*. Descriptions and Lists from the VIDE Database. Australian National University, Canberra.

Chien, A., Edgar, D.B. and Trela, J.M. (1976) Deoxyribonucleic acid polymerase from the extreme thermophile *Thermus aquaticus*. *Journal of Bacteriology* 127, 15.

Clark, M.F. and Adams, A.N. (1977) Characteristics of the microplate method of enzyme-linked immunosorbent assay (ELISA) for the detection of plant viruses. *Journal of General Virology* 34, 475–483.

CMI/AAB (1970–1988) *Descriptions of Plant Viruses, Sets 1–21*. Commonwealth Mycological Institute and The Association of Applied Biologists, Surrey, UK.

Converse, R.H. (ed.) (1987) *Virus Diseases of Small Fruits*. US Department of Agriculture Handbook, Washington, DC, USA.

Dahlen, P. (1987) Detection of biotinylated DNA probes by using Eu-labelled streptavidin and time-resolved fluorometry. *Analytical Biochemistry* 164,78–83.

de Bokx, J.A. (ed.) (1981) *Viruses of Potatoes and Seed-Potato Production*. Pudoc Wageningen, The Netherlands.

Diener, T.O. (1971) Potato spindle tuber 'virus'. IV. A replicating low molecular weight RNA. *Virology* 45, 411–428.

Edwardson, J.R. and Christie, R.G. (eds) (1986) *Viruses Infecting Forage Legumes*, vols I–III. University of Florida, Gainesville.

Feinberg, A.P. and Vogelstein, B. (1983) A technique for radiolabelling DNA restriction endonuclease fragments to high specific activity. *Analytical Biochemistry* 132, 6–13.

Feinberg, A.P. and Vogelstein, B. (1983) A technique for radiolabelling DNA restriction endonuclease fragments to high specific activity, addendum. *Analytical Biochemistry* 137, 266–267.

Forster, A.C., McInnes, J.L., Skingle, D.C. and Symons, R.H. (1985) Non-radioactive hybridization probes prepared by the chemical labelling of DNA and RNA with a novel reagent, photobiotin. *Nucleic Acids Research* 13, 745–761.

Foster, J.A. (1988) Recommendations for improvements in the international exchange of temperate fruit tree germplasm. In: *Conservation and Movement of*

Vegetatively propagated Germplasm: In Vitro *Culture and Disease Aspects.* Report of a Subcommittee Meeting Held at North Carolina State University, USA, 17–19 August 1987. International Board for Plant Genetic Resources, IBPGR Advisory Committee on *in vitro* Storage, IBPGR, Rome, Italy, pp. 27–30.

Francki, R.I.B. (ed.) (1985) *The Plant Viruses. Volume I: Polyhedral Virions with Tripartite Genomes.* Plenum Press, New York, NY, USA.

Francki, R.I.B., Milne, R.G. and Hatta, T. (1985) *Atlas of Plant Viruses. Volumes I and II.* CRC Press, Boca Raton, FL, USA.

Gebeheyu, G., Rao, P.Y., Soo Chan, P., Simms, D.A. and Klevan, L. (1987) Novel biotinylated nucleotide-analogs for labelling and colorimetric detection of DNA. *Nucleic Acids Research* 15, 4513–4534.

Gratzner, H.G. (1982) Monoclonal antibody to 5-bromo- and 5-iododeoxyuridine: a new reagent for detection of DNA replication. *Science* 218, 474–475.

Haber, S., Polston, J.E. and Bird, J. (1987) Use of DNA to diagnose plant diseases caused by single-strand DNA plant viruses. *Canadian Journal of Plant Pathology*, 9, 156–161.

Hodgson, C.P. and Fisk, R.Z. (1987) Hybridization probe size control: optimized 'oligolabelling'. *Nucleic Acids Research* 15, 6295.

Hsu, S.M., Raine, L. and Fanger, H. (1981) Use of avidin–biotin–peroxidase complex (ABC) in immunoperoxidase techniques: a comparison between ABC and unlabelled antibody (PAP) procedures. *The Journal of Histochemistry and Cytochemistry* 29, 577–580.

Hull, R. (1984) Rapid diagnosis of plant virus infections by spot hybridization. *Trends in Biotechnology* 4, 88–91.

Hull, R. and Al-Hakim, A. (1988) Nucleic acid hybridization in plant virus diagnosis and characterization. *Trends in Biotechnology* 6, 213–217.

Kanematsu, S., Hibi, T., Hashimoto, J. and Tsuchizaki, T. (1991) Comparison of non-radioactive cDNA probes for detection of potato spindle tuber viroid by dot-blot hybridization assay. *Journal of Virology Methods* 35, 189–197.

Keese, P. and Symons, R.H. (1987) The structure of viroids and virusoids. In: Semancik, J.S. (ed.), *Viroids and Viroid-like Pathogens.* CRC Press, Boca Raton, FL, USA, pp. 1–47.

Kemp, D.J., Smith, D.B., Foote, S.J., Samaras, N. and Peterson, M.G. (1989) Colorimetric detection of specific DNA segments amplified by polymerase chain reactions. *Proceedings of the National Academy of Sciences, USA* 86, 2423–2427.

Keohavong, P., Kat, A.G., Cariello, N.F. and Thilly, W.G. (1988) DNA amplification in vitro using T4 DNA polymerase. *DNA* 7, 63–70.

Koltunow, A.M. and Rezaian, M.A. (1989) A scheme for viroid classification. *Intervirology* 30, 194–201.

Korschineck, I., Himmler, G., Sagl, R., Steinkellner, H. and Katinger, H.W.D. (1991) A PCR membrane spot assay for the detection of plum pox virus RNA in bark of infected trees. *Journal of Virological Methods* 31, 139–146.

Kurppa, A., and Martin, R.R. (1986) Use of double-stranded RNA for detection and identification of virus diseases of *Rubus* species. *Acta Horticulturae* 186, 51–62.

Lakshman, D.K., Hiruki, C., Wu, X.N. and Leung, W.C. (1986) Use of [^{32}P] RNA probes for the dot-hybridization detection of potato spindle tuber viroid. *Journal of Virological Methods* 14, 309–319.

Langer, P.R., Waldrop, A.A. and Ward, D.C. (1981) Enzymatic synthesis of biotin-labelled polynucleotides: novel nucleic acid affinity probes. *Proceedings of the National Academy of Sciences, USA* 75, 6633–6637.

Langeveld, S.A., Dore, J.M., Memelink, J., Derks, A.F.L.M., van der Vlugt, C.I.M., Asjes, C.J. and Bol, J.F. (1991) Identification of potyviruses using the polymerase chain reaction with degenerate primers. *Journal of General Virology* 72, 1531–1542.

Leary, J.J., Brigati, D.J. and Ward, D.C. (1983) Rapid and sensitive colorimetric method for visualizing biotin-labelled DNA probes hybridized to DNA or RNA immobilized on nitrocellulose: bio-blots. *Proceedings of the National Academy of Sciences, USA* 80, 4045–4049.

Lejour, c. and Kummert, J. (1986) Comparative study of the detection of double-stranded RNA in different plant viral systems. *Annales de Gembloux* 92, 177–188.

Li, P., Medon, P.P., Skingle, D.C., Lanser, J.A. and Symons, R.H. (1987) Enzyme-linked synthetic oligonucleotide probes: non-radioactive detection of enterotoxigenic *Escherichia coli* in faecal specimens. *Nucleic Acids Research* 15, 5275–5287.

Maniatis, T., Fritisch E.F. and Sambrook, J. *Molecular Cloning: A Laboratory Manual.* Cold Spring Harbor Laboratory, Cold Spring Harbor.

Matthews, R.E.F. (1982) *Classification and Nomenclature of Viruses.* Fourth Report of the International Committee on Taxonomy of Viruses. Karger, Basel, Switzerland.

Matthews, R.E.F. (1991) *Plant Virology*, 3rd edn. Academic Press, San Diego, California, USA.

Maxam, A.M. and Gilbert, W. (1980) Sequencing end-labelled DNA with base-specific chemical cleavages. *Methods in Enzymology* 65, 499–560.

McInnes, J.L., Dalton, S., Vize, P.D. and Robins, A.J. (1987) Non-radioactive photobiotin-labeled probes detect single copy genes and low abundance messenger RNA. *Bio/Technology* 5, 269–272.

McInnes, J.L., Habili, N. and Symons, R.H. (1989) Non-radioactive, photobiotin labelled DNA probes for routine diagnosis of viroids in plant extracts. *Journal of Virological Methods* 23, 299–312.

McLaughlin, G.L., Ruth, J.L., Jablonski, E., Steketee, R. and Campbell, G.H. (1987) Use of enzyme-linked synthetic DNA in diagnosis of falciparum malaria. *Lancet* i, 714–716.

Mifflin, T.E., Bowden, J., Lovell, M.A., Bruns, D.E., Hayden, F.G., Groschel, D.H.M. and Savory, J. (1987) Comparison of radioactive (P-32 and S-35) and biotinylated probes for detection of cytomegalovirus DNA. *Clinical Biochemistry* 20, 231–236.

Mills, P.R. (1987) Comparison of cDNA hybridization and other tests for the detection of potato mop-top virus. *Potato Research* 30, 319–328.

Mossop, D.W., Elliott, D.R. and Richards, K.D. (1985) Association of closterovirus-like particles and high molecular weight double-stranded RNA with grapevines affected by leafroll disease. *New Zealand Journal of Agricultural Research* 28, 419–425.

Navot, N., Zeidan, M., Piechersky, E., Zamir, D. and Czosnek, H. (1992) Use of the polymerase chain reaction to amplify tomato yellow leaf curl virus DNA from infected plants and viruliferous whiteflies. *Phytopathology* 82, 1199–1202.

Pinkel, D., Gray, J.W., Trask, B., Van den Engh, G., Fuscoe, J. and Van Dekken, H.

(1986) Cytogenetic analysis by *in situ* hybridization with fluorescently labelled nucleic acid probes. *Cold Spring Harbor Symposia on Quantitative Biology* 51, 151–157.

Reisner, D., Colpan, M., Goodman, T.C., Nagel, L., Schuemaker, J. and Steger, G. (1987) Dynamics and interactions of viroids. *Journal of Biomolecular Structure and Dynamics* 1, 669–687.

Renz, M. and Kurz, C. (1984) A colorimetric method for DNA hybridization. *Nucleic Acids Research* 12, 3435–3444.

Richardson, C.C. (1965) Phosphorylation of nucleic acid by an enzyme from T4 bacteriophage infected *Escherichia coli*. *Proceedings of the National Academy of Sciences, USA* 54, 158–165.

Rigby, P.W.J., Dieckmann, M., Rhodes, C. and Berg, P. (1977) Labelling deoxyribonucleic acid to high specific activity in vitro by nick translation with DNA polymerase I. *Journal of Molecular Biology* 113, 237–251.

Robertson, N.L., French, R. and Gray, S.M. (1991) Use of group specific primers and polymerase chain reaction for the detection and identification of luteoviruses. *Journal of General Virology* 72, 1473–1477.

Rochlitz, C.F., Scott, G.K., Dodson, J.M. and Benz, C.C. (1988) Use of the polymerase chain reaction technique to create base-specific ras oncogene mutations. *DNA* 7, 515–519.

Rybicki, E.P. and Hughes, F.L. (1990) Detection and typing of maize streak virus and other distantly related geminiviruses of grasses by polymerase chain reaction amplification of a conserved viral sequence. *Journal of General Virology* 71, 2519–2526.

Saiki, R.K., Scharf, S., Faloona, F., Mullis, K.B., Horn, G., Erlich, H.A. and Arnheim, N. (1985) Enzymatic amplification of B-globin genomic sequences and restriction site analysis for diagnosis of sickle cell anemia. *Science* 230, 1350–1354.

Saiki, R.K., Gelfand, D.H., Stoffel, S., Scharf, S.J., Higuchi, R., Horn, G.T., Mullis, K.B. and Erlich, H.A. (1988) Primer-directed enzymatic amplification of DNA with a thermostable DNA polymerase. *Science* 239, 487–491.

Sanger, H.L. (1987) Viroid function: viroid replication. In: Diener, T.O. (ed.), *The Viroids*. Plenum Press, New York, NY, USA, pp. 117–166.

Sanger, H.L., Klotz, G., Riesner, D., Gross, H.J. and Kleinschmidt, A.K. (1976) Viroids are single-stranded covalently closed circular RNA molecules existing as highly base paired rod-like structures. *Proceedings of the National Academy of Sciences, USA* 73, 3852–3856.

Schafer R., Zischler, H. and Epplen, J.T. (1998) DNA fingerprint using non-radioactive oligonucleotide probes specific for simple repeats. *Nucleic Acids Research* 16, 9344.

Scharf, S., Horn, G.T. and Erlich, H.A. (1986) Direct cloning and sequence analysis of enzymatically amplified genomic sequences. *Science* 233, 1076–1078.

Schumacher, J., Randles, J.W. and Riesner, D. (1983) Viroid and virusoid detection: an electrophoretic technique with the sensitivity of molecular hybridization. *Analytical Biochemistry* 135, 288–295.

Schumacher, J., Meyer, N., Riesner, D. and Weidemann, H.L. (1986) Diagnostic procedure for detection of viroids and viruses with circular DNA by return-gel electrophoresis. *Journal of Phytopathology* 115, 332–343.

Seriwatana, J., Echeverria, P., Taylor, D.N., Sakuldaipetra, T., Changchawalit, S. and

Chivoratanand, O. (1987) Identification of enterotoxigenic *Escherichia coli* with synthetic alkaline phosphatase-conjugated oligonucleotide DNA probes. *Journal of Clinical Microbiology* 25, 1438–1441.

Shi, S.-R., Itzkowitz, S.H. and Kim, Y.S. (1988) A comparison of three immunoperoxidase techniques for antigen detection in colorectal carcinoma tissues. *Journal of Histochemistry and Cytochemistry* 36, 317–322.

Simpson, D., Crosby, R.M. and Skopek, T.R. (1988) A method for specific cloning and sequencing of human HPRT cDNA for mutation analysis. *Biochemical and Biophysical Research Communications* 151, 487–492.

Singh, R.P. (1982) Evaluation of procedures for the detection of potato spindle tuber viroid by polyacrylamide gel electrophoresis. *Canadian Plant Disease Survey* 62, 41–44.

Sogo, J.M., Koller, T. and Diener, T.O. (1973) Potato spindle tuber viroid. X. Visualization and size determination by electron microscopy. *Virology* 55, 70–80.

Spiegel, S. (1987) Double-stranded RNA in strawberry plants with strawberry mild yellow-edge virus. *Phytopathology* 77, 1492–1494.

Tabares, E. (1987) Detection of DNA viruses by radioactive and non radioactive DNA probes: application to African swine fever virus. *Archives of Virology* 92, 233–242.

Tchen, P., Fuchs, R.P.P., Sage, E. and Leng, M. (1984) Chemically modified nucleic acids as immunodetectable probes in hybridization experiments. *Proceedings of the National Academy of Sciences, USA* 81, 3466–3470.

Valverde, R.A., Dodds, J.A. and Heick, J. (1986) Double-stranded RNA from plants infected with viruses having elongated particles and undivided genomes. *Phytopathology* 76, 459–465.

Varveri, C. Ravelonandro, M. and Dunez, J. (1987) Construction and use of a cloned cDNA probe for the detection of plum pox virus in plants. *Phytopathology* 77, 1221–1224.

Vunsh, R., Rosner, A. and Stein, A. (1990) The use of the polymerase chain reaction (PCR) for the detection of bean yellow mosaic virus in gladiolous. *Annals of Applied Biology* 117, 561–569.

Waterhouse, P.M., Gerlach, W.L. and Miller, W.A. (1986) Serotype-specific and genera luteovirus probes from cloned complementary DNA sequences of barley yellow dwarf virus. *Journal of General Virology* 67, 1273–1282.

Weier, H.U. and Gray, J.W. (1988) A programmable system to perform the polymerase chain reaction. *DNA* 7, 441–447.

Wetzel, T., Candresse, T., Ravelonandro, M. and Dunez, J. (1991) A polymerase chain reaction assay adapted to plum pox potyvirus detection. *Journal of Virological Methods* 33, 355–366.

Wetzel, T., Candresse, T., Macquarie, G., Ravelonandro, M. and Dunez, J. (1992) A highly sensitive immunocapture polymerase chain reaction assay method for plum pox potyvirus detection. *Journal of Virological Methods* 39, 27–37.

Zuckermann, R., Corey, D. and Schultz, P. (1987) Efficient methods for attachment of thiol specific probes to the 3'ends of synthetic oligonucleotides. *Nucleic Acids Research* 15, 5305–5321.

Zwadyk, P., Cookley, R.C. and Thornsberry, C. (1986) Commercial detection methods for biotinylated gene probes: comparison with 32P-labelled DNA probes. *Current Microbiology* 14, 95–100.

Monitoring Safety of Plant Foods: Immunodiagnostics for Mycotoxins and Other Bioactive Compounds

M.R.A. Morgan
Department of Food Molecular Biochemistry,
Institute of Food Research, Norwich Research Park, Colney,
Norwich NR4 7UA, UK.

10.1 Introduction

Food of plant origin contains a chemically diverse group of low molecular mass, biologically active non-nutrient compounds. Traditionally, this group has been regarded as being toxicant in nature and, accordingly, has usually included mycotoxins. The mycotoxins, the secondary metabolite products of molds, are in fact contaminants. However, since these compounds are often inevitable products of crop storage, the association with inherent plant toxicants is not unreasonable and will continue. Research on the mycotoxins has been considerable, in line with the high toxicological potencies of some of the group in both animals and man. Research on the natural toxicants as components of food has by comparison been considerably neglected, apart from elegant plant biochemical studies on structural elucidation and pathways of biosynthesis, and the search for anti-tumour agents. Consequently, it has only recently become apparent that our understanding of *in vivo* biochemical activity following normal ingestion of the non-nutrient food components has been seriously flawed in many cases. Recently, for example, considerable interest in broccoli consumption was generated by studies extending understanding of 'protective' action. The inherent component sulphorophane was reported as having beneficial properties of Phase I enzyme induction in mice (Zhang *et al.*, 1992). Such research has been extended in human systems by the finding of novel transcription factors induced by dietary non-nutrients (Wang and Williamson, 1994). As a result, whether a compound has to be regarded as potentially toxic or potentially beneficial is a question which though clearly of some importance remains difficult to answer in many cases. The term 'non-nutrient' covers our lack of understanding of metabolic behaviour following normal dietary ingestion.

Why is our understanding so poor? Part of the reason lies in the attitude of consumers to 'natural' chemical components of the diet compared to that for man-made chemicals, the permitted additives and residues of production. Research on the added components has, as a result, been considerably out of phase with their actual or potential toxic effects. A further part of the reason is that the analytical resources available in the conventional laboratory carrying out analyses based on chromatography and 'wet' chemistry do not always lend themselves to large sample numbers or small sample volumes. Such limitations are often relevant to animal and human studies where samples containing trace amounts of analyte with no possibility for concentration are the norm. It helps, of course, if the end-point bioactivity is dramatic in nature, such as the acute effects that can be observed in bioassays for certain of the mycotoxins. Sub-acute effects are much more difficult to study, and animal models become much less convincing as predictors of human biochemistry.

Antibody-based methods of analysis are ideally suited to sensitive and specific analysis of large sample numbers. Accordingly, use of immunoassays by regulatory and monitoring bodies, and by the food industry as part of raw and finished product quality control, has become increasingly popular.The demonstrated cost advantages are also important! It can also be shown that the availability of an immunochemically-based assay can generate useful information in order to increase understanding of *in vivo* behaviour following ingestion of particular plant foods. Such an approach ought to refine and improve the regulatory needs. A further most important aspect should be that the immunoassay can also be utilized by the plant breeder in order to increase the efficiency of production of useful new lines. The absence of suitable analytical methodology often means that important factors (potentially deleterious) are not assessed until a late stage in the process of bringing a new line to the market.

In this chapter, the inherent components covered will include the potato glycoalkaloids, the glucosinolates and the phyto-oestrogens; also covered will be the contaminant aflatoxins, trichothecenes and other mycotoxins.

10.1.1 Immunogen production

Molecules such as the mycotoxins or the inherent plant non-nutrients sized below a range of between 1000 and 10,000 Da require covalent conjugation to a carrier molecule that is immunogenic in its own right. A molecule conjugated in this way is known as a hapten. A protein such as bovine serum albumin is a commonly used carrier for this purpose, and when it is administered to an animal with an immune system, the animal responds after a period of time (which can vary between one and 24 months) by production of antibodies – some of which will recognize the protein and some of which will recognize the conjugated molecule (the hapten) when 'free' of the carrier.

The means by which the hapten is coupled to carrier protein is critical in determining the specificity of the resultant antibody. In general, the antibody will be most specific for sites distal to the point of chemical linkage, and least specific for sites at or near to the linkage site. Accordingly, careful thought needs to be given before choosing the conjugation strategy – is it required to analyse for a single compound or are there a group of related compounds which could usefully be determined? Is the assay to be essentially screening in nature, or investigative of a specific compound? Will there be regulatory implications for findings? Are there known structurally related compounds which it is imperative for the antibody not to recognize? On occasion the required strategy might demand synthesis of a new molecule in order to place a functional group (for the linkage to protein) in the right place; on others little if any extra work might be necessary. Specific examples will be given in the subsequent text; see Robins *et al.* (1988) for further relevant discussion.

10.1.2 Polyclonal or monoclonal?

The relative merits of polyclonal and monoclonal antibodies are described elsewhere in this volume. As far as low molecular mass, haptenic compounds are concerned it is not the case, despite the popular conception, that monoclonal antibodies are always more specific – indeed, many examples are available where polyclonal preparations are superior. It is also not always the case that a monoclonal antibody is the only option where diagnostic kits for commercial production are required – large volumes of polyclonal antibody able to be used at high dilutions can be obtained in some instances, and in quantities more than sufficient to meet market requirements. It is important to emphasize that a pool of antiserum should be established and characterized sufficient for foreseeable needs before dissemination of material, whether for commercial or research purposes. Anticipation of future production is a difficult process. Many of the problems relating to polyclonal antibody preparations have occurred because characterization and publication has related to small amounts of material. For certain applications (production of an immunoaffinity column, for example, where extremely high amounts of antibody are required) a monoclonal antibody is virtually essential if a commercial end is in mind.

As a general strategy, it is recommended that the polyclonal route is explored first. In this way, conjugates can be tested, formats explored and application processes initiated. If required, a monoclonal antibody can be pursued subsequently and with greater confidence. Haptenic compounds, it should be remembered, are by far the hardest targets to aim for with an antibody-based approach.

10.1.3 Immunoassay format

The enzyme-linked immunosorbent assay (ELISA), particularly that utilizing microtitration plates, is now virtually standard at the research laboratory level, and provides many advantages for commercial application in addition. There are many variations on the theme, often reflecting a particular company's proprietary strategy, but also different analytical requirements. Thus a quantitative assay requires a rather different approach to that of a yes/no or above/below. Application to single samples is different to batch analysis of many samples.

Whatever the approach adopted, it should always be remembered that immunoassays are based on the immutable law of mass action, and are capable of fully quantitative application. Semi-quantitative assays at their best are quantitative assays with wider bands of precision deliberately put in.

10.2 Low Molecular Mass, Biologically Active Non-nutrients of Food Origin

There are probably more than half a million naturally occurring compounds in fresh plant foods (Morgan and Fenwick, 1990). Of the low molecular mass, biologically active non-nutrient fraction, very few will have been the subject of in-depth research as to their *in vivo* activity, in either man or farm animals, as discussed previously. It is noteworthy that where there is suspicion concerning occurrence, potential toxicity and risk factors, then were they synthetic chemicals, it is difficult to envisage approval by the appropriate government committees. Further, whilst there is considerable public concern expressed at specific gene transfer technology in plants (and animals), the consequences of conventional plant breeding are unpredictable as far as the effect on secondary product pathways are concerned. Recent research has led to the view that there might have to be reappraisal of the effects following ingestion of some of these compounds. At certain levels, it appears that compounds such as the glucosinolates and the phytoestrogens might well be beneficial rather than deleterious in man (Cassidy *et al.*, 1994; Johnson *et al.*, 1994). The application of immunossays can help to clarify the position.

The types of compounds of interest to the food chemist include cyanogenic glycosides, found particularly in cassava (a staple dietary component in many parts of the world), the glucosinolates of cruciferous vegetables such as cabbage and Brussels sprouts, the oestrogenic isoflavones, saponins and lectins from legumes, the pyrrolizidine alkaloids and the potato glycoalkaloids. It is the latter group that is probably best known, as a result of the major scale of potato consumption in many parts of the world. Of the inherent non-nutrients, the potato glycoalkaloids have also been the most widely studied using immunochemical approaches.

10.2.1 *The potato glycoalkaloids*

Figure 10.1 shows the structure of α-solanine and α-chaconine which together make up 95% or more of total glycoalkaloid (TGA) fraction of the common potato destined for human consumption, *Solanum tuberosum* (Paseshnichenko and Guseva, 1956). The two compounds are glycosidic derivatives of the 3-hydroxysteroid alkaloid solanidine and arise biosynthetically from cholesterol. The TGA fraction occurs throughout the plant, with particularly high levels in leaves and flowers. Within the tuber, highest concentrations occur in the tuber peel immediately beneath the skin and in areas of high metabolic activity such as damaged tissue and eye regions (Wolf and Duggar, 1940; Lampitt *et al.*, 1943). Levels are controlled by genetic (Sanford and Sinden, 1972) and environmental factors, such as location of growth, soil type, climate, agricultural practice and postharvest stress (Jadhav *et al.*, 1981). The latter includes physical and chemical damage, exposure to light, storage conditions and age. The function of the glycoalkaloids in the potato is not known, though they do possess both antifungal and antimicrobial activity, and can confer resistance to insect pests. The glycoalkaloid fraction is highly heritable, and it may be that higher levels in new lines would provide just the resistance properties being sought.

The TGA fraction of potatoes has been associated with cases of poisoning in animals and man that have occasionally been fatal (McMillan and Thompson, 1979). Symptoms include gastrointestinal and neurological disturbances, and haemolytic and haemorrhagic damage. Studies on potatoes associated with elevated glycoalkaloid levels (particularly in genetic stocks

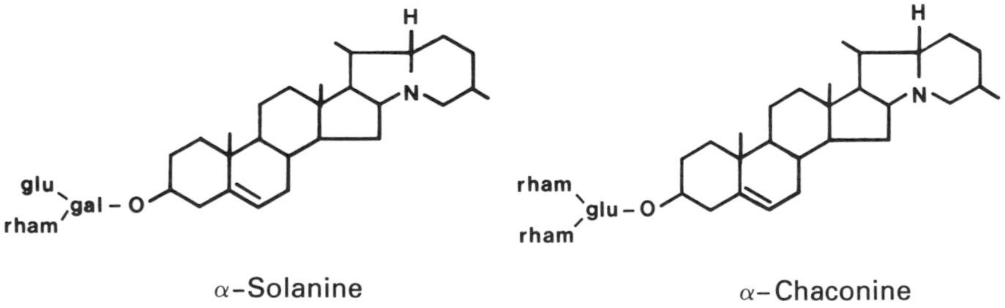

α–Solanine   α–Chaconine

Fig. 10.1. Structures of the predominant components of the potato glycoalkaloid fraction, α-chaconine. Solanidine is the 3-hydroxysteroidal aglycone.

not destined for human consumption) have, unfortunately, not reached the literature. The potato glycoalkaloids can also impart an extreme bitter taste at appropriate levels, regarded by those who have experienced it as particularly nauseous. It is generally accepted that glycoalkaloids in potatoes destined for consumption should not exceed 20 ppb or 200 mg kg^{-1}. Surveillance studies in the UK, USA and Australia of retail supplies show levels between 1 and 20 ppb (Sinden and Webb, 1972; Davies and Blincow, 1984; Morris and Peterman, 1985). For a more detailed discussion on glycoalkaloid properties, see Morgan and Coxon (1987).

The need for an appropriate method of TGA analysis seems clear, both for plant breeders to monitor new varieties, for processors to monitor raw material and finished product, and for regulatory bodies to monitor periodically food material as consumed. The traditional chemical methods of analysis were totally unsuited, involving complex manipulation, reagents that were often highly toxic, and extremely slow rates of sample through-put (for a review, see Coxon, 1984). High performance liquid chromatographic analysis was introduced comparatively recently (Hellenas, 1986) but while improving analytical quality, suffered from high equipment and running costs, and slow rates of sample analysis. Further, few if any of the methods were suitable for application to studies of metabolism of glycoalkaloid compounds following ingestion. Such studies are required in order to fully understand likely effects and to set appropriate safety margins.

Immunochemical approaches

A number of different immunoassays have now been described for determination of the potato TGA fraction (Vallejo and Ercegovitch, 1971; Morgan *et al.*, 1983a; Plhak and Sporns, 1992). The first was a radioimmunoassay, utilizing a tritiated solanidine label. The antibody was produced by forming the solanidine hemisuccinate and conjugating to protein for use as the immunogen. The antibody recognized α-chaconine and α-solanine only poorly, and so potato extracts had to be hydrolysed prior to the immunoassay. The method was cumbersome, the antibody of comparatively poor quality because of the low titre, and the assay does not seem to have been widely used.

The first ELISA (Morgan *et al.*, 1983b) utilized a different approach to antibody production. A Schiff base was generated from α-solanine by periodate cleavage of the sugar residues, which could be directly conjugated to carrier protein to form the immunogen. The polyclonal antiserum generated was of high quality, being produced in large amounts, able to be used at high dilutions, having the desired specificity, and resulting in an assay format that was entirely fitted to the application. Thus, the antibody recognized α-solanine, α-chaconine and other components of the TGA fraction directly and equally well; structurally related steroids did not interact. The ELISA format performed on microtitration plates was able to be used on solubilized potato material without preparation other than dilution in assay buffer. Comparison of the results obtained with other procedures was excellent, both in a cross-laboratory test (Morgan *et al.*, 1985) and in an independent assessment (Hellenas, 1986). This assay, based on a polyclonal antiserum, has been used in laboratories in many countries since 1983 without problem. The applications have been screening of potential new potato lines and surveillance of the retail supply. The normal requirement for TGA analysis is for a quantitative determination. The microtitration plate format was also investigated for above–below assessment (Ward *et al.*, 1988). Probit analysis of results visually scored showed that the use of horseradish peroxidase as marker enzyme together with 2,2-azino-bis (3-ethylbenzthiazoline)-3-sulphonic acid gave greatest discrimination, and that all potato samples at greater than a chosen critical level could be correctly identified without use of a spectrophotometer.

Recently a slightly different approach to immunogen synthesis has been reported (Plhak and Sporns, 1992), whereby solanidine was conjugated to carrier protein through a glycosidic bond. Good quality antibodies were obtained, and correlation of ELISA results with high performance liquid chromatography was high. It was reported that a further disadvantage of the chromatographic assay was that it did not detect some of the minor components of the TGA fraction. The same group has just produced a monoclonal antibody (Plhak and Sporns, 1994) using the same conjugate chemistry, though application to potato samples is awaited. It will then be interesting to compare the properties of the different antibody preparations.

The advantages of the immunochemical approach to TGA determination have been generally accepted to be simplicity (by avoiding major sample preparation and requiring only the essential step of solubilization of the analyte), high rates of sample through-put, stable, non-toxic reagents, reduced cost (both in terms of equipment and labour), the desired assay specificity and excellent precision and accuracy. A further advantage is derived from the use of the immunoassays to quantify glycoalkaloid material in human tissue and body fluid samples. For such application a method must be intrinsically sensitive, because relying on extraction of large sample volumes and concentration of analyte is not an option with most human volunteer donors of samples.

Solanidine has been measured in human serum and saliva by radio-immunoassay (Matthew *et al.*, 1983; Harvey *et al.*, 1985a,b) and glycoalka-loids have also been measured in serum (Harvey *et al.*, 1985b). It was found that all UK samples studied contained potato glycoalkaloid material, that it was present in amounts that paralleled potato intake, and that the half-life in serum was sufficiently long to suggest accumulation (probably in the liver). It is now necessary to try and formulate hypotheses as to the effects of ingested glycoalkaloid material, particularly given evidence of membrane permeabolizing effects as shown with isolated intestinal wall preparations (Gee et al., 1989) and how such activity might relate to absorption of potential allergenic material of high molecular mass (Gee et al., 1994). Future immunochemical strategy might have to be adjusted according to studies suggesting that α-chaconine and α-solanine have significantly different bioactivities (Price et al., 1995).

10.2.2 Glucosinolates

The glucosinolates are sulphur-containing glycosides found in cruciferous vegetables. Considerable efforts have been made over the years to reduce levels of glucosinolates in animal feed crops in particular in order to increase palatability. In human food, hydrolysis products can be responsible for desirable flavour characteristics, for example in mustard, radish and brassica vegetables. Thiocyanate as a cyanide precursor is an undesirable breakdown product. Goitrogenic activity has been identified in some of the group members. In contrast, the ability of glucosinolates to induce synthesis of beneficial enzymes, those able to remove toxic moieties before deleterious interaction with DNA, proteins and membrane components, leads us to suspect that glucosinolates might have properties 'protective' against cancer. Only one immunoassay has been described for any members of this group, for the alkenyl glucosinolates (Hassan et al., 1988). The alkenyl glucosino-lates include sinigrin and gluconapin. The immunogen was produced by forming sinigrin hemisuccinate and linking it to carrier protein. The antibody specificity was towards the alkenyl sidechain and the thiohydroximate moiety; desulphoglucosinolates were not recognized at all. Progoitrin, gluconapin and sinigrin were highly recognized. Analysis of mustard seed by ELISA gave very good agreement with standard high performance liquid chromatographic determination. The ELISA was more sensitive, less expen-sive and faster than any other glucosinolate assay of comparable specificity. Given the considerable current interest in this group of compounds, it seems likely that more immunochemical procedures will be developed. Elucidation of any 'protective' role would seem to be of considerable interest to plant breeders and those involved in diet and health matters.

10.2.3 Phyto-oestrogens

Compounds present in food and animal feedstuffs of animal origin with oestrogenic activity (phyto-oestrogens) have been known for many years, particularly since the oestrogens can cause readily recognized infertility problems in farm animals (Bennetts et al., 1946). The major dietary sources for both animals and plants are the legumes. It is somewhat surprising then to realize that comparatively little is known about phyto-oestrogens in human food in terms of their intake, their metabolism and their contribution to oestrogenic status. A recent report has shown that ingestion of soya (a major dietary source of phyto-oestrogens) led to lengthened menstrual cycles compared to control subjects (Cassidy *et al.*, 1994). Such activity has the potential to be 'protective' against breast cancer, as suggested by epidemiological studies of populations consuming diets rich in phyto-oestrogens (Messina *et al.*, 1994). Regulatory authorities have concerned themselves with residues of veterinary drugs with oestrogenic activity and residues of the mycotoxin, zearalenone, which is also a weak oestrogen – in spite of the oestrogenic potential of the contaminants being considerably less than that of the inherent components of dietary intake. Thus a number of immunoassays have been developed for zearalenone, including ELISAs (Liu *et al.*, 1985; Warner *et al.*, 1986) and an ELISA employing a monoclonal antibody which had specificity and sensitivity advantages over the polyclonal antisera (Dixon *et al.*, 1987). Antibodies were produced against the phyto-oestrogen genistein by covalently linking a carboxylic acid derivative of the oestrogen to a synthetic polypeptide carrier (Bauminger *et al.*, 1969). The antibodies produced recognized genistein (possibly the most active phyto-oestrogen found in many leguminous food plants) but not endogenous circulating oestrogens such as oestradiol. Though use of immunization of livestock as a means of protection was discussed, no application to plant food samples seems to have taken place.

10.2.4 *Miscellaneous plant bioactives*

The pyrrolizidine alkaloids are secondary metabolites present in a number of plant foods, and are deemed to be toxic to animals and man. A class-specific ELISA was developed by producing antibodies against a retronecine–protein conjugate, and using the antibodies to analyse a hydrolysed extract of naturally occurring pyrrolizidine alkaloids in various plants (Roseman *et al.*, 1992). Accordingly, monocrotaline, retrosrine and senicione could all be quantified.

Naringin and limonin are bitter flavour compounds found in citrus fruit. A radioimmunoassay for naringin (Jourdan *et al.*, 1983) and an ELISA for limonin (Jourdan *et al.*, 1984) have been used to study the localization and production of these undesirable compounds, and as quality control assessment of raw materials. The limonin ELISA is commercially available. An ELISA for the bittering agent quinine, a product of the bark of particular tropical trees, has been used for quality control of levels in soft drinks (Ward and Morgan, 1988).

Many immunoassays have been described for caffeine, though few seem to have been applied to food material. In contrast to the position elsewhere, all the work seems to have been applied to serum determinations based on the well-known physiological actions of caffeine. Of particular note is a homogeneous, substrate-labelled fluoroimmunoassay not requiring separation of antibody-bound or free fractions (Pearson *et al.*, 1984).

Glycyrrhizin is the principle component of certain Chinese herbal remedies. It is also found in liquorice at levels not far short of those which would give rise to significant physiological activity, including aldosterone-like activity. A monoclonal antibody has been used in an ELISA to look at levels in sera following drug therapy. There was no reported application to food samples (Mizugaki *et al.*, 1994).

10.3 Mycotoxins

The scientific literature concerning the immunoassay of the mycotoxins is extensive and rapidly expanding. It will be the purpose of this article to try and extract some of the key features of progress over the last 20 years since the first reported attempt to produce antibodies to a mycotoxin (Aalund *et al.*, 1975) and to anticipate future progress. Reviews of the literature are, fortunately, frequent and extensive (Chu, 1992; Wilkinson *et al.*, 1992; Pestka, 1994). The mycotoxins are the secondary metabolite products of moulds. Their biosynthesis will reflect the type of mould, the state of mould growth and nutrition, and environmental factors such as temperature and water activity. Thus *Aspergillus flavus* and *Aspergillus parasiticus* can produce aflatoxins under conditions including high temperatures and humidity typically found in the southern USA and many parts of the developing world. The production of trichothecenes by *Fusaria* spp. or of ochratoxin A by *Aspergillus ochraceous* requires the lower temperatures found in northern Europe. Some moulds possess the genetic capability to biosynthesize toxins but do not seem to do so. Accordingly there has been some interest in elucidating the mechanisms of regulation of such 'sleeping' genes, since if genes in fungi currently present in food material quite safely could be switched on, then the implications would be quite serious (Chang *et al.*, 1992). The use of antibody methods to investigate the enzymes involved is key in such research (Lui *et al.*, 1993).

The interest in the mycotoxins stems from their potential toxicity in both animals and man. Throughout recorded history there have been outbreaks of poisoning and death, as circumstances have contrived to produce the right conditions for toxin synthesis, usually related to stored food material. Gangrenous ergotism was recorded as early as the ninth and tenth centuries and periodically since, and is the result of consuming cereals contaminated with mycotoxin alkaloids produced by *Claviceps purpurea*. Consumption of cereals that had spent the winter under snow is believed to have killed many thousands of people (some have estimated hundreds of thousands) in famine-affected parts of the former Soviet Union after the last war. The cause is generally ascribed to the trichothecenes.

Toxicological studies have indicated that mycotoxin potency – and in particular that of the aflatoxins – can be extreme. Consequently, nearly all nations have regulations governing permitted maximum levels in both human and animal food that are in the very low parts per billion range and occasionally (notably for aflatoxin M_1) down to the parts per trillion level (van Egmond, 1989). Contamination of agricultural and food material is non-uniform. The large majority of particles (such as individual nuts or grains) of a batch might be contaminated at low or undetectable levels, a few particles might contain very high levels. The consequent necessity for appropriate sampling plans generates many additional problems for the analyst.

Immunoassay offers many advantages for mycotoxin analysis. Sensitivity and specificity are clearly highly relevant, particularly the latter in circumstances where compounds of closely related structures are also likely to be present. High rates of sample through-put will be important in cases of large batch sizes where decisions are required quickly. Cost is a factor that will be affected by equipment considerations, associated reagents such as solvents, and the labour costs needed for long sample extraction and purification protocols. The additional benefits provided by rapid and simple screening technology to eliminate large numbers of negative samples should also be noted. The use of immunoassay procedures has seen the development of rapid and straightforward extraction procedures that, because they are essentially for solubilization and not purification of the analyte, probably could not be improved upon. Even so, extraction often remains the most time-consuming and labour-intensive part of the entire assay procedure.

The use of immunoaffinity columns has found considerable favour with aflatoxin analysts in particular. The principle of the approach is to use immobilized antibodies (usually linked covalently to Sepharose beads) to extract the target analyte or analytes from a complex mixture. Extraneous material is then washed away, leaving the adsorbed material of interest, which can be quantified almost *in situ* if it has the right properties of fluorescence (as do the aflatoxins) or can be eluted by appropriate choice of solvent to allow quantification by alternative procedures. The columns are almost always reusable if regenerated correctly. The advantages of such an approach are clear for direct determination (Candlish *et al.*, 1988; Groopman and Donahue, 1988), and for the simple but considerable power of concentration provided for extremely sensitive analysis (Mortimer *et al.*, 1987). For resolution of a group of analytes, using an alternative method of separation and detection, the immunoaffinity approach also has advantages. Thus adsorption of the aflatoxins and individual quantification by high performance liquid chromatography is a popular procedure (Sharman and Gilbert, 1991). Recently, extension of the application has been made to determination of ochratoxin A in cereal samples (Bisson *et al.*, 1994).

10.3.1 Aflatoxins

The aflatoxins (Fig. 10.2) are regarded as highly potent hepatotoxins; aflatoxin B_1 is regarded as the most potent natural carcinogen tested in some animal models and is now generally accepted to be a human carcinogen. Regulations for maximum acceptable levels apply to both human and animal food, and may refer to B_1 alone or to total aflatoxin content (B_1, B_2, G_1 and G_2; van Egmond, 1989). The mammalian hydroxylated metabolite of B_1, aflatoxin M_1, can appear in milk. Regulations concerning M_1 in milk and milk products tend to be addressed to even lower levels than B_1, even though M_1 is likely to be less toxic than B_1; this is because of the higher consumption of milk by the young and because the young are more susceptible to carcinogens. Aflatoxin contamination of cereals, nuts and commodities such as figs means that analysis of raw materials, animal feed, and processed human food materials is essential.

Production of antibodies by synthesizing a carboxymethyloxime derivative of B_1 generates antibodies with a range of desirable specificities, including specificity for aflatoxin B_1 and for total aflatoxin determination

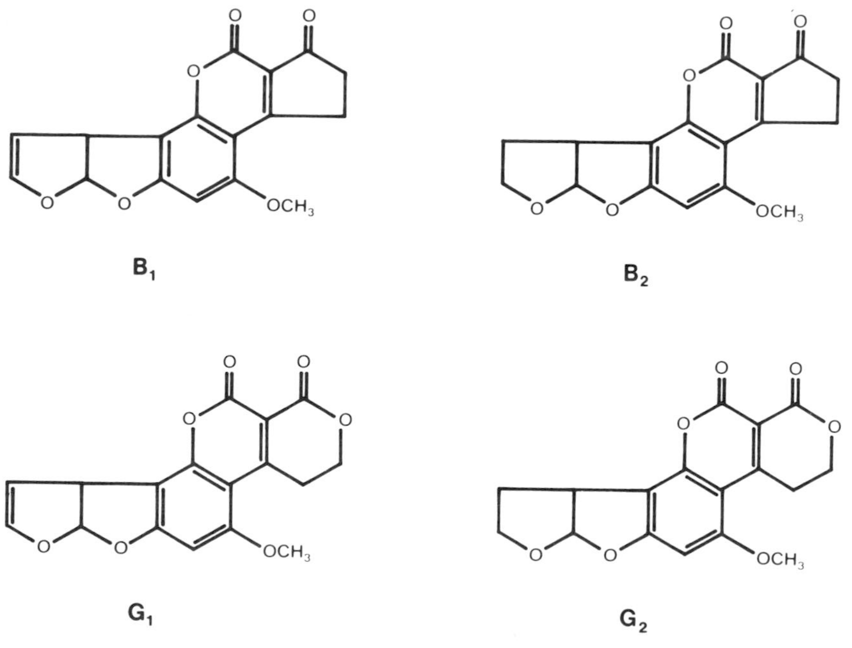

Fig. 10.2. Structures of the aflatoxins. Aflatoxin B_1 is thought to be the most toxic. Aflatoxin M_1 is a hydroxylated mammalian metabolite of B_1, appearing in milk after consumption of contaminated food.

(Morgan *et al.*, 1986a,b,c; Ward *et al.*, 1990), and would appear to be the method of choice for both polyclonal and monoclonal antibody production. Alternative routes have been described, which have produced antibodies of interesting properties but which have not been widely utilized (Lawellin *et al.*, 1977; Sizaret *et al.*, 1980; Lau *et al.*, 1981; Gaur *et al.*, 1981; Fan *et al.*, 1984a,b; Pestka and Chu, 1984; Jackman, 1985; Chu *et al.*, 1985). Both polyclonal and monoclonal preparations appear capable of suitable performance, the major consideration being the use to which the antibody is to be put and the compatibility of the antibody with the solvent of choice. As an example of the latter, a quantitative assay would usually perform better if the solvent able to give maximum and most reproducible extraction of analyte could be employed. Some solvents can – even in trace quantities after dilution – deleteriously affect antibody performance.

A sizeable portion of the worldwide agri-food immunodiagnostic industry (see Appendix 10.1) is involved in kits for aflatoxin analysis (Wilkinson *et al.*, 1990; Pestka, 1994). The formats utilized reflect considerable ingenuity in providing something different from the necessarily limited number of basic assay components. A number of different needs are covered, from fully quantitative determinations suited to batchwise sample analysis to rapid, semiquantitative and simple procedures taking as little as 3 min and appropriate to single samples. In order to speed up the assays it is critical to address the phase separation component of the assay and the assay end-point, particularly since sample preparation is almost invariably a straightforward solubilization step. A number of commercially available assays have received Official First Action Approval from the AOAC and approval from the AOAC Research Institute.

An important consideration in early publications was to correlate results obtained by immunoassay determinations with those obtained by alternative procedures. Nowadays, the fact that properly developed and validated immunoassays can indeed provide results comparable to alternative procedures (albeit quicker, simpler, cheaper) is rather less newsworthy in the mycotoxin area, though still essential. Truckess *et al.* (1990) showed that two commercially available diagnostic kits based on immunoassays were just as effective as long-established AOAC methodology in determining aflatoxin in animal feeds. Dell *et al.* (1990) compared ELISA, high performance liquid chromatography and high performance thin-layer chromatography for the analysis of peanut butter samples. The quantitative analysis of aflatoxin in peanut butter by ELISA was successfully examined by Patey *et al.* (1992).

Applications of immunoassays for aflatoxins to human tissue and fluid samples has been quite extensive in attempts to further elucidate the role of the toxin in disease. Of interest is the interaction of aflatoxin with DNA (and the detoxification procedures employed to combat this which appear to be particularly effective in man compared to certain other animal species), and the subacute effects of aflatoxin which are believed to include immunosuppression. Immunoassays detecting the adducts of aflatoxins to DNA and proteins (and fragments of these) in urine and tissue have been described (Garner *et al.*, 1988; Groopman and Donahue, 1988; Ross *et al.*, 1992), and are clearly important in looking at toxin exposure. Immunoassays of different specificity can look at potential for toxin exposure by picking up aflatoxins prior to derivatization (Wilkinson *et al.*, 1989; Denning *et al.*, 1990a, 1994; Makarananda *et al.*, 1990). Whichever the analytical target, the results are clear on the different magnitude of exposure to aflatoxin experienced by those in developing countries compared to others.

As discussed previously, requirements for determination of aflatoxin M_1 are rather different to those for B_1. ELISA procedures of great specificity, sensitivity (5pg per g dry weight) and rapidity (requiring no clean-up) have been described and used to monitor milk samples for contamination (Okumura *et al.*, 1993; Kawamura *et al.*, 1994).

10.3.2 Trichothecenes

The trichothecenes are an interesting group of compounds (see Fig. 10.3) for which there has been much interest in generating immunoassays. Though

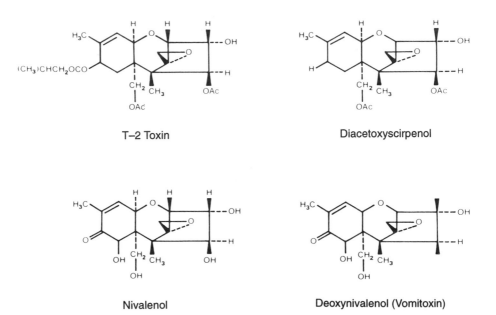

Fig. 10.3. The four trichothecenes generally accepted to be potential contaminants of food. Deoxynivalenol is also known as vomitoxin.

several of the group have been positively identified in food samples and in particular cereals (T-2 toxin, diacetoxyscirpenol, nivalenol and deoxynivale-nol – also known as vomitoxin) it has to be noted that significant interest has been related to their supposed use as agents of biological warfare. The absence of easily measured physicochemical properties (especially when compared to the aflatoxins) means that the trichothecenes are difficult to detect using conventional chromatographic procedures.

Generation of successful immunoassays for trichothecenes has revolved even more than usual around the strategy for antibody production. In general, the polyclonal antisera produced have been of higher specificity than the monoclonals, and have produced standard curves with better limits of detection. In some cases, some very esoteric assay formats have been described in order to improve the performance of the monoclonal antibodies (Warden *et al.*, 1987; Ligler *et al.*, 1987).

For T-2 toxin, the strategy for immunogen synthesis has been to conjugate the toxin to protein via the hydroxyl group, usually after making the hemisuccinate derivative (Pestka *et al.*, 1981; Peters *et al.*, 1982; Gendloff *et al.*, 1984; Fan *et al.*, 1984; Barna-Vetro *et al.*, 1994). Similarly, the chemistry required for diacetoxyscirpenol is straightforward. Linkage through the available hydroxyl group provides an immunogen able to generate specific antibody production (Chu *et al.*, 1984; Mills *et al.*, 1988). Nivalenol has been more difficult to conjugate to protein because of the complexity of the chemistry. Thus far, the less than ideal approach of generating antibodies to the tetraacetate has been employed. During the assay, putative nivalenol in the sample has to be derivatized in order for the antibody to recognize it (Ikebuchi *et al.*, 1990; Teshima *et al.*, 1990).

Deoxynivalenol has provided a challenge to the immunochemist, a challenge readily accepted because of all the trichothecenes this is the one of most commercial interest. Deoxynivalenol occurs frequently and in some years at high levels in cereals, particularly in north America. As for nivalenol, acetylation of hydroxyl groups has provided one option (Xu *et al.*, 1986) for immunogen synthesis. Two differing strategies have allowed the preferred option of direct analysis. Casale *et al.* (1988) used protection of the adjacent hydroxyl groups to allow derivatization of the third. Removal of the protection and conjugation to protein led to production of high specificity antibodies. An alternative procedure was devised by Mills *et al.* (1990). Starting from 3-acetyldeoxynivalenol, the hemiglutarate derivative of the primary hydroxyl group was achieved. An enzyme was then used to remove the 3-acetyl group, prior to conjugation of the hemiglutarate group to the carrier protein. Highly specific antibodies were obtained, which were used to analyse wheat samples. A monoclonal antibody of somewhat broader specificity was obtained rather cleverly by manipulating the reaction conditions between deoxynivalenol and succinic anhydrides to favour production of mono derivatives (Nicol *et al.*, 1993).

Preliminary reports have appeared looking at recombinant methods of antibody production for trichothecene recognition, and specifically for diacetoxyscirpenol (Lee *et al.*, 1995). Considerable advantages might be offered by this approach, including the potential speed of antibody production and the possibility of altering the properties of the recombinant protein. It is too early to predict when the new technology will impact on this field.

10.3.3 Other mycotoxins and related contaminants?

Of the mycotoxins that can be found in agri-food materials other than the aflatoxins and trichothecenes, the most important are probably ochratoxin A and sterigmatocystin. Ochratoxin A is a nephrotoxin, for which claims of carcinogenic effects in humans have been made and in which renewed interest has been demonstrated in the last few years. It is important to monitor animal feed components (Morgan *et al.*, 1983a,b; Lee and Chu, 1984; Candlish *et al.*, 1988a,b) and in certain countries porcine kidneys have been investigated for contamination (Morgan *et al.*, 1986a,b,c; Kawamura *et al.*, 1989; Rousseau and van Peteghem, 1989). A number of laboratories are analysing for ochratoxin A directly in human serum, tissues and body fluids, on the basis that contamination of individual foods might be at a low level but that the exposure of the consumer might be high because of the range and number of foods that contain the toxin (Gareis *et al.*, 1988). The immunochemical strategy is straightforward as far as ochratoxin A is concerned; utilization of the available carboxylic acid function yields antibodies of the desired specificity and affinity.

For sterigmatocystin, little research has been done with immunochemical technology to look at its occurrence in plant food, though antibody production has been described (Li and Chu, 1984; Morgan *et al.*, 1986).

ELISAs for sporidesmin and the indole, diterpenoid and ergot alkaloids of endophytic fungi (problem contaminants of pastoral agriculture which have been extensively researched in New Zealand) have also been described (Garthwaite *et al.*, 1994) and applied to routine monitoring programmes.

Much interest has been generated in the fumonisin group of mycotoxins in recent years as potential animal and human toxins. Antibody-based procedures of analysis have been described by two groups (Azcona-Olivera *et al.*, 1992; Usleber *et al.*, 1994).

10.3.4 Multi-analyte approaches to mycotoxin analysis

It is often required to analyse for a number of mycotoxins in a single sample. There has, in consequence, been significant interest in approaches to multi-analyte determination incorporating antibody technology. One possibility, the use of instrumental techniques to look at immuno-extracts, has been discussed previously. The drawback of returning to expensive and time-consuming equipment might be regarded as a disadvantage. Some researchers have taken alternative approaches, for example, by using simple chromatographic procedures to separate the analytes and individual antibodies to label and quantify. Thus Pestka (1991) has described the ELISAGRAM system to detect aflatoxins and zearalenones. Individual antibodies can be used to extract individual toxins from a mixture, using labelled toxin material for quantification. Thus Schneider *et al.* (1991) have described an elegant dipstick system for detection of aflatoxins, trichothecenes, ochratoxin A and zearalenone. Abouzied and Pestka (1993) have described a computer-assisted multi-analyte system (CAMAS) for simultaneous detection of aflatoxins, zearalenones and fumonisins.

A very different possibility for multi-analyte detection is provided by the use of multivariate statistical methods to assess the contributions of cross-reacting compounds in immunoassays set up to include (rather than more usually to exclude) such binding (Karu *et al.*, 1994). Though this powerful approach has not yet been applied to mycotoxins or toxicants in food, it will become more widely utilized in future.

10.4 Future Trends

In the immunochemistry area, the trends for the future will continue being towards development of faster, simpler assays with cost being at a premium, and the information value of the results being paramount. If the promise of genetically manipulated assays can be realized, then the days of antibody production being the rate-limiting factor will be over (Griffiths *et al.*, 1994; Karu *et al.*, 1994; Lee *et al.*, 1995). In some areas, we will be able to more easily obtain the specificity we desire, rather than that we are given. The opportunities for marrying immunochemistry with other technologies will increase, and maybe the long-awaited disposable biosensors will become a reality for analysts seeking rapid answers in the field.

Over the last decade, immunoassay methods of analysis in the area of mycotoxin determination have found acceptance by analysts in research, and in the quality control laboratories of the industry and the regulatory authorities. Much more remains to be achieved in the area of the biologically active non-nutrient plant food components, however, both by plant breeders and by those with an interest in the metabolism of the non-nutrients once consumed. The opportunities for improvement of crop varieties relative to yield and resistance are increasing rapidly. It will be important to monitor for deleterious factors early in the process, and immunoassay techniques offer an ideal approach. The variety of potato known as *Lenape* had to be withdrawn from the market two years after its introduction to the north American market in 1968 because of elevated glycoalkaloid levels. Are plant breeders well-placed today to study changes in the toxicant secondary metabolites before varieties reach the market? On the beneficial side, enormous opportunities exist for providing enhanced health benefits through dietary manipulation. It seems clear that a major barrier to such advances remains the poor level of understanding of human metabolism of dietary components in the normal state. The paradox is that increased consumption of plant foods is being urged upon us and that agriculture is increasing its use of low input systems – and yet we have little information as to how consumption of biologically active plant components will affect health status in the longer term. The elevated status of herbal remedies with sections of the public (which in the UK last year saw a significant number of cases of human poisoning) at least ensures that the risks do not disappear from view entirely.

Acknowledgements

The author would like to thank Roger Fenwick, Bob Heaney, Andy Wilkinson, Colin Ward and Rebecca Sweet for much valued assistance in the production of this manuscript.

References

Aalund, O., Brundelt, K., Hald, B., Krogh, P. and Poulsen, K. (1975) A radio-immunoassay for ochratoxin A: a preliminary investigation. *Acta Pathologica Microbiologica Scandinavia, Section C* 83, 390–392.

Abouzied, M.M. and Pestka, J.J. (1993) Simultaneous screening of fumonisin B$_1$ aflatoxin B$_1$, and zearalenone by line immunoblot: a computer-assisted multi-analyte system (CAMAS). *Journal of the Association of Official Analytical Chemists International* 77, 495–500.

Azcona-Olivera, J.I., Abouzied, M.M., Plattner, R.D. and Pestka, J.J. (1992) Production of monoclonal antibodies to the mycotoxins fumonisin B$_1$, B$_2$ and B$_3$. *Journal of Agricultural and Food Chemistry* 40, 531–534.

Barna-Vetro, I., Gyongyosi, A. and Solti, L. (1994) Monoclonal antibody-based enzyme-linked immunosorbent assay of *Fusarium* T-2 and zearalenone toxins in cereals. *Applied and Environmental Microbiology* 60, 729–731.

Bauminger, S., Lindner, H.R., Perel, E. and Arnon, R. (1969) Antibodies to a phytoestrogen: antigenicity of genistein coupled to a synthetic polypeptide. *Journal of Endocrinology* 44, 567–578.

Bennetts, H.W., Underwood, E.J. and Shier, F.L. (1946) A specific breeding problem of sheep on subterranean clover pastures in Western Australia. *Australian Veterinary Journal* 22, 2–12.

Bisson, E., Bypass, L., Garner, A. and Garner, R.C. (1994) Analysis of wheat and kidney samples for ochratoxin A using immunoaffinity columns in conjunction with HPLC. *Food and Agricultural Immunology* 6, 331–339.

Candlish, A.A.G., Haynes, C.A. and Stimson, W.H. (1988a) Detection and determination of aflatoxins using affinity chromatography. *International Journal of*

Food Science and Technology 23, 479–485.

Candlish, A.A.G., Stimson, W.H. and Smith, J.E. (1988b) Determination of ochratoxin A by monoclonal antibody-based enzyme immunoassay. *Journal of the Association of Official Analytical Chemists* 71, 961–964.

Casale, W.L., Pestka, J.J. and Hart, L.P. (1988) Enzyme-linked immunosorbent assay employing monoclonal antibody specific for deoxynivalenol (vomitoxin) and several analogues. *Journal of Agricultural and Food Chemistry* 36, 663–668.

Cassidy, A., Bingham, S. and Setchell, K.D.R. (1994) Biological effects of a diet of soy protein rich in isoflavones on the menstrual cycle of pre-menopausal women. *American Journal of Clinical Nutrition* 60, 333–340.

Chang, P.K., Skory, C.D. and Linz, J.E. (1992) Cloning of a gene associated with aflatoxin B_1 biosynthesis in *Aspergillus parasiticus*. *Current Genetics* 21, 231–233.

Chu, F.S. (1992) Recent progress on analytical techniques for mycotoxins in feedstuffs. *Journal of Animal Science* 70, 3950–3963.

Chu, F.S., Chen Liang, M.Y. and Zhang, G.S. (1984) Production and characterisation of antibody against diacetoxyscripenol. *Applied and Environmental Microbiology* 48, 777–780.

Chu, F.S., Steinert, B.W. and Gaur, P.K. (1985) Production and characterisation of antibody against aflatoxin G_1. *Journal of Food Safety* 7, 161–170.

Coxon, D.T. (1984) Methodology for glycoalkaloid analysis. *American Potato Journal* 51, 169–183.

Davies, A.M.C. and Blincow, P.J. (1984) Glycoalkaloid content of potatoes and potato products sold in the U.K. *Journal of the Science of Food and Agriculture* 35, 553–557.

Dell, M.P.K., Haswell, S.J., Roch, O.D., Coker, R.D., Medlock, V.F.P. and Tomkins, K. (1990) Analytical methodology for the determination of aflatoxins in peanut butter: comparison of HPTLC, ELISA and HPLC methods. *Analyst* 115, 1435–1439.

Denning, D.W., Sykes, J.A., Wilkinson, A.P. and Morgan, M.R.A. (1990a) Aflatoxin in Nepalese sera measured by enzyme-linked immunosorbent assay. *Human Experimental Toxicology* 9, 143–146.

Denning, D.W., Allen, R., Wilkinson, A.P. and Morgan, M.R.A. (1990b) Transplacental transfer of aflatoxin in humans. *Carcinogenesis* 11, 1033–1035.

Denning, D.W., Queipo, S.C., Altman, D.G., Makarananda, K., Neal, G.E., Camallere, E.L., Morgan, M.R.A. and Tupasi, T.E. (1995) Aflatoxin and outcome from acute lower respiratory infection in children in the Philippines. *Paediatric Infectious Disease Journal* (in press).

Dixon, D.E., Warner, R.L., Ram, B.P., Hart, L.P. and Pestka, J.J. (1987) Hybridoma cell line production of specific monoclonal antibody to the mycotoxins zearalenone and α-zearalenol. *Journal of Agricultural and Food Chemistry* 35, 122–126.

Fan, T.S.L., Zhang, G.S. and Chu, F.S. (1984a) Production and characterisation of antibody against Q_1. *Applied and Environmental Microbiology* 47, 526–532.

Fan, T.S.L., Zhang, G.S. and Chu, F.S. (1984b) An enzyme-linked immunosorbent assay for T-2 toxin in biological fluids. *Journal of Food Protection* 47, 964–967.

Gareis, M., Martlbauer, E., Bauer, J. and Gedek, B. (1988) Determination of ochratoxin A in human milk. *Zeitschrift fur Lebensmittel Unters Forschung* 186, 114–117.

Garner, R.C., Dvorackova, I. and Tursi, F. (1988) Immunoassay procedures to detect exposure to aflatoxin B_1 and benzo(a) pyrene in animals and man at the DNA level. *International Archives of Occupational and Environmental Health* 60, 145–148.

Garthwaite, I., Sporsen, J., Briggs, L., Collin, R. and Towers, N. (1994) Food quality on the farm: immunological detection of mycotoxins in New Zealand pastoral agriculture. *Food and Agricultural Immunology* 6, 123–130.

Gaur, P.K., Lau, H.P., Pestka, J.J. and Chu, F.S. (1981) Production and characterisation of aflatoxin B_{zd} antiserum. *Applied and Environmental Microbiology* 41, 478–482.

Gee, J.M., Price, K.R., Ridout, C.L., Johnson, I.T. and Fenwick, G.R. (1989) The effects of some purified saponins on trans-mural potential difference in mammalian small intestine. *Toxicology in Vitro* 3, 85–90.

Gee, J.M., Johnson, I.T., Wal, J.M., Atkinson, H.A.C. and Miller, K. (1994) Studies on the relationship between mucosal gut permeability and allergic reactions in the brown Norway rat. *Human Experimental Toxicology* 13, 619.

Gendloff, E.H., Pestka, J.J., Swanson, S.P. and Hart, L.P. (1984) Detection of T-2 toxin in *Fusarium sporotrichioides*-infected corn by enzyme-linked immunosorbent assay. *Applied and Environmental Microbiology* 47, 1161–1163.

Griffiths, A.D., Williams, S.C., Hartley, O., Tomlinson, I.M., Waterhouse, P., Crosby, W.L., Koutermann, R.E., Jones, P.T., Low, M.N., Allison, T.J., Prospero, T.D., Hoogenboom, H.R., Nissim, A., Cox, J.P.L., Harrison, J.L., Zaccolo, M., Gherardi, E. and Winter, G. (1994) Isolation of high affinity human antibodies directly from large synthetic repertoires. *The EMBO Journal* 13, 3245–3260.

Groopman, J.D. and Donahue, K.F. (1988) Aflatoxin, a human carcinogen: determination in foods and biological samples by monoclonal antibody affinity chromatography. *Journal of the Associations of Official Analytical Chemists* 71, 861–867.

Harvey, M.H., McMillan, M., Morgan, M.R.A. and Chan, H.W.S. (1985a) Solanidine is present in sera of healthy individuals and in amounts dependent on their dietary potato consumption. *Human Toxicology* 4, 187–194.

Harvey, M.H., Morris, B.A., McMillan, M. and Marks, V. (1985b) Measurement of potato steroidal alkaloids in human serum and saliva by radioimmunoassay. *Human Toxicology* 4, 503–512.

Hassan, F., Rothnie, N.E., Yeung, S.P. and Palmer, M.V. (1988) Enzyme-linked immunosorbent assays for alkenyl glucosinolates. *Journal of Agricultural and Food Chemistry* 36, 398–403.

Hellenas, K-E. (1986) A simplified procedure for quantification of potato glycoalkaloids in tuber extracts by HPLC: comparison with an ELISA and a colorimetric method. *Journal of the Science of Food and Agriculture* 37, 776–782.

Ikebuchi, H., Teshima, R., Hirai, K., Sato, M., Ichinoe, M. and Terao, T. (1990) Production and characterisation of monoclonal antibodies to nivalenol tetraacetate and their application to enzyme-linked immunoassay of nivalenol. *Zeitschrift Biologische Chemie Hoppe-Seyler* 371–31–36.

Jackman, R. (1985) Determination of aflatoxins by enzyme-linked immunosorbent assay with special reference to aflatoxin M_1 in raw milk. *Journal of the Science of Food and Agriculture* 36, 685–698.

Jadhav, S.J., Sharma, R.P. and Salunkhe, D.K. (1981) Naturally-occurring toxic alkaloids in foods. *CRC Critical Reviews in Toxicology* 9, 21–104.

Johnson, I.T., Williamson, G. and Musk, S.R.R. (1994) Anticarcinogenic factors in plant foods. A new class of nutrients? *Nutrition Research Reviews* 7, 175–204.

Jourdan, P.S., Weiler, E.W. and Mansell, R.L. (1983) Radioimmunossay for naringin and related flavenone 7-neohesperidosides using a titrated tracer. *Journal of Agricultural and Food Chemistry* 31, 1249–1255.

Jourdan, P.S., Mansell, R.L., Oliver, D.G. and Weiler, E.W. (1984) Competitive solid phase enzyme-linked immunoassay for the quantification of limonin in citrus. *Analytical Biochemistry* 138, 19–24.

Karu, A.E., Lin, T.H., Breiman, L., Muldoon, M.T. and Hsu, J. (1994) Use of multivariate statistical methods to identify immunochemical cross-reactants. *Food and Agricultural Immunology* 6, 371–384.

Kawamura, O., Sato, S., Kajii, H., Nagayama, S., Ohtani, K., Chiba, J. and Ueno, Y. (1989) A sensitive enzyme-linked immunosorbent assay of ochratoxin A based on monoclonal antibodies. *Toxicon* 27, 887–897.

Kawamura, O., Wang, D.S., Liang, Y.X., Hasegawa, A., Saga, C., Visconti, A. and Ueno, Y. (1994) Further survey of aflatoxin M₁ in milk powders by ELISA. *Food and Agricultural Immunology* 6, 465–467

Lampitt, L.H., Bushill, J.H., Rooke, H.S. and Jockson, E.M. (1943) Solanine, glycoside of the potato. 2. Distribution in the potato plant. *Journal of the Society of Chemical Industry* London 62, 48–51.

Lau, H.P., Gaur, P.K. and Chu, F.S. (1981) Preparation and characterisation of aflatoxin B$_{z\alpha}$ antiserum. *Applied and Environmental Microbiology* 41, 478–482.

Lawellin, D.W., Grant, D.W. and Joyce, B.K. (1977) Enzyme-linked immunosorbent analysis of aflatoxin B₁. *Applied and Environmental Microbiology* 34, 94–96.

Lee, H.A., Wyatt, G.M., Garrett, S.D., Lacarra, T.G., Alcocer, M.J.C. and Morgan, M.R.A. (1995) Recombinant antibodies directed towards diacetoxyscirpenol and a wheat peptide. *Food and Agricultural Immunology* (in press).

Lee, S.C. and Chu, F.S. (1984) Enzyme-linked immunosorbent assay of ochratoxin A in wheat. *Journal of the Association of Official Analytical Chemists* 67, 45–49.

Li, Y. and Chu, F.S. (1984) Production and characterisation of antibody against sterigmatocystin. *Journal of Food Safety* 6, 119–127.

Ligler, F.S., Bredehorst, R., Talebian, A., Shriver, L.C., Hammer, C.F., Sheridan, J.P., Vogel, C.-W. and Gaber, B.P. (1987) A homogeneous immunoassay for the mycotoxin T-2 using liposomes, monoclonal antibody and complement. *Analytical Biochemistry* 163, 369–375.

Liu, B.H., Keller, N.P., Bhatnager, D., Cleveland, T.E. and Chu, F.S. (1993) Production and characterisation of antibodies against sterigmatocystin O-methyltransferase. *Food and Agricultural Immunology* 5, 155–164.

Liu, M.T., Ram, B.P., Hart, L.P. and Pestka, J.J. (1985) Indirect enzyme-linked immunosorbent assay for the mycotoxin zearalenone. *Applied and Environmental Microbiology* 50, 332–336.

Makarananda, K., Wild, C.P., Jiang, Y.Z. and Neal, G.G. (1990) Possible effect of infection with liver fluke on the monitoring of urine by enzyme-linked immunosorbent assay for human exposure to aflatoxins. In: O'Neill, I.K., Chen, J.S. and Bartch, S. (eds), *Relevance to Human Cancer of N-Nitroso Compounds, Tobacco Smoke and Mycotoxins*. IARC Scientific Publications, No. 105, International Agency for Cancer Research, Lyon.

Matthew, J.A., Morgan, M.R.A., McNerney, R., Chan, H.W.S. and Coxon, D.T. (1983) Determination of solanidine in human plasma by radioimmunoassay. *Food and Chemical Toxicology* 21, 637–640.

McMillan, M. and Thompson, J.C. (1979) An outbreak of suspected solanine poisoning in schoolboys. *Quarterly Journal of Medicine* 48, 227–243.

Messina, M.J., Perzky, V., Setchell, K.D.R. and Barnes, S. (1994) Soy intake and cancer risk: a review of the *in vivo* and *in vitro* data. *Nutrition and Cancer* 21, 113–131.

Mills, E.N.C., Johnston, J.M., Kemp, H.A. and Morgan, M.R.A. (1988) An enzyme-linked immunosorbent assay for diacetoxyscripenol applied to the analysis of wheat. *Journal of the Science of Food and Agriculture* 42, 225–233.

Mills, E.N.C., Alcock, S.M., Lee, H.A. and Morgan, M.R.A. (1990) An enzyme-linked immunosorbent assay for deoxynivalenol in wheat utilising novel hapten derivatisation procedures. *Food and Agricultural Immunology* 2, 109–118.

Mizugaki, M., Itoh, K., Hayasaka, M., Ishiwata, S., Nozaki, S., Nagata, N., Hanadate, K. and Ishida, N. (1994) Monoclonal antibody-based enzyme-linked immunosorbent assay for glycyrrhizin and its aglicon, glycyrrhetic acid. *Journal of Immunoassay* 15, 21–34.

Morgan, M.R.A. and Coxon, D.T. (1987) Tolerances: glycoalkaloids in potatoes. In: Watson, D.H. (ed.), *Natural Toxicants in Food: Progress and Prospects*. Ellis Horwood, Chichester, pp. 221–230.

Morgan, M.R.A. and Fenwick, G.R. (1990) Natural foodborne toxicants. *Lancet* 336, 1492–1495.

Morgan, M.R.A., McNerney, R., Matthew, J.A., Coxon, D.T. and Chan, H.W.S. (1983a) An enzyme-linked immunosorbent assay for total glycoalkaloids in potato tubers. *Journal of the Science of Food and Agriculture* 34, 593–598.

Morgan, M.R.A., McNerney, R. and Chan, H.W.-S. (1983b) Enzyme-linked immunosorbent assay of ochratoxin A in barley. *Journal of the Association of Official Analytical Chemists* 66, 1481–1484.

Morgan, M.R.A., Coxon, D.T., Bramham, S., Chan, H.W.S., van Gelder, W.M.H. and Allison, M.J. (1985) Determination of the glycoalkaloid content of potato tubers by three methods including enzyme-linked immunosorbent assay. *Journal of the Science of Food and Agriculture* 36, 282–288.

Morgan, M.R.A., Kang, A.S. and Chan, H.W-S. (1986a) Aflatoxin determination in peanut butter by enzyme-linked immunosorbent assay. *Journal of the Science of Food and Agriculture* 37, 908–914.

Morgan, M.R.A., McNerney, R., Chan, H.W.-S. and Anderson, P.H. (1986b) Ochratoxin A in pig kidney determined by enzyme-linked immunosorbent assay (ELISA). *Journal of the Science of Food and Agriculture* 37, 475–480.

Morgan, M.R.A., Kang, A.S. and Chan, H.W.-S. (1986c) Production of antisera against sterigmatocystin hemiacetal and its potential for use in an enzyme-linked immunosorbent assay for sterigmatoystin in barley. *Journal of the Science of Food and Agriculture* 37, 873–880.

Morris, S.C. and Peterman, J.B. (1985) Genetic and environmental effects on the levels of glycoalkaloids in cultivars of potato (*Solanum tuberosum*). *Food Chemistry* 18, 271–282.

Mortimer, D.N., Gilbert, J. and Shepherd, M.J. (1987) Rapid and highly sensitive analysis of aflatoxin M, in liquid and powdered milks using an affinity column clean-up. *Journal of Chromatography* 407, 393–398.

Nicol, M.J., Lauren, D.R., Miles, C.O. and Jones, W.T. (1993) Production of a monoclonal antibody with specificity for deoxynivalenol, 3-acetyl-deoxynivalenol and 15-acetyldeoxy-nivalenol. *Food and Agricultural Immunology* 5, 199–210.

Okumura, H., Okimoto, J., Kishimoto, S., Hasegawa, A., Kawamura, O., Nakajima, M., Miyabe, M. and Ueno, Y. (1993) An improved indirect competitive ELISA for aflatoxin M_1 in milk powders using novel monoclonal antibodies. *Food and Agricultural Immunology* 5, 75–84.

Paseschnichenko, V. and Guseva, A.R. (1956) Quantitative determination of potato glycoalkaloids and their preparative separation. *Biochemistry, USSR* 21, 606–611.

Patey, A.L., Sharman, M. and Gilbert, J. (1992) Determination of total aflatoxin levels in peanut butter by enzyme-linked immunosorbent assay: collaborative study. *Journal of the Association of Official Analytical Chemists International* 75, 693–697.

Pearson, S., Smith, J.M. and Marks, V. (1984) Measurement of plasma caffeine concentrations by substrate-labelled fluoroimmunoassay. *Annals of Clinical Biochemistry* 21, 208–212.

Pestka, J.J. (1991) High performance thin layer chromatography ELISAGRAM: application of a multi-hapten immunoassay to analysis of the zearalenone and aflatoxin mycotoxin families. *Journal of Immunological Methods* 136, 177–183.

Pestka, J.J. (1994) Application of immunology to the analysis of toxicity assessment of mycotoxins. *Food and Agricultural Immunology* 6, 219–233.

Pestka, J.J. and Chu, F.S. (1984) Aflatoxin B_1 dihydrodiol antibody: production and specificity. *Applied and Environmental Microbiology* 47, 472–477.

Pestka, J.J., Lee, S.C., Lau, H.P. and Chu, F.S. (1981) Enzyme-linked immunosorbent assay for T-2 toxin. *Journal of the American Oil Chemists Society* 58, 940A–944A.

Peters, I.I., Dierich, M.P. and Dose, K. (1982) Enzyme-linked immunosorbent assay for detection of T-2 toxin. *Hoppe-Seyler's Zeitschift für Physiologische Chemie* 363, 1437–1441.

Plhak, L.C. and Sporns, P. (1992) Enzyme immunoassay for potato glycoalkaloids. *Journal of Agricultural and Food Chemistry* 40, 2533–2540.

Plhak, L.C. and Sporns, P. (1994) Development and production of monoclonal antibodies for the measurement of solanidine potato glycoalkaloids. *American Potato Journal* 71, 297–313.

Price, K.R., Gee, J.M., Ng., K., Wortley, G.M., Johnson, I.T. and Rhodes, M.J.C. (1995) Preliminary studies into the relationship between saponin structure and bioactivity. *Natural Toxins* (in press).

Robins, R.J., Morgan, M.R.A., Rhodes, M.J.C. and Furze, J.M. (1985) Cross-reactions in immunoassays for small molecules: use of specific and non-specific antisera. In: Morris, B.A. and Clifford, M.N. (eds), *Immunoassays in Food Analysis*. Elsevier Applied Science, Barking, UK.

Roseman, D.M., Wu, X.Y., Milco, L.A., Bober, M., Miller, R.B. and Kurth, M.J. (1992) Development of a class-specific competitive enzyme-linked immunosorbent assay for the detection of pyrrolizidine alkaloids *in vitro*. *Journal of Agricultrual and Food Chemistry* 40, 1008–1014.

Ross, R.K., Yuan, J-M., Yu, M.C., Wogan, G.N., Qian, G-S., Tu, J-T., Groopman,

J.D., Gao, Y-T. and Henderson, B.E. (1992) Urinary aflatoxin biomarkers and risk of hepatocellular carcinoma. *Lancet* 339, 943–946.

Rousseau, D.M. and van Peteghem, C.H. (1989) Spontaneous occurrence of ochratoxin residues in porcine kidneys in Belgium. *Bulletin of Environmental Contaminant Toxicology* 42, 181–186.

Sanford, L.L. and Sinden, S.L. (1972) Inheritance of potato glycoalkaloids. *American Potato Journal* 49, 209–217.

Schneider, E., Dietrich, R., Martlbauer, E., Usleber, E. and Terplan, G. (1991) Detection of aflatoxins, trichothecenes, ochratoxin A and zearalenone by test strip immunoassay: a rapid method for screening cereals for mycotoxins. *Food and Agricultural Immunology* 3, 185–193.

Sharman, M. and Gilbert, J. (1991) Automated aflatoxin analysis of foods and animal feeds using immunoaffinity column clean-up and high performance liquid chromatographic determination. *Journal of Chromatography* 543, 220–225.

Sinden, S.L. and Webb, R.E. (1972) Effect of variety and location on the glycoalkaloid content of potatoes. *American Potato Journal* 49, 334–338.

Sizaret, P., Malavielle, C., Montesano, R. and Frayssinet, C. (1980) Detection of aflaxtoxins and related metabolites by radioimmunoassay. *Journal of the National Cancer Institute* 69, 1375–1380.

Teshima, R., Hirai, K., Sato, M., Ikebuchi, H., Ichinoe, M. and Terao, T. (1990) Radioimmunoassay of nivalenol in barley. *Applied and Environmental Microbiology* 56, 764–768.

Trucksess, M.W., Young, K., Donahue, K.F., Morris, D.K. and Lewis, E. (1990) Comparison of two immunochemical methods with thin-layer chromatographic methods for determination of aflatoxins. *Journal of the Association of Official Analytical Chemists* 73, 425–428.

Usleber, E., Straka, M. and Terplan, G. (1994) Enzyme immunoassay for fumonisin B_1 applied to corn-based food. *Journal of Agricultural and Food Chemistry* 42, 1392–1396.

Vallejo, R.P. and Ercegovitch, C.D. (1979) Analysis of potato for glycoalkaloid content by radioimmunoassay (RIA). *NBS Special Publication (U.S.)* 519 (Trace Organic Analysis: New Frontiers in Analytical Chemistry), pp. 330–340.

van Egmond, H.P. (1989) Current situation on regulations for mycotoxins. Overview of tolerances and status of standard methods of sampling and analysis. *Food Additives and Contaminants* 6, 139–188.

Wang, B. and Williamson, G. (1994) Detection of a nuclear protein which binds specifically to the antioxidant responsive element (ARE) of the human NAD(P)H: quinone oxidoreductase gene. *Biochemica et Biophysica Acta* 1219, 645–652.

Ward, C.M. and Morgan, M.R.A. (1988) An immunoassay for determination of quinine in soft drinks. *Food Additives and Contaminants* 5, 555–561.

Ward, C.M., Franklin, J.G. and Morgan, M.R.A. (1988) Investigations into the visual assessment of ELISA end-points: application to determination of potato total glycoalkaloids. *Food Additives and Contaminants* 5, 621–627.

Ward, C.M., Wilkinson, A.P., Bramham, S., Lee, H.A., Chan, H.W.-S., Butcher, G.W., Hutchings, A. and Morgan, M.R.A. (1990) Production and characterisation of polyclonal and monoclonal antibodies against aflatoxin B_1 oxime BSA in an enzyme-linked immunosorbent assay. *Mycotoxin Research* 6, 73–83.

Warden, B.A., Allam, K., Seutissi, A., Cecchini, D.J. and Giese, R.W. (1987) Repetitive hit-and-run fluoroimmunoassay for T-2 toxin. *Analytical Biochemistry* 162, 363–369.

Warner, R., Ram, B.P., Hart, L.P. and Pestka, J.J. (1986) Screening for zearalenone in corn by competitive direct enzyme-linked immunosorbent assay. *Journal of Agricultural and Food Chemistry* 34, 714–717.

Wilkinson, A.P., Denning, D.W. and Morgan, M.R.A. (1989) Immunoassay of aflatoxin in food and human tissue. *Journal of Toxicology, Toxin Reviews* 8, 69–79.

Wilkinson, A.P., Ward, C.M. and Morgan, M.R.A. (1992) Immunological analysis of mycotoxins. In: Linskens, H.F. and Jackson, J.F. (eds), *Modern Methods of Plant Analysis. Vol.* 13. Springer-Verlag, Berlin, pp. 185–225.

Wolf, M.J. and Duggar, B.M. (1940) Solanine in the potato and the effects of some factors on its synthesis and distribution. *American Journal of Botany* 27, Suppl. 20S.

Xu, Y.-C., Zhang, G.S. and Chu, F.S. (1986) Radioimmunoassay of deoxynivalenol in wheat and corn. *Journal of the Association of Official Analytical Chemists* 69, 967–969.

Zhang, Y., Talalay, P., Cho, C.G. and Posner, G.H. (1992) A major inducer of anticarcinogenic protective enzymes from broccoli: isolation and elucidation of structure. *Proceedings of the National Academy of Sciences, USA* 89, 2399–2403.

Appendix 10.1. Commercially available immunodiagnostic kits relevant to non-nutrients in plant foods.

Company	Analytes	Format	Application
United Kingdom	Mycotoxins	A	Q
Biocode Ltd University Road Heslington York YO1 5DE			
CIS (UK) Ltd Unit 5 Lincoln Park Business Centre Lincoln Road High Wycombe Bucks HP12 3RD	Steroids	P/R	Q
Cortecs Diagnostics Techbase 1 Newtech Square Deeside Industrial Park Deeside Clwyd CH5 2NT	Mycotoxins Proteins Pathogens Meat species Antibiotics Steroids Herbicides	P/D	Q/Sq

Appendix 10.1. Continued.

Company	Analytes	Format	Application
Digen Ltd Rectory Mews Crown Road Oxford OX33 1UL	Mycotoxins Drugs Steroids Vitamins	P	Q
Flurochem Ltd Wesley Street Old Glossop Derbyshire SK13 9RY	Mycotoxins Drugs Steroids Pesticides	P	Q
IDEXX UK 46–48 High Street Slough Berkshire SL1 1EL	Drugs Bioactives	P	Sq
Randox Laboratories Ltd Ardmore Diamond Road Crumlin County Antrim Northern Ireland BT29 4QY	Drugs Steroids Pesticides	A/P/D	Q/Sq
Rhône Poulenc Diagnostics Ltd Montrose House 187 George Street Glasgow G1 1YT	Mycotoxins *Salmonella* Proteins	A Ag D	Sq Sq
Sigma Fancy Road Poole Dorset BH17 7NH	Mycotoxins Animal proteins Gliadin Soya		
Transia Ltd Rectory Road Upton Industrial Estate Upton-on-Severn Worcestershire WR8 0XL	Mycotoxins Drugs	P/C C	Sq
Europe			
Chemopol Ltd Kodanská 100 10 Prague-10 Czech Republic	Mycotoxins	R	Q
Holland Biotechnology B.V. Nields Bohrweg 13 PO Box 394 2300 AJ Leiden Netherlands	Solanine	P	Q
Lumac B.V. PO Box 3101 6370 AC Landgraaf Netherlands	*Salmonella* Antibiotics	D	Sq

Appendix 10.1. Continued.

Company	Analytes	Format	Application
Noack Gesellschaft mbH Geylinggasse 27/8 A 1130 Vienna Austria	Mycotoxins Pesticides Bacteria Antibodies Meat specification	C Ag C D	Q/Sq Sq
Novo Food Diagnostics A/S Frydenalsvej 30 1809 Frederiksberg C Copenhagen Denmark	Drugs	P	Q
R-Biopharm GmbH Rosslerstrasse 94 6100 Darmstadt Germany	Mycotoxin	P	Q
Riedel-de-Haën Wunstorfer Str. 40 D 3016 Seelze 1 Germany	Mycotoxins Drugs Steroids Pesticides Hormones	P	Q
Transia Ltd 8 Rue Saint-Jean de Dieu 69007 Lyon France	Mycotoxins	P/C	Q/Sq
Non-European			
Actio Inc 1127 57th Ave Oakland CA 94621 USA	Pesticides	T	Q/Sq
Agri-Diagnostics Associates 2611 Branch Pike Cinnaminson NJ 08077 USA	Mycotoxins Pesticides	P	Q
Bioman Products Inc. 400 Matheson Blud Unit 4 Mississauga Ontario L4Z 1N8 Canada	Drugs Mycotoxins	P/C	Q/Sq
Idetek 1057 Sneath Lane San Bruno CA 94066 USA	Drugs Bioactives	P	Q

Appendix 10.1. Continued.

Company	Analytes	Format	Application
IDEXX 100 Fore Street Portland ME 04101 USA	Drugs Bioactives		Sq
International Diagnostic Systems Corp PO Box 799 2613 Niles Ave St Joseph MI 49085 USA	Mycotoxins Steroids Drugs Antibodies	P	Q
Neogen Corporation 620 Lesher Place Lansing MI 48912 USA	Mycotoxins Drugs	P C	Q Sq
Penicillin Assays Inc 36 Franklin Street Malden MA 02148 USA	Mycotoxins Antibodies Tetracyclines Sulphonomides Aminoglycocides		
Sceti Co Ltd DF Building 2-2-8 Minami Adyama Minato-Ku Tokyo 107 Japan	Antibiotics Sulphonamides Pesticides Mycotoxins Steroids Meat specification		
Unipath Inc 217 Colonade Road Nepean Ontario K23 7K3 Canada	Mycotoxins	A	Sq
Vicam 313 Pleasant Town Watertown MA 02172 USA	Aflatoxin *Listeria*	A	Sq

Abbreviations: A, affinity columns; Ag, agglutination; B, magnetic bread; C, card; D, dipstick; P, plate; Q, quantitative; R, radioimmunoassay; Sq, semiquantitative; T, tube.

Diagnostics for Plant Agrochemicals – a Meeting of Chemistry and Immunoassay

S.J. Gee[1], B.D. Hammock[1] and J.H. Skerritt[2]

[1]University of California, Department of Entomology, Davis, CA 95616, USA; [2]CSIRO, Division of Plant Industry, Canberra, ACT 2601, Australia.

11.1 Role of Chemicals in Plant Agriculture

The role of agrochemicals in crop sciences is changing rather than diminishing. Even in situations where the proportion of crop or harvest product treated is becoming lower, the need for testing may actually increase as sellers and customers want to segregate treated and untreated material. In general, there has been a move away from 'spraying by the calendar', to use of strategically timed and targeted sprays based on regular field ascertation of insect or weed pressure. Minimal-till agricultural methods, introduced to save labour and to conserve topsoil, soil fertility and soil water, rely heavily on chemical weed control. The three groups of agrochemicals of special interest to the crop sciences include herbicides, insecticides and fungicides. Herbicides account for over half of all agrochemical use in developed countries (minor groups such as defoliants and plant hormones are included here). Herbicides are applied either pre-plant to prepare crop beds for sowing, or post plant for on-going weed control. They are also used to clear irrigation channels from being choked by aquatic weeds. Insecticides are not used as widely as herbicides in the field on broadacre grain crops, but are of major importance in horticultural (especially orchard) crops. They are also critical to the viability of certain broadacre crops such as cotton, while postharvest use of insecticides on stored commodities, such as cereal grains, is also significant. Fungicides are applied to growing crops, but a major use is also in postharvest protection of fruit and vegetables during storage and transport.

11.2 Immunoassay as a Pesticide Analytical Method

When faced with an analytical problem, the instrumental method is often the first technology evaluated by the analyst. One of the purposes of this chapter is to provide the analyst with enough information about immunoassay so that it may be included among the methods considered. Immunoassay methods have a variety of advantages as analytical methods. They may be more sensitive than competing instrumental methods. For example, a triazine immunoassay may measure atrazine at the parts-per-trillion level in a water sample, with no prior sample preparation (Wittmann and Hock, 1989; Schneider and Hammock, 1992). The corresponding gas chromatographic method requires an extraction and concentration step prior to analysis.

The antibody determines the selectivity of the assay. Assays can be designated as class selective, selective for parent and metabolites, compound selective or even isomer selective. For example, with the herbicide alachlor, immunoassays have been developed which will detect the parent compound (Sharp et al., 1991), the mercapturic acid metabolite as well as other thioether metabolites (Feng et al., 1990). Others are more reactive to the class of chloroacetanilide herbicides (Feng et al., 1992). Bromacil is another example in which the metabolites cross-react 3% or less (Szurdoki et al., 1992). Antibodies can be sufficiently selective to distinguish between chiral derivatives, as demonstrated by early work for the pyrethroid S-bioallethrin (Wing and Hammock, 1979).

Specific interferences in immunoassay are termed cross-reactivity. These can result from interaction of the antibody with structurally related compounds that may be present with the analyte of interest. Non-specific interferences can result in false positive or false negative results. Because of their selectivity, immunoassays are generally single analyte methods, thus cannot be directly applied to situations in which simultaneous detection of a variety of analytes is required.

For the analysis of agrochemicals, commercially available immunoassay kits are generally more cost effective than having a commercial laboratory run the standard analysis. Using the medical diagnostic field as a model, we would assume that the price per analysis would continue to decline as the technology comes into demand. Since this is a new technology, reagents for all analytes of interest are not readily available. The cost effectiveness is also measured in the speed of analysis. Generally immunoassays require less sample workup, improving sample throughput time. The decreased sample workup time, along with the ability to process a number of samples in parallel, can result in a significant saving of analyst time. For example, the paraquat immunoassay is significantly faster, simpler and more sensitive than the instrumental (gas-liquid chromatography (GLC) or high-performance liquid chromatography (HPLC)) methods (Van Emon et al., 1987). To measure paraquat from a glass fibre filter from an air sampler, 11 sample preparation steps are required prior to GLC analysis. The immunoassay utilized either two or five steps (Van Emon *et al.*, 1985).

Another advantage of immunoassay is its adaptability. The reagents developed for immunochemical analysis can be prepared as field portable test kits that give quick 'yes or no' type answers, or as highly quantitative laboratory tests or even fully automated robotic systems that handle all pipetting steps and data analysis. Field-portable kits are used currently as part of the US Department of Agriculture Food Safety and Inspection Service monitoring programmes (Van Emon and Lopez-Avila, 1992).

The wide applicability of immunoassay could help in the development of a broader range of materials for the enhancement of productivity and profitability in agriculture. A few years ago as modern alternatives to many classical pesticides were sought, it was found that many promising substances could not be analysed by GLC with the detectors commonly available. This presented a severe barrier to the development of these materials. Now we have a variety of new GLC detectors, HPLC systems and particularly new mass spectrophotometric techniques that can analyse compound classes such as acylureas and sulphonylureas. However, we now see a similar analytical barrier to many biological insecticides including products of recombinant DNA research. Fortunately, immunoassay is an even more powerful analytical technique for these materials than it is for classical pesticides (Cheung and Hammock, 1988). Thus, immunochemical methods for analysis can speed the acceptance of modern systems for the control of agricultural pests.

With classical chromatographic methods, the use of the analytical technology requires a major investment in equipment and laboratory facilities and the efforts of a highly trained analytical chemist. In contrast, immunoassays can be formatted so that little or no equipment is needed to obtain analytical data of reasonable quality. Even a very sophisticated automated immunoassay laboratory can be equipped at a fraction of the cost of a classical analytical laboratory. As discussed earlier several hundred samples can be run quantitatively per man day in a field laboratory, needing to be equipped with only water and electricity. Thus, analytical data now can be generated by people who need those data in a timely fashion, largely due to decreased start-up costs.

11.3 Application of Agrochemical Immunoassay to the Crop Sciences

11.3.1 *Generation of data for product registration*

In the past, analytical data were generated largely to satisfy regulatory agencies. The cost of analysis limits the incentive of industry to register pesticides for use on anything other than major crops. The availability of immunochemical analyses should speed product registration on major crops and facilitate expanded registration to minor crops. Many agrochemical companies have in-house programmes of immunoassay development, for provision of analytical data to the scientists involved in product development. Analytical data are often developed largely under regulatory pressure. Since resources are largely directed there, in-house scientists have less access to analytical data, for example, to optimize formulation systems and application methods. Because the data are only available late in the development of a chemical, analytical results from greenhouse and small-scale field trials are seldom available for use in determining which of a set of compounds are best to develop.

11.3.2 Analysis of residues in plant foods

The low cost and ease of analysis of immunoassay will also be attractive to agricultural marketing groups and consumer groups. Numerous studies in developed countries have shown that our food supply is of high quality and generally free of alarming levels of pesticide residues. However, the public is still very concerned about pesticides in food, for both scientific as well as sociological reasons. Our current analytical technologies can allow us to say with a high degree of confidence that the average diet probably contains no dangerous residues of agrochemicals. However, the cost and slowness of current analytical methods means that we lack the technology to say that any specific truckload of produce is free of pesticide residues. In general consumers are not convinced by statistics and want to know that the food that they are purchasing is in fact free of residues. The speed and low cost of immunoassay will permit lot-by-lot screening for residues, with greater opportunities for timely traceback. The most important food analytes are insecticides and fungicides. This is because the greatest use of herbicides is early rather than late in the crop cycle. Detection in foods is complex because of the wide variety of sample matrices (harvested plant materials instead of water and soil); this will lead to obvious variation in extraction efficiency and in matrix effects. There has been little systematic matrix analysis apart from fruit juices and grain matrices, so there remains the need to develop generic extraction methods for residues from plant foods.

11.3.3 Contamination of the cropping environment

One of the major trends in agrochemical use over the last two decades has been the move towards use of compounds with lower environmental persistence. However, some of the compounds with shorter half-lives (and thus notionally lower chronic ecotoxicity) are actually of greater acute ecotoxicity to mammals, fish and desirable insect species. An example is the somewhat greater acute toxicity of endosulphan and organophosphates compared with DDT. Persistence and decay behaviour determined in laboratory studies using clear, neutral water may bear little relevance to field conditions where water may be more turbid, of varying temperature and of non-neutral pH.

Certain herbicides can be a serious concern in crop rotation systems because the residues of the previous year may damage the crop of the existing year, if persistence is unexpectedly high due to unseasonal temperature or precipitation, or if the soil pH or organic content is other than expected. Similar concerns apply to plant-back damage that can be caused by persistence of pre-plant herbicides, such as diuron. Immunoassay will give the farmer the power to ensure that enough of a herbicide is used in the first year of a rotation to control weeds, but that it is sufficiently dissipated to allow planting of the second crop in the rotation with no problems of residual plant toxicity. Special attention must be paid to analysis of sulphonylureas in soils, since they are herbicidally active at ppb levels and persistence is highly dependent on soil pH.

11.3.4 Contamination of irrigation drainage water

Formerly, there was the widespread attitude that much of the compound used would dissipate by hydrolysis or photolysis and that dilution in irrigation water would lead to vanishingly small levels in drainage water and subsequently in rivers. However, the finding that fish and aquatic invertebrates can be severely affected by ppb levels of some major-use compounds is leading to a change in attitudes, and a recognition of the importance of gathering more data on the dissipation of common agrochemicals under field conditions. Contamination of water (thus providing greater potential for environmental toxicity than soil contamination) is most likely in irrigated plant agricultural systems, such as cotton, where pesticides are applied (often aerially) near periods of irrigation or storm rains and rice, where pesticide is sprayed into water-filled paddies. Several aspects of the ecotoxicology of major pesticides are not well understood, and the ability of immunoassay to capture data in a timely and cost-effective manner will be a significant bonus. Stream sediment may act as a sink, increasing persistence of chemicals such as endosulfan.

11.3.5 Monitoring occupational exposure

Every agriculturist has the ethical responsibility to avoid occupational exposure of his workers to pesticides. Immunoassays can facilitate the complex studies of worker exposure that will lead to better protective systems, better application methods and improved worker training. They allow real-time dissipation curves to be developed for re-entry. This will prevent early re-entry that can result in unacceptable worker exposure to pesticides while avoiding delayed re-entry times that can result in the loss of profitability. Finally, solid analytical data on the presence of pesticides in a field and/or exposure of a worker population can protect a legitimate operator from unfounded charges of exposure. A number of assays have been developed for pesticides and their metabolites in urine; in most cases urine is just diluted and added directly to the microwell. Examples include nitrophenols as a marker for parathion and related compounds (Li *et al.*, 1991a), 2,4-D and picloram in urine (Hall *et al.*, 1989) and paraquat (Van Emon *et al.*, 1986). In the latter case air monitors, clothing patches and hand rinses during aerial spraying of cotton, and also plasma and urine were studied.

11.3.6 Role in integrated pest management

Another application of the technology will be to monitor and optimize the use of chemicals. We know that pesticides degrade very differently depending upon the crop and the agricultural situation. Since analytical data are so expensive and time consuming to obtain, we seldom use such hard data to decide how much material to apply. This can result in the use of unnecessarily high levels of agrochemicals to ensure that we maintain levels needed to control pests during the sensitive period of a crop. By having rapid and reliable analytical data provided by immunoassay in support of a cropping or postharvest situation, we will be able to reduce the use of agrochemicals. This clearly helps the consumer by reducing pesticide residues. It helps the producer by reducing costs of agrochemicals, and it helps the agrochemical industry by reducing residues and thus preserving the life of valuable compounds in an often tense regulatory climate. An example of this application is in the postharvest protection of grain. Effective insecticide levels must be maintained or unacceptable crop loss and contamination of grain with insect parts results. The ability to analyse the levels of pesticides in a grain store will assist in determining if any additional pesticide is needed to maintain complete control of pests. Immunochemistry offers the rapid, field-portable analytical technology that can be used to improve formulation and application methods to reduce the build-up of resistance. It can be used to monitor actual levels of pesticide in the field for use in resistance management systems to ensure that optimal pressures are put on the insect population.

11.4 How Agrochemical Immunoassays Work

Immunoassays use antibodies that have been prepared in rabbits, mice or sheep to a particular pesticide or family of pesticides. The pesticide molecules are too small to elicit an immune response by themselves, so much of the art in developing a pesticide immunoassay consists of developing a route for coupling a chemical analogue of the pesticide to a carrier, usually a protein that is foreign to the animal being immunized. The analogue must retain all of the characteristic features of the pesticide, but contain a new chemical group in its structure that can act as a handle for coupling to the protein. Once coupled to the protein, the pesticide–protein 'conjugate' is usually able to evoke antibodies, but for a useful test to be possible, antibodies must bind to the free pesticide. The key steps in development of an antibody test are as follows:

1. Synthesis of a pesticide derivative, coupled to a suitable carrier protein for immunization.
2. Immunization of rabbits, mice and/or other species. Preparation and purification of antibodies.
3. Development of initial immunoassay using pesticide standards. Check assay sensitivity and specificity.
4. Assessment of assay performance with water and soil matrices in laboratory-spiked and field samples.
5. Formatting of methods as prototype kits, stabilization and stability trials on components and prototypes.
6. Field trials of kits and training workshops.

There are many textbooks available which describe the variety of formats in which the immunoassay can exist based mostly on the work of the clinical laboratories (Voller *et al.*, 1978; Tijssen, 1985). The primary format currently used for agrochemical analysis is the competitive enzyme-linked immunosorbent assay (ELISA). In these assays, the analyte competes with a structural analogue of the analyte that is covalently attached to a protein (coating antigen format) or to an enzyme (haptenated enzyme format). Common to each format is a pesticide-specific antibody. An example of the coating antigen format is shown in Fig. 11.1. In other research applications and in some of the commercially available kits, the coating antigen is usually provided already adsorbed to the solid phase (A). The solid phase may either be a 96 well microwell plate, a polystyrene test tube or a polystyrene-coated magnetic bead. The analyte sample is added by pipetting the solution into the well or tube (B). Next the pesticide-specific antibody is added (C). These components are allowed to incubate (D). The pesticide derivative that is part of the coating antigen and the analyte added to the tube compete for the fixed amount of pesticide (analyte)-specific antibody. After the incubation, the unbound reagents are washed out of the well or tube. The analyte-specific antibody is labelled with an enzyme. We can then detect the amount of antibody that binds to the coating antigen on the solid phase by the addition of the substrate for the enzyme (E) and subsequent colour development (F). Since this is a competitive assay, if there is a large amount of pesticide analyte

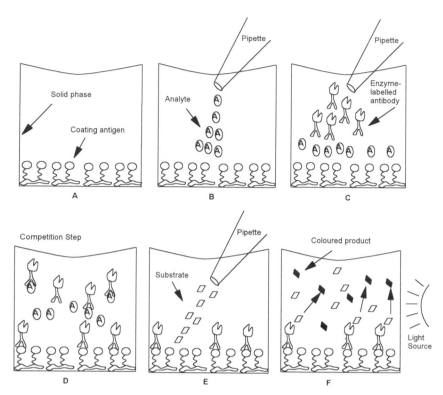

Fig. 11.1. Incubation steps required for determination of agrochemicals by direct ELISA. **A.** Coating of antigen–agrochemical conjugate on to microwell surface. **B.** After blocking non-specific binding, agrochemical analyte is added. **C.** Enzyme-labelled antibody is then added. **D.** The incubation or 'competition step' is performed before unbound components are removed by washing. **E.** Chromogenic substrate is added. **F.** The absorbance of the coloured product is determined.

then the antibody will bind to it and be washed from the well or tube. If there is little pesticide analyte then the enzyme-labelled antibody will bind to the solid phase. The result is that as the concentration of the pesticide analyte increases, the absorbance decreases. Most commercial kits and several research assays use the haptenated enzyme format, in which the pesticide-specific antibody is immobilized. This also provides decreasing absorbance as pesticide analyte concentration increases, but involves only a single incubation and wash step for the assay user.

An example calibration curve is shown in Fig. 11.2. The sigmoidal shaped curve is defined by the four-parameter equation of Rodbard (1981). The 'C' value is the point on the curve where the absorbance has been decreased by 50%. This is often designated the IC_{50} and is a relative measure of the sensitivity of an immunoassay. Another important characteristic of the

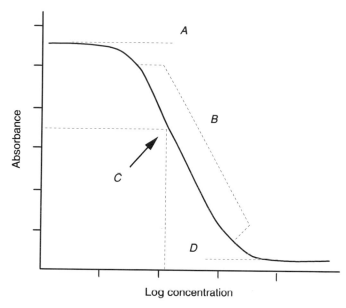

Fig. 11.2. Relationship between assay absorbance and (free) agrochemical concentration observed for most pesticide immunoassays. As pesticide concentration increases, decreases in assay absorbance occur. The curve, which has a sigmoidal shape when plotted on log-linear axes, represents a four-parameter logistic plot, according to the equation:

$$y \text{ (absorbance)} = (A - D) / [1 - (x/C)^B] + D$$

where x represents the free agrochemical concentration; A is the zero dose response (approximating colour produced in the absence of free agrochemical in the sample); B is the slope of the curve; C (the point of inflection) approximates the agrochemical concentration inhibiting colour development by 50%; and D is the infinite dose response, approximating 'background' colour remaining after inhibition by excess free agrochemical.

calibration curve is the lower limit of detectability. This is variously defined, but is the concentration that can be distinguished from background with some degree of confidence. If one back-calculates the zero analyte control from the standard curve, this value averaged over several experiments is an indication of the lower limit of detectability (Harrison *et al.*, 1989a). An often-used rule is that the lower limit of detectability is the concentration that is two or three standard deviations above this calculated background value. When running immunoassays routinely, use of control charts, such as Shewhart control charts (Miller and Miller, 1984) for the IC_{50}, the lower limit of detectability and known positive and negative control sample is a useful quality control tool (Harrison *et al.*, 1989a).

11.5 Advantages and Disadvantages of Immunoassay Compared with Other Analytical Methods

The following are some useful questions to ask when considering immunoassay as the method of analysis:

1. Are the antibodies available? One of the disadvantages of immunoassay is that the antibody must be made for each analyte, although availability of antibodies has increased tremendously in the last few years (Tables 11.1 to 11.4).

2. What is the level of sensitivity desired? The level of detectability will depend upon the analyte and the matrix being studied. A complex matrix may require more sample workup that could lead to an increase in the lower limit of detectability.

3. What is the sample load? Immunoassays are more cost effective for large sample loads, particularly if less sample workup is utilized and/or less analysis time. There is also a time saving due to the ability to process samples

Table 11.1. Commercial kits available for detection of agrochemical residues.

r-Biopharm (Darmstadt, Germany – fax +49 6151 81 0220)
– Triazines (in development – urea herbicides)

Riedel-de Haen (Hannover, Germany – fax +49 5137 707123)
– Triazines

Idetek (Sunnyvale, CA, USA – fax +1 408 829 2625)
– Metolachlor, alachlor, atrazine, triazine, imazaquin, benomyl/carbendazim, chlorothalonil, trifluralin

Randox (Crumlin, Northern Ireland, UK – fax +44 184 94 52912)
– Triazines, 2,4-D and urea herbicides

Transia (Lyon, France – fax +33 72 73 43 34)
– *plate and tube kits:* Triazines

Millipore (Bedford, MA, USA – fax + 1 617 275 5550)
– *plate kits:* Triazine, aldicarb, paraquat, 2,4-D, benomyl/carbendazim, alachlor, cyclodienes, methoprene, metalaxyl, fenitrothion, chlorpyrifos, chlorpyrifos-methyl, bioresmethrin, pirimiphos-methyl, diazinon, polychlorinated biphenyls, pentachlorophenol, procymidone, isoproturon, chlortoluron/urea herbicides, parathion (ethyl/methyl), chlorsulfuron, triasulfuron, metsulfuron-methyl.
– *tube kits:* Triazines, aldicarb, 2,4-D, carbofuran, cyclodienes, alachlor, benomyl, DDT, fenitrothion, chlorpyrifos-methyl

Ohmicron (Newton, PA, USA – fax + 1 215 860 5213)
– Alachlor, aldicarb, atrazine, BTEX (benzene/toluene/ethyl xylenes), benomyl/carbendazim, captan, carbaryl, carbofuran, chlorothalonil, chlorpyrifos, cyanazine, 2,4-D, metolachlor, paraquat, pentachlorophenol, polycyclic aromatic hydrocarbons, polychlorinated biphenyls, procymidone

Biocode (York, UK – fax +44 1904 43 0495)
(in development – microcystin, dioxins)

Table 11.2. Enzyme immunoassays developed for insecticides.

Organophosphates	Parathion	Al-Rubae, 1978
		Vallejo *et al.*, 1982
		Duquette *et al.*, 1988
	Chlorpyrifos	Skerritt *et al.*, 1992b
	Chlorpyrifos-methyl	Skerritt *et al.*, 1992b
		Edward *et al.*, 1993a,b
	Fenitrothion	Hill *et al.*, 1992
		Beasley *et al.*, 1993
	Diazinon	Ferguson *et al.*, 1993b
	Pirimiphos-methyl	Skerritt *et al.*, 1992b
Carbamates	Aldicarb	Brady *et al.*, 1989
		Itak *et al.*, 1992
	Carbaryl	Marco *et al.*, 1993
	Methomyl	Charlton *et al.*, 1991
Organochlorines	Cyclodienes	Bushway *et al.*, 1988a
		Dreher and Podratzki, 1988
		Karu *et al.*, 1990
		Stanker *et al.*, 1991
	DDT	Burgisser, 1990
Pyrethroids	S-Bioallethrin	Wing and Hammock, 1979
	Bioresmethrin	Hill *et al.*, 1993
	Cypermethrin	Wraith *et al.*, 1986
	Deltamethrin	Demoute *et al.*, 1986
	Permethrin and Phenothrin	Stanker *et al.*, 1989
		Skerritt *et al.*, 1992a
Insect growth regulators	Diflubenzuron	Wie and Hammock, 1982,
	Methoprene	1984
		Mei *et al.*, 1991
		Hill *et al.*, 1991
		Heckman *et al.*, 1992
Toxins	*Bacillus thuringiensis*	Wie *et al.*, 1984
		Hofte *et al.*, 1988
		Bekheit *et al.*, 1993

in parallel. Since immunoassays are amenable to automation, high volume, rapid sample processing is possible if the matrix is well defined.

4. Is a portable method needed? Immunoassays are among the few field-portable analytical methods currently available.

5. Is broad screening or compound-specific analysis needed? The ability of many pesticide immunoassays to detect members of a group of structurally related pesticides is both a strength and a weakness. While assays have been described for some individual members of the triazine family of herbicides, many antibodies show cross-reaction. If such cross-reaction is limited, it has been proved possible to use a small panel of antibodies of differing

Table 11.3. Enzyme immunoassays developed for herbicides.

1. *Applied to foliage*		
(a) *Non-selective*		
Quarternary ammonium	Paraquat	Niewola *et al.*, 1986
		Van Emon *et al.*, 1987
Aminotriazole	Amitrole	Jung *et al.*, 1991
(b) *Selective for broad-leafed weeds in cereals*		
Phenoxyacetic acids	2,4-D; 2,4,5-T	Fleeker, 1987
		Hall *et al.*, 1989
		Newsome and Collins, 1989
		Fialova and Franek, 1990
Others	Picloram	Hall *et al.*, 1989
	Bentazon	Li *et al.*, 1991b
(c) *Selective for grasses*		
'Fops'	Diclofop-methyl	Schwalbe *et al.*, 1984
2. *Soil incorporated, often before planting crop*		
Thiolcarbamates	Molinate	Gee *et al.*, 1988
		Harrison *et al.*, 1989a
	Thiobencarb	Karu, 1993
Chloracetamides	Alachlor	Feng *et al.*, 1990
		Lawruk *et al.*, 1992
		Fitzpatrick *et al.*, 1992
		Baker *et al.*, 1993
	Metolachlor	Schlaeppi *et al.*, 1991
		Fitzpatrick *et al.*, 1992
		Hall *et al.*, 1992
		Feng *et al.*, 1992
		Lawruk *et al.*, 1993b
	Metazaclor	Scholz and Hock, 1991
	Amidochlor	Feng *et al.*, 1992
	Butachlor	Feng *et al.*, 1992
Oxazolidinone	Clomazone	Koppatschek *et al.*, 1990
		Dargar *et al.*, 1991
Dinitroanilines	Trifluralin	Riggle, 1991
	Surflan	Kuniyuki and McCarthy, 1986
3. *Foliar and soil action*		
(a) *Photosynthesis inhibitors*		
Urea herbicides	Diuron	Newsome and Collins, 1990
		Schneider *et al.*, 1994b
		Karu *et al.*, 1994
	Chlortoluron	Aherne, 1991
	Isoproturon	Aherne, 1991
		Liegeois *et al.*, 1992
	Methabenzthiazuron	Kreissig and Hock, 1991

Triazines	Broad specificity	Bushway *et al.*, 1988b, 1989
		Dunbar *et al.*, 1990
		Goodrow *et al.*, 1990
		Harrison *et al.*, 1991
		Weil *et al.*, 1991
		Aherne, 1992
		Schneider and Hammock, 1992
	2,4-D compounds	Giersch and Hock, 1990
		Giersch *et al.*, 1993
		Muldoon *et al.*, 1993
	Atrazine (specific)	Huber, 1985
		Fitzpatrick *et al.*, 1992
		Giersch *et al.*, 1993
	Atrazine metabolites	Schlaeppi *et al.*, 1989
		Wittmann and Hock, 1991
		Lucas *et al.*, 1993
	Simazine metabolites	Lucas *et al.*, 1993
	Terbutryn	Huber and Hock, 1985
	Terbuthylazine	Giersch *et al.*, 1993
	Cyanazine	Lawruk *et al.*, 1993a
Uracils	Bromacil	Szudorki *et al.*, 1992
Others	Norflurazon	Riggle and Dunbar, 1990
(b) *Branched-chain amino acid synthesis inhibitors*		
Sulfonylureas	Chlorsulfuron	Kelley *et al.*, 1985
		Chigrin *et al.*, 1989
		Charlton *et al.*, 1991
	Triasulfuron	Schlaeppi *et al.*, 1992
	Chlorimuron ethyl	Charlton *et al.*, 1991
	Metsulfuron-methyl	Rothnie, unpublished.
Imidazolinones	Imazamethabenz-methyl	Newsome and Collins, 1991
	Imazethpyr	Anis *et al.*, 1993
	Imazaquin	Wong and Ahmed, 1992
(c) Others	Maleic hydrazide	Harrison *et al.*, 1989b

specificities and by treatment of the data obtained use simultaneous equations that take into account differences in cross-reactivity between antibodies for related compounds. This approach has been successfully applied to analysis of atrazine, simazine, cyanazine and total *s*-triazines (Muldoon *et al.*, 1993). While interfacing of ELISA readers with computers is routine and automates this approach, it may still involve many assumptions and the increase in complexity of the method must be weighed against performing a single multi-residue analytical run using an alternative, instrumental method of analysis. Another approach could be to utilize

Table 11.4. Immunoassays developed for fungicides.

Non-systemic organic fungicides		
Organosulphur	Captan	Itak *et al.*, 1991
		Newsome *et al.*, 1993
Chlorinated aromatic	Chlorothalonil	Quantix, unpublished
Dicarboximides	Iprodione and vinclozolin	Newsome, 1987
	Procymidone	Ferguson *et al.*, 1993b
Systemic fungicides		
Benzimidazoles	Benomyl	Newsome and Collins, 1987
	and carbendazim	Bushway *et al.*, 1992
		Charlton *et al.*, 1991
	Thiabendazole	Newsome and Collins, 1987
		Brandon *et al.*, 1993
Morpholines	Fenpropimorph	Jung *et al.*, 1989
Triazoles	Triadimefon	Newsome, 1986
	Propiconazole	Forlani *et al.*, 1992
Phenylamides	Metalaxyl	Newsome, 1985

chromatographic separation, and detection and identification by immu-noassay (Krämer *et al.*, 1994).

If the 'variable group' in a family of related agrochemicals can be presented to the immune system by conjugation to a point near to the common groups in the molecule, it may be possible to develop a general method for production of separate antibodies specific for invidual members of a pesticide class. For example, this has been possible for organophosphates (Skerritt *et al.*, 1992b) and acetanilide herbicides (Feng *et al.*, 1992). In some cases, cross-reaction of an immunoassay can lead to new information on the disposition of a particular compound. A commercial immunoassay fre-quently provided apparent 'false positives' when used to screen well water samples for alachlor. Further examination of the false positives revealed that they contained an ethanesulphonic acid metabolite of alachlor, which had not been previously found in water (Baker *et al.*, 1993).

11.6 Sample Preparation for Immunoassay

11.6.1 Integrating sample preparation with analysis

Given that the analyst has chosen to use immunoassay and that the reagents exist, the next question is how to apply the technology to the given matrix. The analyst may use a decision tree such as the one shown in Fig. 11.3. If the sample is water miscible, then it may be a simple direct analysis. If this does not produce desired recoveries, then simple modifications to the aqueous

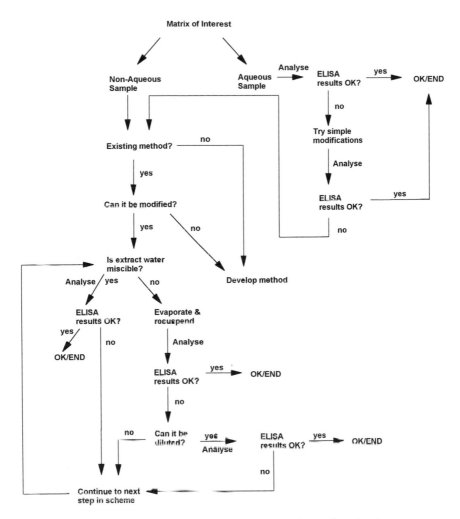

Fig. 11.3. A decision sequence for determination of agrochemical residues by immunoassay.

sample may solve the problem. Suggestions include filtration, adding buffer to neutralize pH and provide constant salt concentration, or addition of chelating agents to bind interfering metals. If the assay is sensitive enough it may be desirable simply to dilute out the interfering material, or to perform a single organic partition step. If the sample is non-aqueous or direct analysis of the aqueous sample did not produce the desired results then we fall back on the existing extraction method. It is desirable to limit the number of sample preparation steps, thus at each step it is important to ascertain whether the solvent used to extract is water miscible, if it can be run directly in the ELISA, or if it needs to be modified to eliminate interferences. An example of this approach is described for molinate (Gee *et al.*, 1988; Li *et al.*, 1989). Coloured matrices rarely interfere in the ELISA.

If it becomes necessary to develop a clean-up method, the analyst should consider first the simplest alternatives, such as solid phase extraction (Lucas *et al.*, 1994). Supercritical fluid extraction may also be a good alternative; the extracting solvent is easily vaporized and if the trapping solvent is water miscible, the samples may be run directly as demonstrated for parathion (Wong *et al.*, 1991). If a lengthy clean-up cannot be avoided, immunoassay may still be useful as a selective 'down-line' detector for HPLC (Krämer *et al.*, 1994) or as a second confirmatory technique. For example, if the compound co-chromatographs on HPLC with authentic standard and analysis of these peak fractions by immunoassay is positive, this may serve as a confirmation step. Similarly, if the HPLC detector cannot resolve the analyte of interest from interfering materials, using the immunoassay as a post-column detector may provide the needed detection. Other means of integrating immunoassay with HPLC can be found in a review by de Frutos and Regnier (1993). Finally, immunoassay may be useful to pre-screen samples in order to rank them for analysis or in order to screen out negative samples, thus helping to reduce total analytical costs.

11.6.2 *Analysis of specific matrices*

The obvious matrices for environmental analysis are soil, silt and water. Analysis of pesticides in water may seem to be a facile procedure but field water samples containing incurred residues can often be quite complex. In instrumental analysis, pesticide from water samples is invariably concentrated by use of either liquid–liquid extraction (into a volatile, immiscible solvent) or by adsorption on to a solid phase, followed by elution with a small volume of polar organic solvent. Apart from concentration, these steps also serve to remove interfering substances such as ions and humic acids. Immunoassay technology enables, for the first time, analysis of pesticides directly in water samples at or below ppb levels. This strength, however, brings with it several methodological challenges, which may only manifest when the correlations between data obtained with a set of water samples for ELISA and an instrumental method are less than perfect.

Does the immunoassay detect only pesticide in true solution or is both dissolved and particle-bound compound measured? The common use of detergents in ELISA buffers may enable pesticide to be freed from the particles for measurement during the assay incubations. Antibodies may extract an analyte from a micelle or solid phase by mass action. In contrast, rapid shaking of a turbid water sample with a solvent such as dichloromethane does not necessarily mean that pesticide dissociation from the particles will be favoured. So the two methods could conceivably be measuring different parameters. Simple filtration of the water sample to remove particles may also be fraught with artefacts due to binding of dissolved pesticide to the filter. Pesticide at low ppb levels may also bind to the vessel used for sampling. Many of these problems are common to both immunoassay and instrumental analyses.

Much of the validation has been done using spikes rather than samples containing incurred residues. Another potential problem is that it is more valid to compare ELISA and GLC when they have both been done on the same extract (even when subsequently cleaned up in different ways). Often ELISA is performed on a methanol extract while GLC uses an extract of a less polar, water-immiscible solvent. The latter may provide cleaner backgrounds for GLC analysis but may actually be a poorer extractant. The simplest approach to analysis of possible pesticide residues in soil samples by immunoassay would be to extract the soil in a water-miscible solvent, followed by direct dilution of the solvent in the assay buffer. Instrumental analysis usually uses an immiscible solvent such as hexane in mixture with acetone or dichloromethane. These are often less efficient extractants than methanol, but are favoured because of the relative lack of co-extractives, which can interfere with instrumental analyses unless the sample is cleaned up extensively.

11.7 Assays Available for Specific Compounds

The assays for compounds of relevance to crop scientists are listed in Tables 11.2 to 11.4. In the late 1970s and early 1980s some assays were developed as radioimmunoassays, but generally we have only listed enzyme immunoassays. While several groups have developed assays for metabolites of pesticides (especially useful in applicator screening and environmental fate studies), with the exception of triazines, we have also only listed assays for parent compounds. Some assays are quite specific for the compounds listed, others are of broader specificity.

11.7.1 Insecticides

Although many of the major insecticides have been in commercial use for several decades, most of the immunoassays for these compounds were developed rather recently compared with immunoassays for herbicides and fungicides. Assays for several organophosphates have been developed (Table 11.2), but attempts to generate antibodies capable of broad organophosphate detection have had limited success (Sudi and Heschen, 1988) due to the small size and aliphatic nature of the common phosphate esters and possibly the lack of difference from chemical groupings that naturally occur *in vivo*.

However, as an alternative, a cholinesterase enzyme inhibition assay, rather than an immunoassay, can be used to screen for most organophosphates and some carbamates. The insecticides pose some difficult chemical challenges as few possess groups that can readily be derivatized for coupling to protein for subsequent antibody production. One reason for the slower development of good assays for organochlorines is their chemical inertness, while the 'designed in' instability of pyrethroids means that many direct derivatization strategies cannot be used for synthesis of pesticide conjugates.

With the exception of grain and grain-based foods (organophosphates, pyrethroids and methoprene), meat (organochlorines and pyrethroids) and tobacco (methoprene), there has only been limited work with other crop and food matrices. Environmental matrices (soil and water) are also important for these compounds, due to effects on both vertebrate (fish) and invertebrate biota. Although DDT has been banned from agricultural use in developed countries for one or two decades, its persistence means that screening for soil residues remains important. Residues can still enter the food chain through grazing and growth of certain crops which translocate organochlorines.

11.7.2 Herbicides

Most immunoassays have been developed for herbicides, with an emphasis on extensively used compounds with moderate to high persistence, such as atrazine. The triazines have been the subject of greatest study because of the interest in performing widespread screening in ground and surface water. The potential for reduced clean-up in soil and foods has also stimulated a reasonable degree of effort with these matrices. Depending mainly on the chemistry used for hapten–protein conjugation for antibody production, the assays developed for triazines have varying degrees of selectivity, either detecting several members of the group or one to three compounds. Both types of assays can be useful. Antibodies have also been developed to triazine metabolites to study biodegradation more easily (Schlaeppi et al., 1989; Lucas et al., 1993). Other compounds such as the sulphonylureas are of interest because of the difficulty in obtaining sufficiently sensitive detection of these very potent compounds using instrumental analyses; here the main focus is on soil, since variable persistence can cause large yield losses. The majority of the work in this area has been performed by US scientists thus logically there has been most assay development for compounds that are used widely on American crops, such as triazines, acetanilides and phenoxyacetic acids.

After water, the important matrix for residue analysis of herbicides is soil. In assay development, it is important to account thoroughly for potential matrix effects. Since clay content, organic matter, sand/silt and pH of particular soils can vary, it is important to assess the performance of particular assays in soils of the same type. A useful approach is to modify or simplify methods that have previously been developed for instrumental analysis. A common matrix effect observed with soils is satisfactory recovery of high-level spikes but inhibition of ELISA colour development with pesticide-free soil; this can lead to false positives, unless the extracts are diluted appropriately before assay. Fortunately dilution may not diminish the usefulness of the assay as the soil detection levels required are usually one to two orders of magnitude higher than needed for water. Some example of approaches used to imunoassay residues in soil include:

1. Clomazone. In the approach of Koppatschek *et al.* (1990), soils were extracted with acetonitrile, the extract dried, then redissolved for assay. Soil components decreased the sensitivity of the assay, possibly by binding the herbicide. Such binding was not reversed under the assay conditions. Dargar *et al.* (1991) used a simple water extract for soil analysis of the same compound. They demonstrated a correlation between clomazone levels measured by ELISA and injury to a susceptible crop, wheat. Such studies are important if we are to demonstrate that the residual herbicide, measured by ELISA, is in a 'free' form that is toxic to rotation crops.

2. Chlorsulfuron. Kelley *et al.* (1985) found that standard curves for chlorsulfuron spikes differed according to the soil type used in the assay – there was greatest sensitivity in a clay soil. In developing an immunoassay for triasulfuron, Schlaeppi *et al.* (1992) compared three extraction methods. Some inhibition with pesticide-free soil extracts was noted if a buffer or methanol/buffer extract was used. Clean-up by ion pair partition removed this effect and enabled detection of triasulfuron down to 0.1 ppb in soil, but obviously increased sample preparation time.

3. Paraquat requires extensive clean-up for its analysis by 'conventional methods'. Niewola *et al.* (1986) extracted paraquat from soil by refluxing in concentrated sulphuric acid. The neutralized soil extracts contained high levels of sodium sulphate so this was added to the standard curve to compensate. This is also an example where a commonly used method of analysis (spectrophotometry) detects another compound (diquat), while the ELISA is paraquat specific.

4. Various polar organic solvents, containing differing proportions of water, have been used to extract triazines from soil. Cyanazine has been extracted in 75% methanol (Lawruk *et al.*, 1993a), while atrazine was extracted in 40% methanol (Goh *et al.*, 1991). Both 90% acetonitrile and methanol gave overestimates (Goh *et al.*, 1990) for atrazine compared with the instrumental assay. Atrazine has been extracted with a variety of solvent ratios: 90% acetonitrile (Bushway *et al.*, 1988b; Leavitt *et al.*, 1991), neat methanol (Fitzpatrick *et al.*, 1992; they also used this for alachlor and metolachlor), as well as pure water (Schneider *et al.*, 1992).

5. Metolachlor has been extracted in neat methanol but samples were read with respect to a standard curve prepared in methanol soil extract. This

approach is of limited use because of variation in soil composition between sites (Hall *et al.*, 1992). Lawruk *et al.* (1993b) used 75% methanol for metolachlor.

There have been several studies on determination of herbicide residues in foods, presumably aimed to detect contamination following uptake from the soil in which the plant crop was grown. Examples include 2,4-D in cereals and fruit (Newsome and Collins, 1989); diclofop-methyl in wheat, soy and sugarbeet (Schwalbe *et al.*, 1984), and imazamethabenz in cereal grains (Newsome and Collins, 1991). Atrazine has been determined in a range of foods both by Bushway *et al.* (1989) and by Wittmann and Hock (1993).

11.7.3 *Fungicides*

The major fungicide that has been studied by immunoassay is benomyl. This reflects its use on a wide number of crops and, more recently perhaps, its notoriety due to possible phytotoxicity of a contaminant in the formulation. Since benomyl rapidly degrades to methyl 2-benzimidazole carbamate (MBC), the ELISAs generally determine benomyl as MBC. Both polyclonal and monoclonal antibody assays have been developed (Charlton *et al.*, 1991). There has been a reasonable amount of work done with spiked fruits and some vegetables, and juices containing incurred residues (Bushway *et al.*, 1990). In the assay developed by Newsome and Collins (1987), the samples are refluxed with alkaline ethyl acetate and an aliquot evaporated off then redissolved in methanol for immunoassay. Thus no clean-up was needed and quantitative recovery was obtained. Determination of thiabendazole is usually more sensitive than benomyl. Bushway *et al.* (1992) used another assay for MBC to analyse incurred residues in blueberries. They analysed both direct methanol extracts of this matrix as well as partitioning the extract, after addition of saline, with methylene chloride. The latter step improved the correlation between immunoassay and GLC data, yet is still far simpler than most sample handling steps required for instrumental analysis.

Assays for several other fungicides have been developed, often by Newsome and co-workers in Canada. The assays include the following:

1. Metalaxyl (Newsome, 1985) – assayed in vegetables after extraction by direct homogenization in methanol. The immunoassays for metalaxyl usually cross-react with the herbicide metolachlor as they are both acetanilides. Possibly conjugation *para-* to the aniline nitrogen (which was not attempted) would have yielded a more specific assay. However, it would be unlikely to find the herbicide in most matrices of interest for metalaxyl.
2. Triadimefon (Newsome, 1986) – assayed in fruits after homogenization in methanol.
3. Other triazole fungicides (Forlani *et al.*, 1992) – no matrix application was reported.

4. Iprodione (Newsome, 1987) – assayed in fruits and vegetables after extraction in ethyl acetate, evaporation to dryness and resuspension in dimethylsulphoxide.

5. Procymidone (Ferguson *et al.*, 1993a) – assayed in wine following addition of a protein/detergent to the diluent to minimize interference from phenolic pigments in red wine. This is an example of the development of an assay by workers in a country (US) to screen for a compound that is not in use there but is of regulatory concern in imports to that country, in this case from Europe.

6. Fenpropimorph (Jung *et al.*, 1989) – assayed in soil percolation water (assayed directly) since it is used in field crops, loss into water is of concern.

Like most pesticide ELISAs, the assays often have detection ranges between 1 and 100 ppb in buffer. Since action levels for most fungicides in fruit and vegetable matrices are of the order of 1–20 ppm, even after extraction and dilution of the extract of the food there is more than adequate sensitivity. In contrast, many assays for water analysis of herbicides are required to satisfy the EC detection limit of 0.1 ppb.

11.7.4 Commercial test kits

Commercial test kits have been available for some compounds since the mid–late 1980s. The most commonly developed assays have been for analytes such as herbicides that appear in ground water, e.g. triazines and alachlor. In contrast to developed countries, which have the greatest regulatory pressure and capacity to pay for pesticide analyses, assays for a number of pesticides of significant concern to developing countries (e.g. lindane) have not been developed. Assays for some analytes are available from several companies. Before committing to a particular kit it is important to evaluate the assay on several criteria – price per assay, sensitivity of the assay, cross reaction and, importantly, availability of published validation, especially by independent groups and presence of local technical support. It is important to determine whether the assay has been validated for the particular matrix for which that analysis is required. For example, a kit and method that is intended for water may give quite spurious results with fruit purée (this comment applies equally well to gas chromatography and HPLC). Commonly, the assay development company will be interested in actively collaborating with groups wishing to extend major assays to new matrices.

11.8 Statistical Considerations

One of the biggest misconceptions about immunoassay is that it is not as accurate or precise as other methods. The accuracy and precision of any immunoassay will vary depending upon the antibody. Thus in some cases, the accuracy and precision will be equivalent to the 'model' method, whereas in other cases, it may be better or worse. This comparison can be made between any analytical technique, regardless of the type of detection system it may utilize.

Variability in the immunoassay may be manifested at several points in the analytical procedure:

1. *Well-to-well* variability in absorbance is a measure of the consistency of the coating reagents and the pipetting precision throughout the assay.

2. *Plate-to-plate* variability in absorbance or in the standard curve can be a function of the inherent plate characteristics, coating reagents, pipetting precision and environmental variables such as time of incubation and temperature.

3. The *day-to-day* variability in the standard curve is a measure of the reproducibility of the assay reagents and assay environmental parameters. Day-to-day variations can be tolerated since all data are compared to standard curves prepared that day and preferably run on the same plate as the samples.

4. A *curve that fits* the analysis model well will provide a back-calculation of the standard curve that is within some predetermined confidence limit. However, the data generated from an immunoassay result in a sigmoidal calibration curve. This curve has been well studied and is defined by the four-parameter logistic fit of Rodbard (1981). The error on the curve can be defined. This error is usually larger at the upper and lower asymptotes of the standard curve as might be expected when the concentrations approach either zero or are exceedingly large. Sample absorbances are extrapolated from within the central portion of the calibration curve. Unlike other methods of analysis, the calibration curve is based on a log-linear scale, rather than a linear scale, thus in a calibration curve with a steep slope, a small change in absorbance results in a large change in sample concentration. General rules that guide the preparation of calibration curves (Wernimont, 1985) are also valid for the immunoassay calibration curve.

5. *Matrix effects* may also be responsible for variations in data from sample to sample. Most often matrix effects are manifested as false positives, because the effect is usually a reduction in colour development. It is very important to test a 'blank' matrix or to run recovery studies to determine the potential for matrix interferences. If this is not possible, use of the method of standard additions (Miller and Miller, 1984) can provide some control for matrix effects.

11.9 Basic Troubleshooting

There are some references in the clinical literature (Perlstein, 1987) and some written for the agrochemical immunoassay (Gee *et al.*, 1994; Schneider *et al.*, 1995) which describe troubleshooting procedures in some detail. The most frequently encountered problems are described here. When analysing the absorbance data derived from the immunoassay, poor precision among plate microwell replicates is sometimes observed. The largest contributors to poor precision are incomplete washing steps and poor pipetting technique. If residual reagents remain in the well, particularly unbound enzyme label, then substrate conversion will occur even though the enzyme is not bound to the plate. Thorough washing is necessary and tapping the plate or tube inverted on to absorbent paper to remove residual droplets is recommended. If pipetting of reagents is not uniform, then some wells may have more reagent than others, which would also result in spurious colour development. When pipetting from well to well, it is also important to be wary of potential places where sample carryover may contribute to variability. Pipettes should be checked for cleanliness and calibrated on a regular basis.

Another contributor to poor precision among well replicates is temperature. The reactions that are occurring on the plate are based on the Law of Mass Action. These are thus physical assays, but are equilibrium reactions and are sensitive to temperature. Reagents should be used at room temperature, and during analysis plates should be protected from wide fluctuations in temperature (i.e. try to perform all steps of the assay at the same ambient temperature each time). With the 96-well plates, there have been reports that the outer wells reach temperature sooner than the inner wells, which then has an effect on the equilibrium reactions. Variations in final absorbances are generally manifested in what is called an 'edge effect'. Conducting incubations in a forced-air incubator may eliminate problems due to temperature fluctuations. Temperature-related effects on equilibrium are more likely to be seen in assays whose incubation times are very short.

An immunoassay in which the colour development is lacking or very low is most likely due to a reagent failure. The most common reagent to fail is the enzyme label. There are two potential problems: first, that the enzyme has lost activity; second, that the conjugate has degraded and can no longer bind efficiently to the antibody. In the second case, in which the conjugate can no longer bind to the antibody, the only remedy is to replace the reagent. Low colour development may also be a result of a sample matrix effect. Matrices should always be evaluated carefully in the immunoassay prior to analysis of samples. As mentioned above, testing of a 'blank' matrix, careful recovery studies or use of the method of standard addition can indicate a matrix effect. Another method for determining a matrix effect is to plot the dilution curve of the sample with the calibration curve. If the slope of the two curves is similar, than there is probably not a matrix effect. However, if the curves are not parallel, this usually indicates an interference due to the matrix (Perlstein, 1987).

The type of 96-well plate used is also an important factor in obtaining precision among replicates. Some plates will bind antigens differently, and some have greater variability in binding capacity from well to well which would contribute to variability. Changes in the calibration curve parameters may also be indicative of an impending assay failure. For the microwell assay, if a standard curve is run on each assay plate, then the extrapolation of data for samples on that plate is normalized for effects specific to that plate. For example, if one plate is allowed to incubate longer than another one the colour development may be greater on the plate that incubated longer. Sample concentration extrapolation from a standard curve on the same plate is desirable in order to normalize for these potential plate-to-plate differences.

11.10 Field Tests for Agrochemicals

A major advantage of immunoassay methods is their potential for residue analysis under field or small laboratory conditions, since all of the reagents can be carried with the operator in the form of a small test kit. Examples of field testing relevant to crop science include the following:

1. Checking for a suspected pesticide spill, or escape during a storm or flood.
2. On-site monitoring of pesticide residues in commodities such as stored grain, where it is important to check that pesticide treatment has been adequate for appropriate insect control.
3. Before sale or shipment of produce, checking that the pesticide levels in the product do not exceed maximum residue limits.
4. Monitoring of water for recycling on herbicide-treated field before planting for phytotoxic residues.

Apart from the speed of the residue test itself, use of a pesticide field test allows many documentation and sample handling and shipping exercises to be eliminated. Field tests usually are a modification of an assay format developed for the conventional laboratory immunoassay. Most commonly, instead of using microwells, polystyrene test tubes are used as the solid phase. Other formats, such as dipsticks, have also been described (Giersch, 1993). Running a field test usually involves two brief incubation steps (sample and enzyme conjugate followed by substrate/ chromogen) separated by a single wash step. The overall test time may be as low as 6 min. Reagents are often added using dropper bottles rather than laboratory automatic pipettes, and washing steps performed using tap water instead of the use of buffers and detergent-containing solutions.

Because field immunoassays will often be used by individuals with limited professional experience in analytical chemistry, there are several potential traps in the gathering and application of pesticide data. The following points should be noted:

1. What is the relevance of screening for particular compounds? Some reasons for measuring pesticides include to assess for each compound that is used in an area, what its fate is (in quantitative, temporal and spatial terms), and to establish the knowledge base for chemical use and monitoring management decisions.
2. What breakdown products of the compound under study should be measured?
3. How quantitative are the data required to be? Field immunoassay tests can be run in three ways:
    (a) with no calibrators, only a zero-pesticide control tube;
    (b) with a calibrator set at the appropriate action tolerance level (and possibly a zero-pesticide control as well);
    (c) with a set of four or so calibrators, enabling development of a standard curve.
Just which approach is selected will obviously relate to the original reason for performing the test.
4. What do the results mean? One aspect that we need to be aware of is that the ability to test for residues can be placed in uninformed hands, so education on proper interpretation of results is very important. For example, 2 ppb molinate in a drain leading from a rice field is not an environmental disaster, while a similar level of endosulfan in major river systems could be a disaster! Also, users need to be aware of possible sources of interference in any test.

11.11 Conclusion

It will not be important to develop tests for each of the hundreds of registered farm chemicals, instead the targets fit into one or more of the following categories, each of which relates to industry need:

1. Major use chemical in world plant agriculture.
2. Of significant toxicity to fish, bees, birds, aquatic organisms, wildlife.
3. Having the ability to damage rotation crops.
4. Associated with occupational safety concerns.
5. May produce residues in foods derived from the crop of concern.

The ability of the tests to be performed cheaply and in 'low-tech' situations would also make them suitable for use in developing countries, where toxicity to farm workers and the public of a number of pesticides (many of which are banned in developed countries) is a serious health problem.

Acknowledgements

The work of B.D.H. and S.J.G. was supported in part by NIEHS Superfund 2P42 ES04699, US Environmental Protection Agency CR 81 9047–01–0, EPA Center for Ecological Health Research CR 81 9658, the NIEHS Center for Environmental Health Sciences (at UCD) 1P30 ES05707 and the US Forest Service NAPIAP USDA/PSW 5–93–25. B.D.H. is a Burroughs Welcome Toxicology Scholar. Although some of the information in this document has been funded wholly or partly by the United States Environmental Protection Agency, it may not necessarily reflect the views of the Agency and no official endorsement should be inferred. The work of J.H.S. and colleagues was supported in part by the Grains R&D Corporation of Australia, the Land and Water Resources R&D Corporation of Australia and the Cotton R&D Corporation of Australia.

References

Aherne, G.W. (1991) Immunoassays for the measurement and detection of pesticides in the environment. *British Crop Protection Council Monograph Molecular Biology – Crop Protection* 48, 59–77.

Aherne, G.W. (1992) The use of ELISAs for the detection and measurement of pollutants in water. *Journal of Chemical Technology and Biotechnology* 54, 189–191.

Al-Rubae, A.Y. (1978) The ELISA: a new method for the analysis of pesticide residues. PhD thesis, Pennsylvania State University, *Dissertation Abstracts International* 39, 4723–4724.

Anis, N.A., Eldefrawi, M.E. and Wong, R.B. (1993) Reusable fibre optic immunosensor for rapid detection of imazethapyr herbicide. *Journal of Agricultural and Food Chemistry* 41, 843–848.

Baker, D.B., Bushway, R.J., Adams, S.A. and Macomber, C. (1993) Immunoassay screens for alachlor in rural wells: false positives and an alachlor soil metabolite. *Environmental Science and Technology* 27, 562–564.

Beasley, H.L., Skerritt, J.H., Hill, A.S. and Desmarchelier, J.M. (1993) Rapid field tests for the organophosphates, fenitrothion and pirimiphos-methyl – reliable estimates of pesticide residues in stored grain. *Journal of Stored Products Research* 29, 357–369.

Bekheit, H.K.M., Lucas, A.D., Gee, S.J., Harrison, R.O. and Hammock, B.D. (1993) Development of an enzyme-linked immunosorbent assay for the β-exotoxin of *Bacillus thuringiensis*. *Journal of Agricultural and Food Chemistry* 41, 1530–1536.

Brady, J.F., Fleeker, J.R., Wilson, R.A. and Mumma, R.O. (1989) Enzyme immunoassay for aldicarb. In: Wang, R.G.M., Franklin, C.A., Honeycutt, R.C. and Reinert, R.C. (eds), *Biological Monitoring for Pesticide Exposure.* American Chemical Society Symposium Series 382, Washington, DC, pp. 262–284.

Brandon, D.L., Binder, R.G., Wilson, R.E. and Montague, W.C. (1993) Analysis of thiabendazole in potatoes and apples by ELISA using monoclonal antibodies. *Journal of Agricultural and Food Chemistry* 41, 996–999.

Burgisser, D. (1990) Preparation and characterization of polyclonal and monoclonal antibodies against the insecticide, DDT. *Biochemical and Biophysical Research Communications* 166, 1228–1236.

Bushway, R.J., Pask, W.M., King, J., Perkins, B. and Ferguson, B.S. (1988a) Determination of chlordane in soil by enzyme immunoassay. *Field Screening Methods for Hazardous Waste Site Investigations*, First International Symposium, pp. 433–437.

Bushway, R.J., Perkins, L.B., Savage, S.A., Lekousi, S.J. and Ferguson, B.S. (1988b) Determination of atrazine residues in water and soil by enzyme-immunoassay. *Bulletin of Environmental Contamination and Toxicology* 40, 647–654.

Bushway, R.J., Perkins, B., Savage, S.A., Lekousi, S.J. and Ferguson, B.S. (1989) Determination of atrazine residues in food by enzyme-immunoassay. *Bulletin of Environmental Contamination and Toxicology* 42, 899–904.

Bushway, R.J., Savage, S.A. and Ferguson, B.S. (1990) Determination of methyl-2-benzimidazole in fruit juices by immunoassay. *Food Chemistry* 35, 51–58.

Bushway, R.J., Kugabalasooriar, J., Perkins, L.B., Harrison, R.O., Young, B.E.S. and Ferguson, B.S. (1992) Determination of methyl 2-benzimidazole in blueberries by competitive inhibition immunoassay. *Journal of AOAC International* 75, 323–328.

Charlton, R.R., Majaran, W.R., Carlson, C.W., Carski, T.H., Liberatore, F.A., Kenyon, L.H., Allison, D.A., Steele, W.J. and Fraser, A. (1991) *Environmental Monitoring of Crop Protection Chemicals by Immunoassay.* Abstract 105th AOAC meeting, Phoenix, Abstract 326.

Cheung, P.Y.K. and Hammock, B.D. (1988) Monitoring BT in the environment with ELISA. In: Hedin, P.A., Menn, J.J. and Hollingworth R.M. (eds), *Biotechnology for Crop Protection. American Chemical Society Symposium Series.* American Chemical Society, Washington, DC, pp. 298–305.

Chigrin, A.V., Ummov, A.M., Sokolova, G.D., Khokhov, P.S. and Chkanikov, D.I. (1989) Determination of chlorsulfuron by immunoenzyme analysis. *Agrokhimiya* 8, 119–123.

Dargar, R.V., Tymonoko, J.M. and van der Werk, P. (1991) Clomazone measurement by ELISA. *Journal of Agricultural and Food Chemistry* 39, 813–819.

de Frutos, M. and Regnier, F.E. (1993) Tandem chromatographic–immunological analyses. *Analytical Chemistry* 65, 17A–25A.

Demoute, J.P., Touer, G. and Mouren, M. (1986) Preparation of pyrethroidal radioactive iodine-labelled amino acid derivatives for radioimmunoassay of deltamethrin. French Patent number 2593503.

Dreher, R.M. and Podratzki, B. (1988) Development of an enzyme-immunoassay for endosulfan and its degradation products. *Journal of Agricultural and Food Chemistry* 36, 1072–1075.

Dunbar, B., Riggle, B. and Niswender, G. (1990) Development of an enzyme-immunoassay for the detection of triazine herbicides. *Journal of Agricultural and Food Chemistry* 38, 433–437.

Duquette, P.H., Guire, P.E. and Swanson, M.J. (1988) Fieldable enzyme-immunoassay kits for pesticides. *Field Screening Methods for Hazardous Waste Site Investigations*, Proceedings of the First International Symposium, pp. 239–242.

Edward, S.L., Hill, A.S., Ashworth, P., Matt, J.L. and Skerritt, J.H. (1993a) Analysis of the grain protectants, chlorpyrifos-methyl and methoprene with a 15-minute immunoassay for field or elevator use. *Cereal Chemistry* 70, 748–752.

Edward, S.L., Skerritt, J.H., Hill, A.S. and McAdam, D.P. (1993b) An improved immunoassay for chlorpyrifos-methyl (Reldan ®) in grain. *Food and Agricultural Immunology* 5, 129–144.

Feng, P.C.C., Wratten, S.J., Sharp, C.R. and Logusch, E.W. (1990) Development of an ELISA for alachlor and its application to the analysis of environmental water samples. *Journal of Agricultural and Food Chemistry* 38, 159–163.

Feng, P.C.C., Horton, S.R. and Sharp, C.R. (1992) A general method for developing immunoassays to chloracetanilide herbicides. *Journal of Agricultural and Food Chemistry* 40, 211–214.

Ferguson, B.S., Kelsey, D., Fan, T.S. and Bushway, R.J. (1993a) Pesticide testing by enzyme immunoassay at trace levels in environmental and agricultural samples. *The Science of the Total Environment* 132, 415–428.

Ferguson, B.S., Larkin, K.A., Skerritt, J.H. and Beasley, H.L. (1993b) The detection of diazinon in lanolin using a rapid and simple enzyme-immunoassay. *Association of Official Analytical Chemists*, 107th Annual Meeting, Washington, DC, Abstract 241.

Fialova, K. and Franek, M. (1990) Development of an ELISA for 2,4-D MCPA and chloridazon herbicides. Symposium on Bioanalytical Methods, Prague, Czechoslovakia, September 1990, Abstract 130.

Fitzpatrick, D.A., Stocker, D.R., Rittenburg, J.L. and Grothaus, G.D. (1992) Determination of herbicides in agricultural samples by immunoassay. *Association of Official Analytical Chemists*, 106th Annual Meeting, Cincinnati, OH, Abstract 265.

Fleeker, J.R. (1987) Two enzyme-immunoassays to screen for 2,4-dichlorophenoxyacetic acid in water. *Journal of the Association of Official Analytical Chemists* 70, 874–878.

Forlani, F., Arnoldi, A. and Pagani, S. (1992) Development of an enzyme-linked immunosorbent assay for triazole fungicides. *Journal of Agricultural and Food Chemistry* 40, 328–331.

Gee, S.J., Miyamoto, T., Goodrow, M.H., Buster, D. and Hammock, B.D. (1988) Development of an ELISA for the analysis of the thiocarbamate herbicide, molinate. *Journal of Agricultural and Food Chemistry* 36, 863–870.

Gee, S.J., Hammock, B.D. and Van Emon, J.M. (1994) *A User's Guide to Environmental Immunochemical Analysis* EPA Publication 540/R-G4/509. US Environmental Protection Agency, Washington, DC.

Giersch, T. (1993) A new monoclonal antibody for the sensitive detection of atrazine with immunoassay in microtiter plate and dipstick format. *Journal of Agricultural and Food Chemistry* 41, 1006–1011.

Giersch, T. and Hock, B. (1990) Monoclonal antibodies for the determination of s-triazines with enzyme-immunoassay. *Food and Agricultural Immunology* 2, 85–97.

Giersch, T., Kramer, D. and Hock, B. (1993) Optimization of a monoclonal antibody-based enzyme-immunoassay for the detection of terbuthylazine. *The Science of the Total Environment* 132, 435–448.

Goh, K.S., Hernandez, J., Powell, S.J. and Greene, C.D. (1990) Atrazine soil residue analysis by enzyme-immunoassay: solvent effect and extraction efficiency. *Bulletin of Environmental Contamination and Toxicology* 45, 208–215.

Goh, K.S., Hernandex, J., Powell, S.J., Garretson, C., Trolano, J., Ray, M. and Greene, C.D. (1991) Enzyme-immunoassay for the determination of atrazine residues in soil. *Bulletin of Environmental Contamination and Toxicology* 46, 30–36.

Goodrow, M.H., Harrison, R.O. and Hammock, B.D. (1990) Hapten synthesis, antibody development and competitive inhibition enzyme-immunoassay for s-triazine herbicides. *Journal of Agricultural and Food Chemistry* 38, 159–163.

Hall, J.C., Deschamps, R.J.A. and Kreig, K.K. (1989) Immunoassays for the detection of 2, 4-D and picloram in river water and urine. *Journal of Agricultural and Food Chemistry* 37, 981–984.

Hall, J.C., Wilson, L.K. and Chapman, R.A. (1992) An immunoassay for metolachlor detection in river water and soil. *Journal of Environmental Science and Health* 27, 523–544.

Harrison, R.O., Braun, A., Gee, S.J., O'Brien, D.J. and Hammock, B.D. (1989a) Evaluation of an ELISA for the direct analysis of molinate (Ordram ®) in rice field water. *Food and Agricultural Immunology* 1, 37–52.

Harrison, R.O., Brimfield, A.A. and Nelson, J.O. (1989b) Development of a monoclonal antibody-based enzyme-immunoassay method for analysis of maleic hydrazide. *Journal of Agricultural and Food Chemistry* 37, 958–964.

Harrison, R.O., Goodrow, M.H. and Hammock, B.D. (1991) Competitive inhibition ELISA for the s-triazine herbicides: assay optimization and antibody characterization. *Journal of Agricultural and Food Chemistry* 39, 122–128.

Heckman, R.A., Ferguson, B.S., Fan, T.S., Mei, J.V., Yin, C.-M., Conner, T.R., Addington, V.W. and Benezet, H.J. (1992) Validation of an enzyme immunoassay for analysis of methoprene residues on tobacco. *Journal of Agricultural and Food Chemistry* 40, 2530–2532.

Hill, A.S., Mei, J.V., Yin, C.-M, Ferguson, B.S. and Skerritt, J.H. (1991) Determination of the insect growth regulator, methoprene, in wheat grain and milling fractions using enzyme-immunoassay. *Journal of Agricultural and Food Chemistry* 39, 1882–1886.

Hill, A.S., Beasley, H.L., McAdam, D.P. and Skerritt, J.H. (1992) Mono- and polyclonal antibodies to the organophosphate, fenitrothion. 2. Antibody specificity and assay performance. *Journal of Agricultural and Food Chemistry* 40, 1471–1474.

Hill, A.S., McAdam, D.P., Edward, S.L. and Skerritt, J.H. (1993) Quantitation of bioresmethrin, a synthetic pyrethroid grain protectant by enzyme-immunoassay. *Journal of Agricultural and Food Chemistry* 41, 2011–2018.

Hofte, H., van Rie, J., Jansens, S., van Houtven, A., Vanderbruggen H. and Vaeck M. (1988) Monoclonal antibody analysis and insecticidal analysis and insecticidal spectrum of three types of lepidopteran-specific insecticidal crystal proteins of *Bacillus thuringiensis*. *Applied and Environmental Microbiology* 54, 2010–2017.

Huber, S.J. (1985) Improved solid-phase enzyme-immunoassay systems in the ppt range for atrazine in fresh water. *Chemosphere*, 14, 1795–1803.

Huber, S.J. and Hock, B. (1985) A solid-phase enzyme-immunoassay for quantitative determination of the herbicide, terbutryn. *Journal of Plant Disease and Protection* 92, 147–156.

Itak, J.A., Selisker, M.Y., Herzog, D.P., Dautlick, J.X. and Fleeker, J.R. (1991) Rapid method for the determination of captan residues. Proceedings of the 105th AOAC meeting, Phoenix, Abstract 2104.

Itak, J.A., Selisker, M.Y. and Herzog, D.P. (1992) Development and evaluation of a magnetic particle-based enzyme immunoassay for aldicarb, aldicarb sulfone and aldicarb sulfoxide. *Chemosphere* 24, 11–21.

Jung, F., Meyer, H.H.D. and Hamm, R.T. (1989) Development of a sensitive ELISA for the fungicide, fenpropimorph. *Journal of Agricultural and Food Chemistry* 37, 1183–1187.

Jung, F., Székás, A. Li, Q.X. and Hammock, B.D. (1991) Immunochemical approach to the detection of aminotriazoles using selective amino group protection by chromophores. *Journal of Agricultural and Food Chemistry* 39, 129–136.

Karu, A.E. (1993) Monoclonal antibodies and their use in measurement of environmental contaminants. In: Saxena, J. (ed.), *Hazard Assessment of Chemicals, vol. 8*. Taylor and Francis, Washington, DC, pp. 205–321.

Karu, A.E., Lilental, J.E., Schmidt, D.J., Lim, A.K., Carlson, R.E., Swanson, T.A., Buirge, A.W. and Chamerlik, M. (1990) Monoclonal antibody-based immunoassay of cyclodienes. Proceedings of the 6th Annual EPA Waste Testing and Quality Assurance Symposium, Washington, DC, pp. I-237–I-245.

Karu, A.E., Goodrow, M.H., Schmidt, D.J. Hammock, B.D. and Bigelow, M.W. (1994) Synthesis of haptens and derivation of monoclonal antibodies for immunoassay of the phenylurea herbicide, diuron. *Journal of Agricultural and Food Chemistry*, 42, 301–309.

Kelley, M.M., Zahnow, E.W., Petersen, W.C. and Toy, S.T. (1985) Chlorsulfuron determination in soil extracts by enzyme-immunoassay. *Journal of Agricultural and Food Chemistry* 33, 962–965.

Koppatschek, F.K., Liebl, R.A., Kriz, A.L. and Melhado, L.L. (1990) Development of an ELISA for the detection of the herbicide, clomazone. *Journal of Agricultural and Food Chemistry* 38, 1519–1522.

Krämer, P.M., Li, Q.X. and Hammock, B.D. (1994) Integration of LC with immunoassay: an approach combining the potential of both methods. *Journal of the AOAC International* 77, 1275–1287.

Kreissig, S. and Hock, B. (1991) Development of an enzyme immunoassay for the determination of methabenzthiazuron. *Zeitschrift für Wasser und Abwasser Forschung*, 24, 10–12.

Kuniyuki, A.H. and McCarthy, S. (1986) A novel method for the development of monoclonal antibodies to surflan. In: Greenhalgh, R. and Roberts, T.R. (eds), *Pesticide Science and Biotechnology*, Proceedings of the 6th International Congress of Pesticide Chemistry, Ottawa, Canada. Blackwell Scientific, Oxford, Abstract 5c–11.

Lawruk, T.S., Hottenstein, C.S., Herzog, D.P. and Rubio, F.M. (1992) Quantification of alachlor in water by a novel magnetic particle-based ELISA. *Bulletin of Environmental Contamination and Toxicology* 48, 643–650.

Lawruk, T.S., Lachman, C.E., Jourdan, S.W., Fleeker, J.R., Herzog, D.P. and Rubio, F.M. (1993a) Quantification of cyanazine in water and soil by a rapid magnetic particle-based ELISA. *Journal of Agricultural and Food Chemistry* 41, 747–752.

Lawruk, T.S., Lachman, C.E., Jourdan, S.W., Fleeker, J.R., Herzog, D.P. and Rubio, F.M. (1993b) Determination of metolachlor in water and soil by a rapid magnetic particle-based ELISA. *Journal of Agricultural and Food Chemistry* 41, 1426–1431.

Leavitt, R.A., Kells, J.J., Bunkelmann, J.R. and Hollingworth, R.M. (1991) Assessing atrazine persistence in soil following a severe drought. *Bulletin of Environmental Contamination and Toxicology* 46, 22–29.

Li, Q.X., Gee, S.J., McChesney, M.M., Hammock, B.D. and Seiber, J.N. (1989) Comparison of ELISA and gas chromatographic procedures for the analysis of molinate residues. *Analytical Chemistry* 61, 819–823.

Li, Q.X., Zhao, M.S., Gee, S.J., Kurth, M., Seiber, J.N. and Hammock, B.D. (1991a) Development of enzyme-immunosorbent assays for 4-nitrophenol and substituted 4-nitrophenols. *Journal of Agricultural and Food Chemistry* 39, 1685–1692.

Li, Q.X., Hammock, B.D. and Seiber, J.N. (1991b) Development of an ELISA for the herbicide, bentazon. *Journal of Agricultural and Food Chemistry* 39, 1537–1544.

Liegeois, E., Dehon, Y., de Brabant, B., Perry, P., Portelle, D. and Copin, A. (1992) ELISA test, a new method to detect and quantify isoproturon in soil. *The Science of the Total Environment* 123/124, 17–28.

Lucas, A.D., Bekheit, H.K.M., Goodrow, M.H., Jones, A.D., Kullman, S., Matsumura, F., Woodrow, J.E., Seiber, J.N. and Hammock, B.D. (1993) Development of antibodies against hydroxyatrazine and hydroxysimazine: application to environmental samples. *Journal of Agricultural and Food Chemistry* 41, 1523–1529.

Lucas, A.D., Gee, S.J., Hammock, B.D. and Seiber, J.N. (1994) The integration of immunochemical methods with analytical chemistry for pesticide residue determination. *Journal of the AOAC International* 78(3).

Marco, M-P., Gee, S.J., Cheng, H.M., Liang, Z.Y. and Hammock, B.D. (1993) Development of an enzyme-linked immunosorbent assay for carbaryl. *Journal of Agricultural and Food Chemistry* 41, 423–430.

Mei, J.V., Yin, C.M. and Carpino, L.A. (1991) Hapten synthesis and competitive ELISA for methoprene. *Journal of Agricultural and Food Chemistry* 39, 2083–2090.

Miller, J.C. and Miller, J.N. (1984) *Statistics for Analytical Chemistry*. Ellis Horwood, Chichester.

Muldoon, M.T., Fries, G.F. and Nelson, J.O. (1993) Evaluation of ELISA for the multivariate analysis of s-triazines in pesticide waste and rinsate. *Journal of Agricultural and Food Chemistry* 41, 322–328.

Newsome, W.H. (1985) An ELISA for metalaxyl in foods. *Journal of Agricultural and Food Chemistry* 33, 528–530.

Newsome, W.H. (1986) Development of an enzyme-linked immunosorbent assay for triadimefon in foods. *Bulletin of Environmental Contamination and Toxicology* 36, 9–14.

Newsome, W.H. (1987) Determination of Iprodione in foods by ELISA. In:

Greenhalgh, R. and Roberts, T.R. (eds), *Pesticide Science and Biotechnology.* Proceedings of the 6th International Congress of Pesticide Chemistry, Ottawa. Blackwell Scientific, Oxford, pp. 349–352.

Newsome, W.H. and Collins, P. (1987) ELISA of benomyl and thiabendazole in some foods. *Journal of the Association of Official Analytical Chemists* 70, 1025–1027.

Newsome, W.H. and Collins, P.G. (1989) Determination of 2,4-D in foods by ELISA. *Food and Agricultural Immunology* 1, 203–210.

Newsome, W.H. and Collins, P.G. (1990) Development of an ELISA for urea herbicides in foods. *Food and Agricultural Immunology* 2, 75–84.

Newsome, W.H. and Collins, P.G. (1991) Determination of imazamethabenz in cereal grain by ELISA. *Bulletin of Environmental Contamination and Toxicology* 47, 211–216.

Newsome, W.H., Yeung, J.M. and Collins, P.G. (1993) Development of enzyme-immunoassay for captan and its degradation product, tetrahydrophthalimide in foods. *Journal of AOAC International* 76, 381–386.

Niewola, Z., Benner, J.P. and Swaine, H. (1986) Determination of paraquat residues in soil by an ELISA. *Analyst* 111, 399–406.

Perlstein, M.T. (1987) In: Chand D.W. and Perlstein, M.T. (eds), *Immunoassays: a Practical Guide.* Academic Press, New York,

Riggle, B. (1991) Development of a preliminary ELISA for the herbicide, trifluralin. *Bulletin of Environmental Contamination and Toxicology* 46, 404–409.

Riggle, B. and Dunbar, B. (1990) Development of an enzyme-immunoassay for the detection of the herbicide, norflurazon. *Journal of Agricultural and Food Chemistry* 38, 1922–1925.

Rodbard, D. (1981) Mathematics and statistics of ligand assays: an illustrated guide. In: Langan, J. and Clapp, J.J. (eds), *Ligand Assay: Analysis of International Developments on Isotopic and Nonisotopic Immunoassay.* Masson Publishing, New York, pp. 45–99.

Schlaeppi, J.-M., Fory, W. and Ramsteiner, K. (1989) Hydroxyatrazine and atrazine determination in soil and water by ELISA using specific monoclonal antibodies. *Journal of Agricultural and Food Chemistry* 37, 1532–1538.

Schlaeppi, J.-M., Moser, H. and Ramsteiner, K. (1991) Determination of metolachlor by competitive enzyme immunoassay using a specific monoclonal antibody. *Journal of Agricultural and Food Chemistry* 39, 1533–1536.

Schlaeppi, J.-M.A., Meyer, W. and Ramsteiner, K.A. (1992) Determination of triasulfuron in soil by monoclonal antibody-based enzyme-immunoassay. *Journal of Agricultural and Food Chemistry* 40, 1093–1098.

Schneider, P. and Hammock, B.D. (1992) Influence of the ELISA format and the hapten–enzyme conjugate on the sensitivity of an immunoassay for s-triazine herbicides using monoclonal antibodies. *Journal of Agricultural and Food Chemistry* 40, 525–530.

Schneider, P., Goodrow, M.H., Gee, S.J. and Hammock, B.D. (1994) A highly sensitive and rapid ELISA for the urea herbicides, diuron, monuron and linuron. *Journal of Agricultural and Food Chemistry*, 42, 413–422.

Schneider, P., Gee, S.J., Kreissing, S., Harris, A.S., Krämer, P., Marco, M.P., Lucas, A.D. and Hammock, B.D. (1995) An introduction to troubleshooting during the development and use of immunoassays for environmental monitoring. In: Kurtz, D.A., Skerritt, J.H. and Stanker, L.H. (eds), *Agrochemical Immunoanal-*

ysis '94. AOAC International, Arlington, VA, USA.

Schneider, R.J., Weil, L. and Neissner, R. (1992) Screening and monitoring of herbicides behaviour in soils by enzyme-immunoassays. *International Journal of Environmental Analytical Chemistry* 46, 129–140.

Scholz, H.M. and Hock, B. (1991) Development of an enzyme-immunoassay for metazachlor. *Analytical Letters* 24, 413–427.

Schwalbe, M., Dorn, E. and Beyermann, K. (1984) Enzyme-immunoassay and fluoroimmunoassay for the herbicide, diclofop-methyl. *Journal of Agricultural and Food Chemistry* 32, 734–741.

Sharp, C.R., Feng, P.C.C., Horton, S.R. and Logusch, E.W. (1991) Development of highly specific antibodies to alachlor by use of a carboxy-alachlor protein conjugate. In: Tweedy, B.G., Dishburger, H.J., Ballantine, L.G. and McCarthy, J. (eds), *Pesticide Residues and Food Safety*. American Chemical Society, Washington, DC, pp. 87–95.

Skerritt, J.H., Hill, A.S., McAdam, D.P. and Stanker, L.H. (1992a) Analysis of the synthetic pyrethroids, permethrin and 1R-phenothrin in grain using a monoclonal antibody-based test. *Journal of Agricultural and Food Chemistry* 40, 1287–1292.

Skerritt, J.H., Hill, A.S., Thwaites, H.L., Edward, S.L. and McAdam, D.P. (1992b) Enzyme-immunoassay for quantitation of the organophosphate pesticides, fenitrothion, chlorpyrifos-methyl and pirimiphos-methyl in wheat grain and flour-milling fractions. *Journal of the Association of Official Analytical Chemists* 73, 519–528.

Stanker, L.H., Bigbee, C., Van Emon, J., Watkins, B., Jensen, R.H., Morris, C. and Vanderlaan, M. (1989) An immunoassay for pyrethroids; detection of permethrin in meat. *Journal of Agricultural and Food Chemistry* 37, 834–839.

Stanker, L.H., Watkins, B., Vanderlaan, M., Ellis, R. and Rajan, J. (1991) Analysis of heptachlor and related cyclodiene insecticides in food products. In: Vanderlaan, M., Stanker, L.H., Watkins, B.E. and Roberts, D.W. (eds), *Immunoassays for Trace Chemical Analysis*. American Chemical Society Symposium Series 451, Washington, DC, pp. 108–123.

Sudi, J. and Hesschen, W. (1988) Studies on the development of an immuno assay for the group-specific detection of the diethyl ester of phosphates, thiophosphates, dithiophosphates and phosphonates. *Kieler Milchwirtschaftliche Forschungberichte* 40, 179–203.

Szurdoki, F., Bekheit, H., Marco, M-P., Goodrow, M.H. and Hammock, B.D. (1992) Synthesis of haptens and conjugates for an enzyme-immunoassay for analysis of the herbicide bromacil. *Journal of Agricultural and Food Chemistry* 40, 1459–1465.

Tijssen, P. (1985) *Practice and Theory of Enzyme Immunoassays*. Elsevier, Amsterdam.

Vallejo, R.P., Bogus, E.R. and Mumma, R.O. (1982) Effects of hapten structure and bridging groups on antisera specificity in parathion immunoassay development. *Journal of Agricultural and Food Chemistry* 30, 572–580.

Van Emon, J.M. and Lopez-Avila, V. (1992) Immunochemical methods for environmental analysis. *Analytical Chemistry* 64, 79A–88A.

Van Emon, J., Seiber, J.N. and Hammock, B.D. (1985) Applications of immunoassay to paraquat and other pesticides. In: Hedin, P.A. (ed.), *Bioregulators for Pest*

Control. ACS Publishers, Washington, DC, pp. 307–316.

Van Emon, J., Hammock, B.D. and Seiber, J.N. (1986) Enzyme-immunoassay for paraquat and its application to exposure analysis. *Analytical Chemistry* 58, 1866–1873.

Van Emon, J., Seiber, J.N. and Hammock, B.D. (1987) Application of an ELISA to determine paraquat residues in milk, beef and potatoes. *Bulletin of Environmental Contamination and Toxicology* 39, 490–497.

Voller, A., Barlett, A. and Bidwell, D.E. (1978) Enzyme immunoassays with special reference to ELISA techniques. *Journal of Clinical Pathology* 31, 507–520.

Weil, L., Schneider, R.J., Schaffer, O., Ulrich, P., Weller, M., Ruppert, T. and Niessner, R. (1991) A heterogeneous immunoassay for the determination of triazine herbicides in water. *Fresenius Journal of Analytical Chemistry* 339, 468–469.

Wernimont, G.T. (1985) *Use of Statistics to Develop and Evaluate Analytical Methods.* Association of Official Analytical Chemists, Arlington, VA.

Wie, S.I. and Hammock, B.D. (1982) Development of ELISAs for residue analysis of diflubenzuron and Bay Sir 8514. *Journal of Agricultural and Food Chemistry* 30, 949–957.

Wie, S.I. and Hammock, B.D. (1984) Comparison of coating and immunizing antigen structure on the sensitivity and specificity of immunoassays for benzoylphenylurea insecticides. *Journal of Agricultural and Food Chemistry* 32, 1294–1301.

Wie, S.I., Hammock, B.D., Gill, S.S., Grate, E., Andrews, R.E., Faust, R.M., Bulla, L.A. and Schaefer, C.H. (1984) An improved ELISA for the detection and quantification of the entomocidal paraporal crystal proteins of *Bacillus thuringinsis* ssp. *kurstaki* and *israelensis. Journal of Applied Bacteriology* 57, 447–454.

Wing, K.D. and Hammock, B.D. (1979) Stereoselectivity of a radioimmunoassay for the insecticide S-bioallethrin. *Experientia* 35, 1619–1620.

Wittmann, C. and Hock, B. (1989) Improved enzyme-immunoassay for the detection of s-triazines in water samples. *Food and Agricultural Immunology* 1, 211–224.

Wittmann, C. and Hock, B. (1991) Development of an ELISA for the analysis of atrazine metabolites, deethylatrazine and deisopropylatrazine. *Journal of Agricultural and Food Chemistry*, 39, 1194–1200.

Wittmann, C. and Hock, B. (1993) Analysis of atrazine residues in food by an enzyme immunoassay. *Journal of Agricultural and Food Chemistry* 41, 1421–1425.

Wong, J.M., Li, Q.X., Hammock, B.D. and Seiber, J.N. (1991) Method for the analysis of 4-nitrophenol and parathion in soil using supercritical fluid extraction and immunoassay. *Journal of Agricultural and Food Chemistry* 39, 1802–1807.

Wong, R.B. and Ahmed, Z.H. (1992) Development of an enzyme-linked immunosorbent assay for imazaquin herbicide. *Journal of Agricultural and Food Chemistry* 40, 811–816.

Wraith, M.J., Hitchings, E.J., Cole, E.R., Cole, D., Woodbridge, A.P. and Roberts, T.R. (1986) Development of immunoassay methods for pyrethroid insecticides. *Sixth International Congress on Pesticide Chemistry*, IUPAC, Ottawa, Canada, Abstract 5C-10.

Measurement of Polysaccharide-degrading Enzymes in Plants Using Chromogenic and Colorimetric Substrates* 12

B.V. McCleary

Megazyme (Australia) Pty Ltd, 2/11 Ponderosa Parade,
Warriewood, NSW 2102 Australia.

12.1 Introduction

Enzymatic degradation of carbohydrates is of major significance in the industrial processing of cereals and fruits. In the production of beer, barley is germinated under well-defined conditions (malting) to induce maximum enzyme synthesis with minimum respiration of reserve carbohydrates. The grains are dried and then extracted with water under controlled conditions. The amylolytic enzymes synthesized during malting, as well as those present in the original barley, convert the starch reserves to fermentable sugars. Other enzymes act on the cell wall polysaccharides, mixed-linkage β-glucan and arabinoxylan, reducing the viscosity and thus aiding filtration, and reducing the possibility of subsequent precipitation of polymeric material (Bamforth, 1982). In baking, β-amylase and α-amylase give controlled degradation of starch to fermentable sugars so as to sustain yeast growth and gas production. Excess quantities of α-amylase in the flour result in excessive degradation of starch during baking which in turn gives a sticky crumb texture and subsequent problems with bread slicing. Juice yield from fruit pulp is significantly improved if cell-wall-degrading enzymes are used to destroy the three-dimensional structure and water-binding capacity of the pectic polysaccharide components of the cell walls. Problems of routine and reliable assay of carbohydrate-degrading enzymes in the presence of high

*Published in part in McCleary, B.V. (1991) *Chemistry in Australia*, pp. 398–401. Royal Australian Chemical Institute, Melbourne.

levels of sugar compounds are experienced with such industrial processes.

Enzyme activities present in cereal and fruit products, or added during processing, are ideally assayed using the natural substrate and under conditions which simulate processing conditions. Polysaccharide *endo*-hydrolases aid processing by catalysing a viscosity reduction, and the ideal assay format would follow the reduction in viscosity with a natural substrate (Hardie and Manners, 1974; Bathgate, 1979). However, viscosity measurements are tedious and require considerable skill. Another routine assay for polysaccharide-degrading enzymes is the measurement of the increase in the reducing sugar levels as the substrate is hydrolysed by the appropriate enzyme. The methods to estimate reducing sugars include dinitrosalycilic acid (DNSA) (Bailey, 1988), Nelson-Somogyi (Somogyi, 1960), *p*-hydroxybenzoic acid hydrazide (PAHBAH) (Lever, 1972) and ferricyanide procedures (Park and Johnson, 1949). However, with cereal and fruit extracts these procedures cannot be used to measure enzymatic activity because of the very high levels of reducing sugars present.

To overcome these problems, assays which exploit a specific reaction characteristic or a solubility property of a particular polysaccharide have been developed. For example, the reaction of starch with iodine (to give a purple colour), and the decrease in this colour as the starch is depolymerized by α-amylase, is commonly employed to measure the activity of this enzyme. The reaction of mixed-linkage β-glucan with Congo Red stain and Calcofluor have been used to assay mixed-linkage β-glucanase activity in malt (malt β-glucanase) (Martin and Bamforth, 1983).

12.2 Chromogenic Substrates

Many of the problems experienced in the assay of polysaccharide *endo*-hydrolases can be resolved by the use of chromogenic or dye-labelled substrates. Such substrates may be soluble (Babson *et al.*, 1970; McCleary, 1978) or rendered insoluble (Ceska *et al.*, 1969) through covalent crosslinks, and have the following major advantages:

1. They allow measurement of enzyme activity in extracts containing high levels of reducing sugars, e.g. the measurement of α-amylase, limit-dextrinase and malt β-glucanase in malted barley extracts.
2. They allow specific measurement of polysaccharide *endo*-hydrolases in the presence of high concentrations of *exo*-acting enzymes, e.g. the measurement of α-amylase in the presence of β-amylase in cereal flours, or the measurement of α-amylase in the presence of amyloglucosidase in industrial microbial enzyme preparations.
3. They form the basis of assays which are simple, quantitative and reproducible. In a Royal Australian Chemical Institute evaluation of azo-

barley glucan for the assay of malt β-glucanase, which involved 18 laboratories and five malt samples, inter-laboratory coefficient of variation values of less than 6% for each of the samples were obtained (Buch, 1986).

4. They relate directly to assays based on viscosity reduction and thus more accurately reflect the likely significance of a given concentration of a particular type of enzyme. This means that these assays can be used to compare directly the activity of the same enzyme from different sources (i.e. different microbial sources) in a particular industrial application. One such example of this is a comparison of the different β-glucanase enzymes which are used in the brewing industry to destroy the viscosity of β-glucan. Bacterial and fungal β-glucanases are employed, but for a given activity based on a reducing-sugar assay, bacterial enzymes are far more effective in viscosity reduction than are the fungal enzymes and this is reflected in assays based on the use of soluble-dye-labelled barley β-glucan (azo-barley glucan) (McCleary and Shameer, 1987). Similar observations have been made in the evaluation of β-glucanases as supplements in chicken-feed diets (Rotter *et al.*, 1990) containing high levels of barley (and thus of β-glucan).

5. They can be used in assays over a wide range of temperature and pH conditions.

The single major limitation with these substrates is that the chemistry of dyeing of the polysaccharides cannot be accurately controlled, thus each production batch of the substrate must be standardized with a particular enzyme before use.

12.2.1 Soluble chromogenic substrates

Soluble chromogenic substrates are produced by the controlled dyeing of highly purified soluble polysaccharides. Dyed polysaccharide is usually separated from unreacted dye by precipitation or by gel permeation chromatography. Assay formats employing these substrates generally involve the incubation of an aliquot of soluble dyed substrate with enzyme solution under defined conditions of pH and temperature, with termination of the reaction and precipitation of non-depolymerized substrate by addition of an organic solvent or an organic solvent-salt solution (Friend and Chang, 1982) (Fig. 12.1). Precipitated material is removed by centrifugation and the colour in the supernatant solution measured. This colour can be directly related to the amount of enzyme in the assay mixture by reference to a standard curve.

The major requirements in the production and use of soluble chromogenic substrates are the following:

1. Availability of highly purified, soluble polysaccharides of moderate viscosity (i.e. a viscosity which allows preparation of solutions of at least 1% (w/v) of polysaccharide in water).

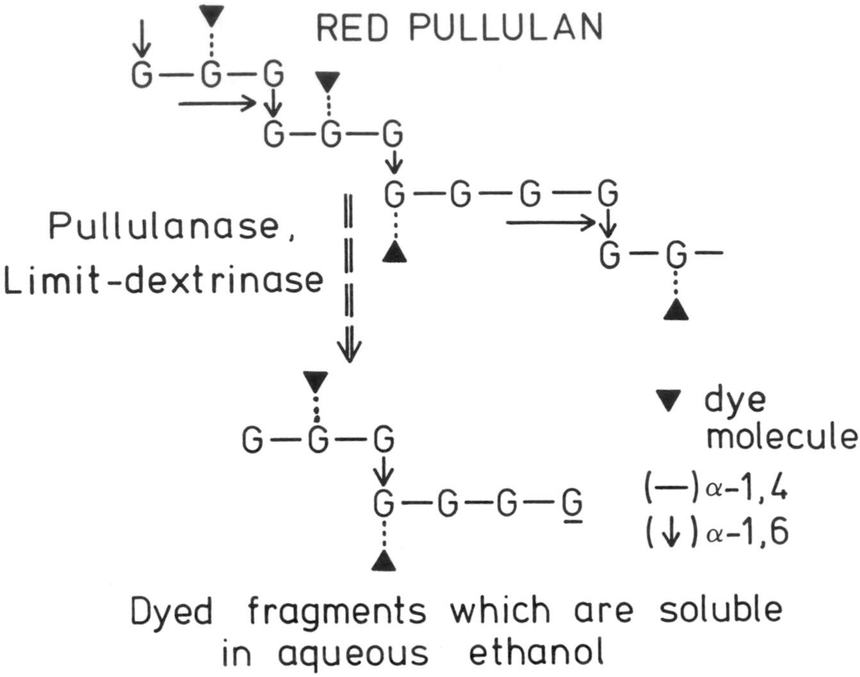

RED PULLULAN

G—G̈—G
G—G̈—G
G—G—G—G
G—G—
Pullulanase,
Limit-dextrinase

G—G̈—G
G—G—G—G̲

▼ dye
  molecule
(—)α-1,4
(↓)α-1,6

Dyed fragments which are soluble in aqueous ethanol

Fig. 12.1. Assay of limit dextrinase using dyed pullulan.

2. Physical, chemical and microbiological stability of the polysaccharide in solution over extended periods of storage.

3. A simple assay format and an effective procedure to quantitatively separate hydrolysed and non-hydrolysed dyed polysaccharide fragments.

4. Adequate dyeing of polysaccharide substrate to give linear, or near-linear, release of dye-labelled fragments over the absorbance range 0.1–1.0 absorbance units.

5. A final dye-labelled polysaccharide substrate which is very susceptible to enzyme attack.

12.2.2 *Insoluble chromogenic substrates*

The substrates referred to here as insoluble chromogenic substrates are those which have been produced from soluble polysaccharides or dye-labelled, soluble polysaccharides and rendered insoluble by covalent crosslinking with a particular crosslinking agent such as epichlorohydrin. These substrates are far superior to, and should not be confused with, substrates which are prepared by dyeing insoluble polysaccharides such as cellulose. These latter substrates are very resistant to enzyme attack and have little analytical use. With covalently crosslinked substrates, individual polysaccharide molecules are 'locked' into a three-dimensional conformation as a result of the crosslinking. The susceptibility of such substrates to enzyme attack is influenced by the concentration of the crosslinks, the size (or length) of the crosslinking agent, the concentration of the dye molecules attached to the polysaccharide and the degree of natural substitution of the native poly-saccharide. The major requirements for useful and effective insoluble (covalently crosslinked), dye-labelled substrates are very similar to those outlined for soluble dye-labelled substrates.

Insoluble substrates have the inherent disadvantage that they must be weighed accurately into each assay tube, whereas soluble substrates are readily and accurately dispensed with commercially available liquid handling equipment such as the Eppendorf Multipette®. The problem of dispensing the insoluble substrate can, however, be overcome by providing the substrate in a tablet form. Such a substrate is available commercially for the assay of α-amylase (Phadebas tablets; Pharmacia Diagnostics) and for the assay of a range of polysaccharide *endo*-hydrolases, including α-amylase, limit-dextrinase, xylanase and *endo*-arabinanase, from Megazyme (Australia) Pty Ltd.

In the preparation of such tablets there are several essential require-ments:

1. The tablet must be sufficiently rigid to cope with normal handling and shipping requirements.
2. On addition to buffer or buffered enzyme solution, it must disintegrate rapidly (i.e. within 20s) and preferably without mechanical agitation (which leaves a varying percentage of the substrate attached to the walls of the test tube).
3. Finally, bulking agents used in conjunction with the active component must not interfere in the assay, i.e. by acting as alternative substrates.

With the tablet substrates produced by Megazyme, the basic assay format involves the addition of the tablet to an aliquot (1ml) of pre-equilibrated and correctly buffered enzyme solution. The tablet is designed to completely disintegrate without agitation within 20s. The reaction is terminated after 10min by addition of 10ml of an alkaline solution (Trizma base, Sigma Chemical Co.) with stirring; the slurry is filtered and the absorbance of the filtrate (at 590nm) is measured. Enzyme activity is determined by reference to a standard curve.

The major advantage of dyed, crosslinked substrates in tablet form is that they are stable indefinitely and, unlike liquid substrates, there is little chance of back-contamination of the bulk substrate preparation. Another advantage of these substrates is that, if prepared correctly and used in a good assay format, they are considerably more sensitive than their soluble counterparts.

All reactions with the insoluble substrates are performed in aqueous solutions, whereas with soluble substrates, the reaction is generally terminated by the addition of an organic solvent to precipitate unhydrolysed material, and this format gives a different molecular size cut-off and a diminished sensitivity. For example, in the measurement of β-xylanase, assays based on the use of Xylazyme tablets (Megazyme Pty Ltd), containing crosslinked and dyed birchwood xylan, are about ten times more sensitive than assays employing soluble dyed arabinoxylan from wheat or soluble dyed oat-spelt xylan.

12.3 Colorimetric Substrates

Colorimetric substrates are based on the use of a defined oligosaccharide which is covalently linked to *p*-nitrophenol through the reducing D-glucosyl residue of the oligosaccharide. At present, such substrates are available for the assay of α-amylase, β-amylase and amyloglucosidase. The action of these enzymes and of limit-dextrinase on starch is shown in Fig. 12.2.

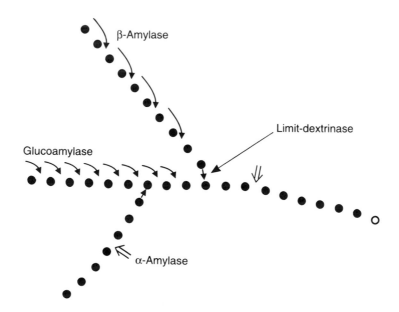

Fig. 12.2. The action of starch-degrading enzymes on amylopectin. ●●, α-1,4-linked D-glucosyl residues; ●⌒●, α-1,6-linked D-glucosyl residue; ○, terminal, reducing D-glucosyl residue.

12.3.1 α-Amylase

For the measurement of α-amylase, the substrate employed is 'non-reducing-end blocked *p*-nitrophenyl maltoheptaoside' (BPNPG7) (Blair, 1989). This substrate is used in conjunction with excess quantities of amyloglucosidase and α-glucosidase (which have no action on the native substrate due to the presence of the 'blocking group'). On hydrolysis of the oligosaccharide by *endo*-acting α-amylase, the excess quantities of amyloglucosidase and α-glucosidase which are present give essentially instantaneous and quantitative hydrolysis of the *p*-nitrophenyl maltosaccharide fragment to glucose and free *p*-nitrophenol (McCleary and Sheehan, 1987). The assay format is shown in Fig. 12.3. A further advantage of this type of substrate is that, with minor modifications, it can be used to assay for different forms of the same enzyme in a mixture. A particular application of this technology is the ability to measure relative proportions of fungal and cereal α-amylases in bread improver mixtures.

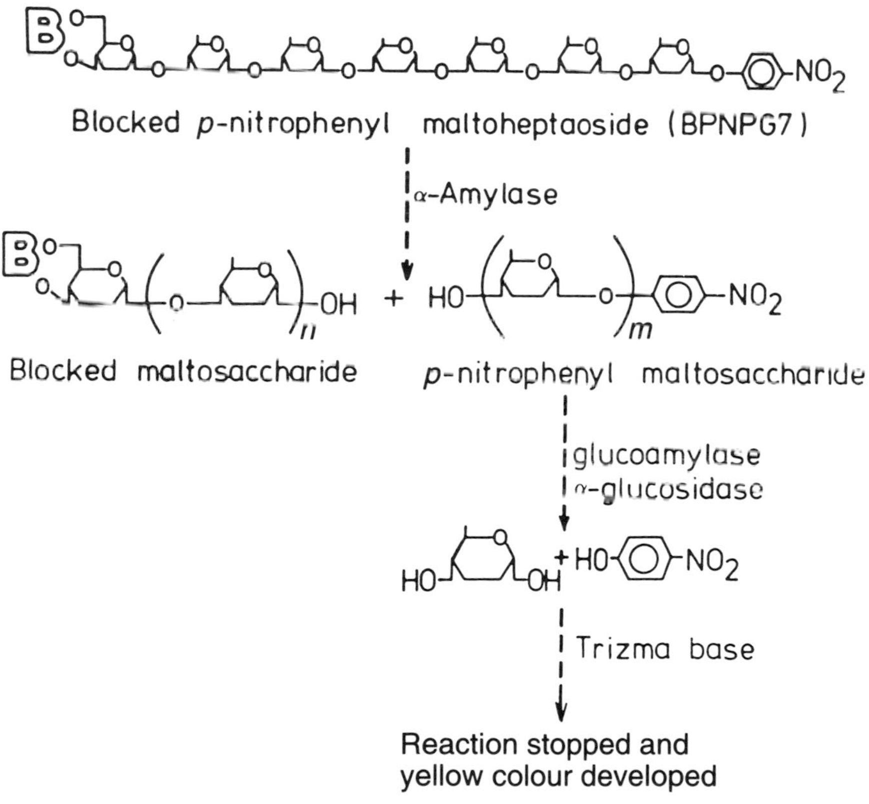

Blocked *p*-nitrophenyl maltoheptaoside (BPNPG7)

α-Amylase

Blocked maltosaccharide + *p*-nitrophenyl maltosaccharide

glucoamylase
α-glucosidase

Trizma base

Reaction stopped and
yellow colour developed

Fig. 12.3. Assay of α-amylase using blocked *p*-nitrophenyl maltoheptaoside.

Cereal α-amylase rapidly hydrolyses blocked *p*-nitrophenyl maltoheptaose, but hydrolysis of the pentasaccharide substrate (blocked *p*-nitrophenyl maltopentaose) is slow. In contrast, fungal α-amylase hydrolyses both substrates rapidly, as a consequence of its different subsite binding requirement at the active site of the enzyme (Megazyme data sheet). Fungal α-amylase is added to bread-making flours to assist the hydrolysis of gelatinized starch to fermentable sugars during the bread proofing stage which assists in improving loaf volume and keeping quality. The enzyme is inactivated at temperatures above 50°C, so it is effectively inactivated during the bread-baking process and thus cannot cause excessive degradation of starch at this stage of the process (which would result in the production of a sticky crumb, and subsequent bread-handling problems). Cereal α-amylases, being more thermostable, would have deleterious effects on bread quality if present in excess quantities in dough.

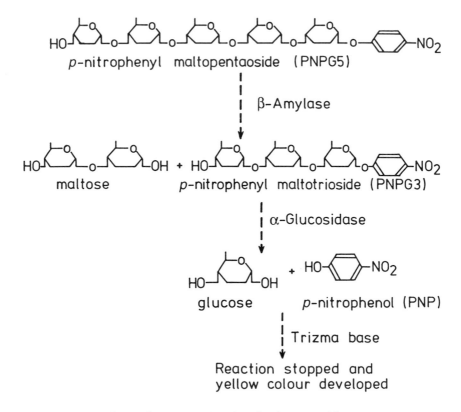

Fig. 12.4. Assay of β-amylase using *p*-nitrophenyl maltopentaoside.

12.3.2 β-Amylase

β-Amylase is an *exo*-acting enzyme which sequentially releases maltose units from the non-reducing terminus of amylose and amylopectin molecules. This enzyme can be effectively assayed in cereal extracts using *p*-nitrophenyl maltopentaose in the presence of yeast α-glucosidase (Fig. 12.4) (Mathewson and Seabourn, 1983; McCleary and Codd, 1989). This assay takes advantage of the fact that cereal α-amylases have very limited action on maltopentaose (oligosaccharides containing at least seven 1,4-α-linked D-glucosyl residues are required for rapid hydrolysis). The assay also takes advantage of the fact that yeast α-glucosidase rapidly hydrolyses maltosaccharides having two or three D-glucosyl residues, but acts extremely slowly on maltopentaose. Thus, when β-amylase removes the terminal maltosyl unit from *p*-nitrophenyl maltopentaose, the resultant *p*-nitrophenyl maltotriose is rapidly hydrolysed by the yeast α-glucosidase, releasing free *p*-nitrophenol.

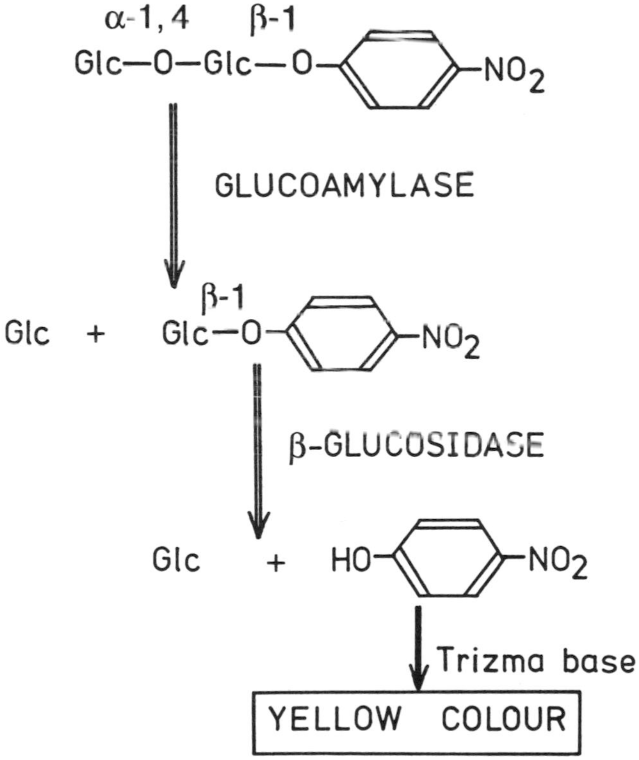

Fig. 12.5. Assay of amyloglucosidase using *p*-nitrophenyl β-maltoside.

12.3.3 Amyloglucosidase

Another *exo* enzyme active on starch fractions is the fungal enzyme amyloglucosidase which finds widespread industrial application. This enzyme is usually assayed with maltose as substrate, rather than starch, because the latter substrate is also rapidly hydrolysed by fungal α-amylase (which occurs at significant but varying levels in amyloglucosidase preparations). As an alternative to maltose, we have employed the substrate *p*-nitrophenyl β-maltoside in the presence of excess quantities of β-glucosidase (Fig. 12.5). (McCleary *et al.*, 1991). When the terminal α-linked D-glucosyl residue is removed by amyloglucosidase, the resultant *p*-nitrophenyl β-glucoside is instantaneously cleaved to glucose and free *p*-nitrophenol by the β-glucosidase. This assay format allows the measurement of amyloglycosidase in 10 min instead of 60 min required with conventional assay procedures.

12.4 Substrate Preparation and Some Applications of Chromogenic and Colorimetric Substrates

12.4.1 α-Amylase

Because of its industrial significance in the brewing and baking industries, numerous methods have been developed for the assay of α-amylase.

Colorimetric substrates

α-Amylase assay procedures employing the colorimetric substrate end-blocked *p*-nitrophenyl maltoheptaose (McCleary and Sheehan, 1987; Sheehan and McCleary, 1988) have many advantages over most other α-amylase assay formats. The substrate is available commercially in a ready-to-use form as Ceralpha α-amylase assay reagent (Megazyme Pty Ltd). The assay simply involves the incubation of an aliquot (0.2 ml) of enzyme preparation with 0.2 ml of the substrate mixture for 10 min at 40°C, termination of the reaction and colour formation by the addition of an alkaline solution (1%, w/v, Trizma base), and measurement of absorbance at 410 nm. In the clinical diagnostics field, this assay format is rapidly replacing most other procedures. However, in some laboratory situations (e.g. in flour mills, in industrial enzyme production facilities and in factories involved in enzyme blending) problems of contamination of the stock substrate solution with trace quantities of α-amylase (possibly from flour dust or aerosols) may be experienced.

Chromogenic substrates

Chromogenic, crosslinked substrates form the basis of standard procedures for the assay of α-amylase in Australia (Barnes and Blakeney, 1974), USA (AACC Method 22–06) and Europe (EBC Method 4.12.3). In Europe (E.B.C. Method 4.12.3) and Australia, Phadebas tablets (Pharmacia Diagnostics AG) are used, whereas in the USA (AACC Method 22–06), special unbuffered Amylochrome tablets (Roche Diagnostics) were employed. These latter tablets are no longer available but can be substituted by Amylazyme tablets (Megazyme). In AACC Method 22–06, the test tablet is added directly to a flour–buffer slurry with filtration after a defined incubation period. In the Australian standard method (Barnes and Blakeney, 1974), the flour is first extracted with buffer and an aliquot of this extract is incubated with the tablet under defined conditions.

Two quite distinct assay formats have been developed for measurement of α-amylase using Amylazyme tablets. With preparations containing high levels of enzyme activity, Format 1 is employed. In this procedure, an aliquot (1.0 ml) of suitably diluted extract is pre-equilibrated at 40°C for 5 min. An

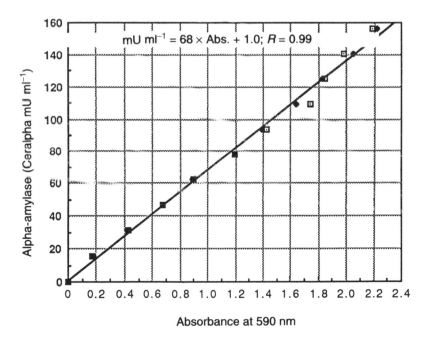

Fig. 12.6. Amylazyme (Lot 11201) standard curve for malt flour α-amylase with assay Format 1. Amylazyme absorbance values (at 590 nm) are converted via a standard curve to α-amylase activity in Ceralpha milliunits (mU) ml^{-1} (or per assay).

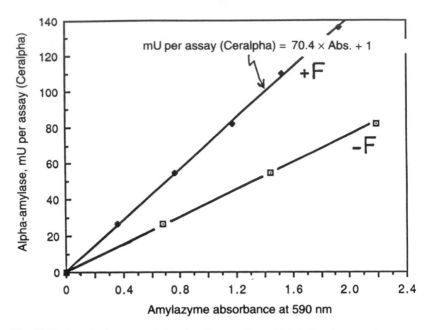

Fig. 12.7. Standard curves relating Amylazyme (Lot 11201) absorbance values to Ceralpha mU per assay (i.e. per 5 ml). An Amylazyme tablet was added to 5 ml of buffer containing Ceralpha 0–140 mU of α-amylase in the presence (+F) or absence (−F) of wheat flour (devoid of additional α-amylase). Stirring and incubation were performed according to assay Format 2.

Amylazyme tablet is added (without stirring) and this rapidly hydrates and absorbs most of the sample volume (i.e. 1.0 ml). After 10 min the reaction is terminated by the addition of Trizma base (a mild alkaline solution) and the slurry is filtered. The absorbance of the filtrate is measured and converted to α-amylase activity via a standard curve (Fig. 12.6). A different assay format (Assay Format 2) is employed for flour samples containing very low levels of α-amylase, e.g. flours from weather-damaged wheat. In this procedure, flour (0.5 g) is suspended in buffer at 55°C and equilibrated with continual stirring. After 5 min, an Amylazyme tablet is added to the slurry and stirring is continued for a further 5 min (exactly). The reaction is terminated with Trizma base and the slurry is filtered. The absorbance of the filtrate is measured at 590 nm. This assay format is about ten times more sensitive than the Ceralpha method and allows the accurate measurement of α-amylase in cereal flour with falling number values as high as 500. Under identical assay conditions, the Amylazyme test tablets are about two to three times more sensitive than Phadebas tablets (Pharmacia Diagnostics AG).

Extraction of α-amylase from cereal flours is likely to be affected by several factors including: the time required to wet the sample; the nature of the buffer employed; affinity binding of α-amylase to starch granules and, possibly, physical entrapment of α-amylase within starch granules. Activity values will be influenced by each of the above factors as well as by the level of starch in the assay mixture (which may act either as an alternative substrate or may physically bind the α-amylase and potentially render it inactive towards the added substrate). Several of these problems are experienced in the direct assay of α-amylase in flour slurries (Format 2). The effect of wheat flour on the measured α-amylase activity using Amylazyme tablets is shown in Fig. 12.7. Enzyme solution (5 ml) was incubated with an Amylazyme tablet in the absence or presence of wheat flour (0.5 g; essentially devoid of endogenous α-amylase). It is apparent that addition of flour to the incubation mixture (+F) results in a reduced rate of hydrolysis of the Amylazyme substrate (reduced rate of increase in absorbance at 590 nm). This reduction in rate is probably due to a combination of affinity binding of α-amylase to granular starch as well as to substrate competition effects (by soluble starch in the flour sample). Thus to obtain an estimate of the absolute

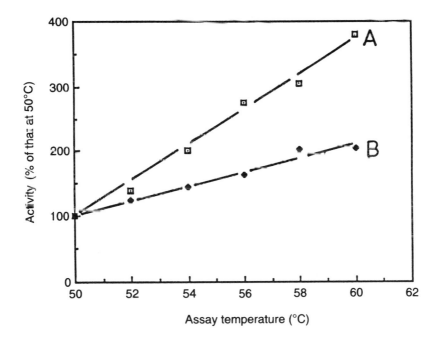

Fig. 12.8. Effect of incubation temperature on Amylazyme values using assay Format 2 with barley α-amylase. Incubations were performed on an aliquot (5 ml) of enzyme extract (B) or a slurry (5 ml) of whole milled barley (A) for 10 min. Activity is presented as a percentage of that at 50°C.

level of α-amylase in flour samples, standard curves should be prepared in the presence of added flour.

A second interesting complication experienced in the measurement of α-amylase, according to assay Format 2, in flour slurries, is the effect of incubation temperature on measured activity. Temperature has a much greater than expected effect on the Amylazyme absorbance values for whole flour slurries; i.e. there is a 3- to 3.5-fold increase in measured activity in the temperature range 50–60°C (Fig. 12.8) (McCleary, 1993). This compares to an increase of just 1.5- to 2-fold for enzyme extracts. This phenomenon may be due to a physical entrapment of the enzyme within starch granules. Heating the granules to temperatures approaching those of starch gelatinization appears to allow the enzyme to diffuse from the granule into the solution, and thus have access to the Amylazyme tablet.

12.4.2 β-Xylanase

Xylanase (1,4-β-D-xylan xylanohydrolase; EC 3.2.1.8) is finding widespread application in chicken feed supplements, in bread improver mixtures and in research on the enzymatic bleaching of wood pulp (Nissen et al., 1992). Historically, xylanase activity has been measured with the DNS reducing sugar assay (Bailey, 1988) with a purified xylan from oat-spelts, larchwood or birchwood as substrate. These assays are limited by their lack of sensitivity and linearity and the fact that they cannot be used to assay activity in materials containing high levels of reducing sugars, e.g. chicken feeds. Alternative assay procedures involve the use of high viscosity wheat-flour arabinoxylan in viscometric assays or of soluble or insoluble, dye-labelled, xylan-based substrates. Commercially available soluble substrates include Azo-xylan (birchwood (McCleary, 1992a), beechwood (Sigma M5019) (Bieley et al., 1985) and oat (McCleary, 1992a)) and Azo-wheat arabinoxylan (McCleary, 1992a). These substrates are prepared by dyeing the poly-saccharide with Remazo Brilliant Blue. Of these substrates, those of most practical use are Azo-xylan (birchwood) and Azo-wheat arabinoxylan as they dissolve rapidly in water and are stable in solution (both chemically and physically) for extended periods at room temperature and 4°C and give a linear standard curve over the absorbance range of 0.1 to 1.0 absorbance units (590 nm).

The only commercially available insoluble (crosslinked) substrates for the measurement of endo-xylanase are AZCL-xylan (birchwood) and AZCL-wheat arabinoxylan which are supplied in powder form or as tablets (Xylazyme or Xylazyme AX). The powdered substrates are useful for incorporation in agar plates or gels for location of xylanolytic microbial cultures and xylanase activity in electrophoretic gels. Xylazyme and Xyla-zyme AX tablets are used in test-tube assays for the quantitative measure-ment of xylanase in microbial fermentation broths and in various feed materials (e.g. chicken feed pellets). Assays employing Xylazyme tablets have ten times the sensitivity of soluble chromogenic substrates (McCleary, 1992a) and about the same sensitivity as viscometric assays employing wheat arabinoxylan as substrate.

Xylanase and cellulase-type enzymes are added to wheat- and barley-based chicken and pig diets to depolymerize arabinoxylan and β-glucan. This results in a decrease in viscosity of the gut contents and an improved absorption of nutrients, with resulting increases in feed conversion ratios. In studies of enzymatic methods for use as predictors of *in vivo* response to enzyme supplementation of barley-based diets when fed to young chicks, Rotter *et al.* (1990) found that the most reliable assay was based on the use of Azo-barley glucan (dyed barley β-glucan). Similar studies have not, as yet, been performed on xylanase assay procedures. However, in preliminary studies we have shown that Xylazyme tablets have the required sensitivity (refer to the assay described in Fig. 12.9). With this assay it has been possible

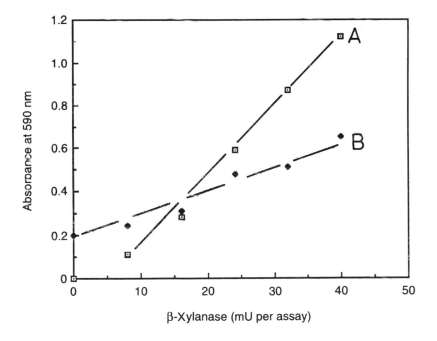

Fig. 12.9. Effect of feed material on the assay of *Trichoderma* sp. xylanase using Xylazyme tablets. Enzyme solution (2.0 ml, 0–50 mU) in 25 mM solium acetate buffer (pH 4.7) was incubated with stirring at 50°C in the presence (B) or absence (A) of feed material (0.2 g) for 10 min. The reaction was terminated after 10 min by the addition of 6 ml of 1% Trizma base. The solution was filtered and the absorbance at 590 nm measured. The absorbance of ~0.2 for the feed sample with no exogenous xylanase added in the assay procedure was due to the presence of xylanase in this feed sample.

to demonstrate some of the problems associated with the measurement of trace levels of enzyme activities in feed samples. The relationship between xylanase concentration (*Trichoderma* sp.) and absorbance, in the presence (B) and absence (A) of added feed sample (0.2g) is shown. It is evident that in the presence of the feed material, the slope of the standard curve is significantly reduced. This could be due either to non-specific binding of the enzyme to feed components or alternatively to the competitive action of alternative substrate material in the feed mixture. It is essential that the nature of this effect is identified and, if possible, that assay strategies to remove it be developed. The need for accurate and reliable assays for the measurement of xylanase, cellulase, β-glucanase and α-amylase in animal feeds pre- and post-pelleting is essential to allow the control of processing conditions and for regulatory purposes.

12.4.3 Limit-dextrinase

Limit-dextrinase, also called debranching enzyme or pullulanase, is the enzyme which cleaves the 1,6-α-branch linkages in the amylopectin fraction of starch. This enzyme is thought to be the limiting activity in malt flour which dictates the degree of conversion of starch to fermentable sugars. Yeast cells readily absorb glucose, maltose and maltotriose, but are unable to absorb branched maltosaccharides. It is thought that the fermentability of wort (the malt extract) may be directly related to the level of limit-dextrinase in the malt. Microbial pullulanases, in combination with β-amylase, are also finding increasing application in the starch-syrup industry in the production of high maltose syrups.

Measurement of limit-dextrinase in malt extracts using reducing sugar methods is difficult, time consuming and at best semi-quantitative. The enzyme has to be partially purified before it can be assayed. However, with dye-labelled pullulan as substrate, activity in malt extracts can be assayed directly (Serre and Lauriere, 1990; McCleary, 1992b). A soluble, red-dyed pullulan is finding widespread use in this area. A linear reaction curve in the range 0.1–1.4 absorbance units (at 510 nm) can be obtained, but the substrate is not as stable as would be preferred. An alternative substrate is dyed and crosslinked pullulan in tablet form (Limit-Dextrizyme tablets) which is both more stable and more sensitive than red pullulan (McCleary, 1992b).

12.4.4 Malt β-glucanase

Malt β-glucanase is a specific *endo* (1–3) (1–4)-β-D-glucanase which cleaves the 1,4-β-glycosidic bond of the 3-linked D-glucosyl units in mixed-linkage barley β-glucan. This enzyme is synthesized by barley during the malting process, but is very susceptible to thermal inactivation during kilning. Malt β-glucanase level is considered to be an important quality parameter of malt.

With judicious mashing temperature profiles, this enzyme can be employed to partially depolymerize the mixed linkage β-glucan extracted from the malt (which causes filtration problems). This enzyme has traditionally been assayed by a viscometric procedure using barley β-glucan of intermediate viscosity as the substrate (Bourne and Pierce, 1970). In recent years a method based on the use of the soluble Azo-barley glucan (McCleary and Shameer, 1987), has gained favour and is routinely used in many breweries and malt houses and by industrial enzyme manufacturers. Dye-labelled, crosslinked barley β-glucan in tablet form is now available (Megazyme). This substrate is more convenient to use than the soluble substrate, Azo-barley glucan, but inter-laboratory evaluations have shown that it is not quite as reliable.

12.4.5 Endo-*arabinanase*

To obtain maximum juice extraction from apple and pear pulp, it is common practice to employ pectinase enzymes to destroy the gelatinous, water-binding properties of the pulp pectic polysaccharides. These pectic polysaccharides have a modified polygalacturonic acid backbone with highly substituted regions to which 1,4-β-D-galactans and α-L-arabinans (or L-arabans) are attached (Schols *et al.*, 1990). The arabinans have a 1,5-α-linked L-arabinan backbone to which single unit L-arabinosyl residues are linked α-1,3 and α-1,2. Pectinase enzyme mixtures are usually deficient in *endo*-1,5-α-L-arabinanase (EC 3.2.1.99) and thus high concentrations (about 3–5 mg ml^{-1}) of linear 1,5-α-L-arabinan can occur in the juice concentrate. This arabinan self-associates and crystallizes in the juice, leading to haze formation. The problem is most readily resolved by ensuring that adequate levels of *endo*-arabinanase (which depolymerizes the linear arabinan) are present in the enzyme preparation employed. However, this has been complicated by the lack of a specific assay procedure for *endo*-arabinanase and even the non-availability of the basic polysaccharide substrate. This problem was partly resolved by major enzyme manufacturers by removing the linear arabinan from 'hazy' apple or pear juice concentrates by filtration, and using the arabinan (after purification) as a substrate for *endo*-arabinanase.

An alternative solution, and that adopted by us (McCleary, 1989), was to purify arabinan from sugarbeet pulp and treat this with α-L-arabinofuranosidase (EC 3.2.1.55) to remove all the 1,3- and 1,2-α-linked arabinofuranosyl residues, leaving linear 1,5-α-L-arabinan (Fig. 12.10). This polysaccharide still contains a small percentage of galacturonic acid, galactose and rhamnose (6%, 4% and 2%, respectively), but is resistant to attack by polygalacturonanase and *endo*-1,4-β-D-galactanase (EC 3.2.1.89). This 'linear' or 'debranched' arabinan was dyed with Reactive Red 120 to produce a soluble, red-dyed 1,5-α-L-arabinan, or was dyed and crosslinked to produce AZCL-arabinan (a black-coloured, crosslinked and

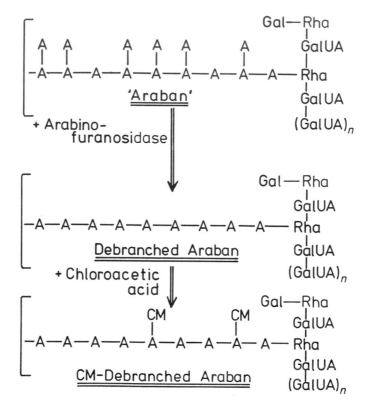

Fig. 12.10. Preparation of debranched araban for use in *endo*-arabinase assays.

debranched arabinan). The latter substrate has been incorporated into tablets (Arabinazyme).

Arabinazyme tablets form the basis of a simple, very specific and highly sensitive assay for the measurement of *endo*-arabinanase in crude enzyme mixtures. This substrate is in routine use in several major enzyme manufacturing companies.

12.4.6 *Cellulase (endo-1,4-β-D-glucanase)*

Cellulase (EC 3.2.1.4) catalyses the *endo*-depolymerization of cellulose (1,4-β-D-glucan). Cellulose, and dyed celluloses, are highly insoluble. Solubility and thus enzyme susceptibility can be greatly increased by chemical modifications which reduce the degree of self-association between cellulose chains, e.g. carboxymethylation or hydroxyethylation. However, the degree of modification of the glucan chains must be sufficient to produce solubility without interfering with the ability of cellulase to bind to, and hydrolyse, the polymer.

A range of chemically modified celluloses are commercially available. Of these, carboxymethylcellulose (CMC) has the greatest solubility with the minimum degree of substitution. CMC with a degree of substitution of 0.7 is completely soluble, but with this degree of substitution it is resistant to cellulase hydrolysis. CMC of degree of substitution 0.4 is only partially hydrated in water. This substrate is readily hydrolysed by cellulases, and RBB-CMC (Azo-CMC) (McCleary, 1980) produced on the dyeing of this with Remazo Brilliant Blue is a useful substrate for the specific assay of *endo*-cellulase in crude microbial mixtures, with a linear standard curve in the absorbance range 0.1–1.4 absorbance units. The substrate is highly specific for *endo*-1,4-β-glucanase (cellulase) and is totally resistant to interference by β-glucosidase (microbial or plant). Assays employing Azo-CMC (Megazyme Pty Ltd) are approximately 50 times more sensitive than those employing Cellulose Azure (cellulose dyed with Remazo Brilliant Blue) (Sigma C8647) and about 15 times more sensitive than TNP-cellulose [O-(2,4,6-trinitrophenyl) carboxymethyl cellulose] (an insoluble dyed substrate) (Huang and Tang, 1976).

High sensitivity, insoluble dyed substrates can be prepared by dyeing and crosslinking soluble cellulose derivatives. In this case, hydroxyethyl propyl cellulose (a neutral polysaccharide) was used because the carboxymethyl groups in CMC interfere with the crosslinking reaction. Alternatively, a cellulase substrate can be prepared by dyeing and crosslinking xyloglucan from tamarind seed. Xyloglucan has a 1,4-β-D-glucan backbone which is partially and regularly substituted by D-xylose residues linked α-1,6. Some of the xylose residues are further substituted. AZCL-HE-Cellulose is readily hydrolysed by a range of cellulase enzymes, whereas AZCL-Xyloglucan (tamarind) is attacked by only selected cellulases. *Trichoderma* sp. *endo*-cellulases act on AZCL-Xyloglucan, but *Aspergillus niger* cellulases do not. In tablet form, AZCL-HE-Cellulose (Cellazyme C) is useful for the routine assay of cellulase in microbial enzyme preparations (Megazyme data sheet EZC 7/92). The standard curve is relatively linear in the absorbance range 0.2–2.6 absorbance units, and the sensitivity of the assay is twice that based on Azo-CMC.

12.4.7 β-Mannanase

β-Mannanase (EC 3.2.1.78) catalyses *endo*-hydrolysis of 1,4-β-D-mannans, galactomannans and glucomannans. β-Mannanase activity in mixtures can be specifically assayed viscometrically using carob galactomannan (23% D-galactose). In most cases glucomannan has been employed as substrate, but this substrate is also depolymerized by cellulase (endo-1,4-β-glucanase). Reducing sugar methods employing 1,4-β-D-mannan and carob galactomannan as substrate have been described, but with these assay formats it is necessary to assay separately for β-mannosidase (EC 3.2.1.25) and/or α-galactosidase (EC 3.2.1.22).

An alternative assay for β-mannanase employs carob galactomannan dyed with Remazol Brilliant Blue as substrate (McCleary, 1978). The assay system measures the hydrolysis of RBB-carob galactomannan in terms of the rate of release of fractions soluble in 66% (v/v) aqueous ethanol. The rate of release of dyed fragments on hydrolysis of RBB-carob galactomannan is a function of the dye to anhydrohexose ratio. Maximal sensitivity in the assay system is obtained with a dye to anhydrohexose ratio in RBB-carob galactomannan of 1 : 15 to 1 : 50. Standard curves for the conversion of absorbance values (at 590nm) to enzyme units (on carob galactomannan at 40°C) for RBB-mannan, RBB-carob galactomannan and RBB-guar galacto-mannan are shown in Fig. 12.11. The low rate of hydrolysis of RBB-mannan is due to the highly insoluble nature of this substrate. The low relative rate of hydrolysis of RBB-guar galactomannan (a soluble substrate) is due to a high degree of D-galactose substitution (38%) of this galactomannan.

In the use of RBB-carob galactomannan to measure β-mannanase in crude enzyme mixtures, an enzyme which could potentially interfere is α-galactosidase. There are two possible mechanisms by which α-galactosidase might increase the rate of release of dyed fragments from RBB-carob galactomannan: the enzyme either might release galactosyl units dyed with RBB or, by releasing galactosyl residues, might make the substrate more susceptible to β-mannanase hydrolysis (McCleary, 1978). Figure 12.12 shows the effect that the release of D-galactose (by purified lucerne

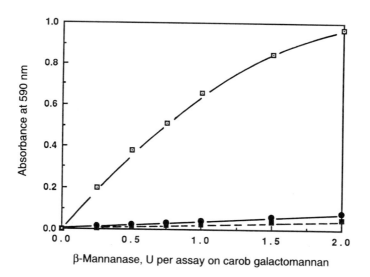

Fig. 12.11. Standard curves relating β-D-mannanase enzyme units on carob galactomannan to absorbance increase (at 590nm) on hydrolysis of RBB-carob galactomannan (1 : 50) □; RBB-guar galactomannan (1 : 50) ●; and RBB-mannan (1 : 35) ■.

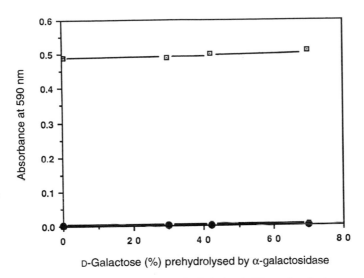

Fig. 12.12. Release of dyed fragments on hydrolysis of RBB-carob galactomannan by α-D-galactosidase (●); and on hydrolysis of α-D-galactosidase-prehydrolysed substrate by β-D-mannanase (□).

α-galactosidase A) from RBB-carob galactomannan (1 : 50) has on absorbance increase at 590 nm, and the effect it has on the susceptibility of the substrate to β-mannanases hydrolysis. Since 70% of the D-galactosyl residues were removed with no release of ethanol-soluble, dyed fragments, it can be concluded that α-galactosidase is unable to remove D-galactosyl residues substituted with Remazo Brilliant Blue dye. Furthermore, removal of up to 70% of the D-galactosyl residues gave only a very slight (~4%) increase in the susceptibility of the substrate to β-mannanase hydrolysis. In agreement with this, Marshall (1970) has shown that although the *exo*-hydrolase, amyloglucosidase, can remove D-glucose from Cibacron-Blue-dyed amylose, it cannot concurrently release Cibacron-Blue-dyed fragments.

Thus, RBB-carob galactomannan forms the basis of a simple, reliable, relatively sensitive and highly specific assay for β-mannanase. Alternative substrates for the assay of β-mannanase have been prepared by dyeing and crosslinking carob galactomannan to produce AZCL-galactomannan. This substrate in tablet form (β-Mannazyme) is currently under evaluation.

12.4.8 Endo-1,4-β-D-galactanase

Endo-galactanase (EC 3.2.1.89) is one of several enzymes involved in the depolymerization of pectic polysaccharides. Galactan in the pectic molecule is attached to rhamnosyl residues in the highly branched ('hairy') rhamnogalacturonosyl regions. The contribution of the galactan chains to the physiochemical properties or suspensions of solutions of pectin is not known; however, it would appear that galactose substitution interferes with hydrolysis of the rhamnogalacturonan regions of pectins by the enzyme rhamnogalacturonanase (Schols *et al.*, 1990).

Endo-galactanase is usually assayed with a partially purified galactan from potato fibre with a reducing sugar method (Rombouts *et al.*, 1988). However, galactan also contains high levels of arabinofuranosyl residues and thus can be hydrolysed by endo-1,5-α-L-arabinanase and arabinofuranosidase.

AZCL-Galactan (dyed, crosslinked) is a sensitive and specific substrate for *endo*-galactanase. The source of the galactan employed can be either potato fibre or lupin-seed fibre. The galactan from lupin-seed fibre contains 12% arabinose, but because this is present as short chains, it is not susceptible to hydrolysis by endo-arabinanase. The sensitivity and specificity of AZCL-galactan (lupin) is not altered by pretreatment of the galactan with a mixture of *endo*-arabinanase and arabinofuranosidase (to reduce the arabinose content from 12% to 2%) before dyeing and crosslinking (B.V. McCleary, unpublished). This finding is consistent with the arabinan existing as multiple, short chains.

12.4.9 Endo-*1,3-β-glucanase*

Endo-1,3-β-glucanase (laminarinase; EC 3.2.1.39) is one of the enzymes involved in degradation of fungal cell-wall material. It has been assayed with laminaran, pachyman and curdlan, but a more convenient substrate is carboxymethyl pachyman (which is soluble) (Clarke and Stone, 1962). Endo-1,3-β-glucanases can also be assayed with AZCL-curdlan or -pachyman (Megazyme data sheet). The latter substrate, available in tablet form as 1,3-beta-glucazyme substrate, is extremely sensitive (assay range 4–24 milliunits (mU) of 1,3-β-glucanase/assay) and highly specific for endo-1,3-β-glucanase. Exo-1,3-β-glucanase (EC 3.2.1.58) from *Trichoderma* sp. is unable to release any soluble dyed fragments from this substrate.

12.4.10 Endo-*polygalacturonanase*

A dyed substrate for the measurement of *endo*-pectinase (*endo*-polygalacturonanase) was prepared from polygalacturonic acid by coupling the dye N{1-[4-(3,6-disulpho-1-naphthyl)-azo] naphthyl}ethylenediamine (DISANED) using a water-soluble carbodimide (Friend and Chang, 1982). Pectin methylesterase does not release measurable levels of DISANED from the substrate.

Conclusion

With continued pressure on cereal and fruit processors for increased throughput, improved yields and simpler processing formats, the efficient and effective exploitation of both endogenous and industrial enzymes is imperative. This is only possible if rapid, reliable and specific enzyme assay procedures are available. Many of the procedures currently in use in the cereals and industrial microbiology industries are based on outdated and tedious technologies. There is an immediate requirement for further research, development and education in this area.

It is interesting to note that with the advent of genetic engineering, the technology available for the large-scale production of specific enzymes has actually outpaced the development of technologies for the accurate and reliable measurement of the enzyme activity.

References

American Association of Cereal Chemists (1988) *AACC Approved Methods*, Method 22-06, St Paul, MN.

Babson, A.L., Tenney, S.A. and Megrew, R.E. (1970) New amylase substrate and assay procedure. *Clinical Chemistry* 16, 39–43.

Bailey, M.J. (1988) A note on the use of dinitrosalicylic acid for determining the products of enzymatic reactions. *Applied Microbiology and Biotechnology* 29, 494–496.

Bamforth, C.W. (1982) Barley β-glucans; their role in malting and brewing. *Brewers' Digest* 57, 22–35.

Barnes, W.C. and Blakeney, A.B. (1974) Determination of cereal alpha-amylase using a commercially available dye labelled substrate. *Starch* 26, 193–197.

Bathgate, G.N. (1979) The determination of endo-β-glucanase activity in malt. *Journal of the Institute of Brewing* 85, 92–94.

Bieley, P., Mislovičová, D. and Toman, R. (1985) Soluble chromogenic substrates for the assay of endo-1,4-β-xylanases and endo-1,4-β-glucanases. *Analytical Biochemistry* 144, 142–146.

Blair, H.E. (1989) Alpha-amylase assay. US Patent No. 4,794,078.

Bourne, D.T. and Pierce, J.S. (1970) β-Glucan and β-glucanase in brewing. *Journal of the Institute of Brewing* 76, 328–338.

Buch, G.J. (1986) Malt β-glucanase: a collaborative test on a new rapid assay. *Journal of the Institute of Brewing* 92, 513–514.

Ceska, M., Hultman, E. and Ingelman, B.G.-A. (1969) A new method for determination of α-amylase. *Experientia* 15, 555–556.

Ceska, M. (1971) Hydrolysis of a water-insoluble substrate incorporated into solidified medium by enzyme α-amylase contained in normal human urine. *Clinica Chemica Acta* 26, 437–444.

Clarke, A.E. and Stone, B.A. (1962) β-1,3-Glucan hydrolases from the grape vine (*Vitis vinifera*) and other plants. *Phytochemistry* 1, 175–188.

European Brewing Convention (1987) *Analytica-EBC*, 4th edn. Method 4.12.3.

Friend, D.R. and Chang, G.W. (1982) Simple dye release assay for determining endopectinase activity. *Journal of Agriculture and Food Chemistry* 30, 982–985.

Hardie, D.G. and Manners, D.J. (1974) A viscometric assay for pullulanase-type, debranching enzymes. *Carbohydrate Research* 36, 207–210.

Huang, J.S. and Tang, J. (1976) Sensitive assay for cellulase and dextranase. *Analytical Biochemistry* 73, 369–377.

Lever, M. (1972) A new reaction for colorimetric determination of carbohydrates. *Analytical Biochemistry* 47, 273–279.

Marshall, J.J. (1970) Action of amylolytic enzymes on a chromogenic substrate. *Analytical Biochemistry* 37, 466–470.

Martin, H.L. and Bamforth, C.W. (1983) Application of a radial diffusion assay for the measurement of β-glucanase in malt. *Journal of the Institute of Brewing* 89, 34–37.

Mathewson, P.R. and Seabourn, B.W. (1983) A new procedure for specific determination of β-amylase in cereals. *Journal of Agriculture and Food Chemistry* 31, 1322–1326.

McCleary, B.V. (1978) A simple assay procedure for β-D-mannanase. *Carbohydrate Research* 67, 213–221.

McCleary, B.V. (1980) New chromogenic substrates for the assay of α-amylase and β-1,4-glucanase. *Carbohydrate Research* 86, 97–104.

McCleary, B.V. (1989) Novel and selective substrates for the array of endo-Arabinase. In: Phillips, G.O., Wedlock, D.J. and Williams, P.A. (eds), *Gums and Stabilisers for the Food Industry*, vol. 5. IRL Press, Oxford, pp. 291–300.

McCleary, B.V. (1992a) Measurement of endo-1,4-β-D-xylanase. In: Visser, J., Beldman, B., Kusters-van-Someren, M.A. and Voragen, A.G.J. (eds), *Xylans and Xylanases. Progress in Biotechnology*, vol. 7. Elsevier, Amsterdam, pp. 161–170.

McCleary, B.V. (1992b) Measurement of the content of limit-dextrinase in cereal flours. *Carbohydrate Research* 227, 257–268.

McCleary, B.V. (1993) Measurement of α-amylase in weather damaged wheat and barley. *Chemistry in Australia* 60, 485.

McCleary, B.V. and Codd, R. (1989) Measurement of β-amylase in cereal flours and commercial enzyme preparations. *Journal of Cereal Science* 9, 17–33.

McCleary, B.V. and Shameer, I. (1987) Assay of malt β-glucanase using azo-barley glucan: an improved precipitant. *Journal of the Institute of Brewing* 93, 87–90.

McCleary, B.V. and Sheehan, H. (1987) Measurement of cereal α-amylase: a new assay procedure. *Journal of Cereal Science* 6, 237–251.

McCleary, B.V., Bouhet, F. and Driguez, H. (1991) Measurement of amyloglucosidase using *p*-nitrophenyl β-maltoside as substrate. *Biotechnology Techniques* 5, 255–258.

Nissen, A.M., Anker, L., Munk, N. and Krebs Lange, N. (1992) Xylanases for the pulp and paper industry. In: Visser, J., Beldman, B., Kusters-van Someren, M.A. and Voragen, A.G.J. (eds), *Xylans and Xylanases. Progress in Biotechnology*, vol. 7. Elsevier, Amsterdam, pp. 325–338.

Park, J.T. and Johnson, M.J. (1949) A submicrodetermination of glucose. *Journal of Biological Chemistry* 181, 149–151.

Rombouts, F.M., Voragen, A.G.J., Searle-van Leeuwen, M.F., Geraeds, C.J.M., Schols, H.A. and Pilnik, W. (1988) The arabinanases of *Aspergillis niger* – purification and characterisation of two α-L-arabinofuranosidases and an endo-

1,5-α-L-arabinanase. *Carbohydrate Polymers* 9, 25–47.

Rotter, B.A., Marquardt, R.R., Guenter, W. and Crow, G.H. (1990) Evaluation of three enzymic methods as predictors of *in vivo* response to enzyme supplementation of barley-based diets when fed to young chicks. *Journal of the Science of Food and Agriculture* 50, 19–27.

Schols, H.A., Geraeds, C.J.M., Searle-van Leeuwen, F., Kromelink, F.J.M. and Voragen, A.G.J. (1990) Rhamnogalacturonanase: a novel enzyme that degrades the hairy regions of pectins. *Carbohydrate Research* 206, 105–115.

Serre, L. and Lauriere, C. (1990) Specific assay of α-D-dextrin 6-glucanohydrolase using labelled pullulan. *Analytical Biochemistry* 186, 312–315.

Sheehan, H. and McCleary, B.V. (1988) A new procedure for the measurement of fungal and bacterial α-amylase. *Biotechnology Techniques* 2, 289–292.

Somogyi, M. (1960) Modifications of two methods for the assay of amylase. *Clinical Chemistry* 6, 23–35.

Isozyme Variation and Analysis in Agriculturally Important Plants

T. Konishi

Faculty of Agriculture, Kyushu University, Fukuoka 812, Japan.

13.1 Introduction

During the past three decades, isozymes have contributed to the study of systematics, evolutionary dynamics, population genetics, cytogenetics, breeding and the conservation of genetic resources in plants. In 473 plant species examined, an average of 16.5 isozyme loci per species were determined, approximately half of which were found to be polymorphic with 1.96 alleles per isozyme locus (Hamrick and Godt, 1990). In tomato (*Lycopersicon esculentum*), for example, most cultivars and land races have the same genotype for isozyme loci. Among distant relatives, however, isozyme differences can be expected at ten or more of the loci (Tanksley and Rick, 1980) and can be used in genetic studies. In interspecific hybrids between cultivated tomato and its wild species (*L. hirsutum*), 28 of the 32 isozyme loci detected were genetically mapped on 10 of the 12 chromosomes of *Lycopersicon* (Vallejos and Tanksley, 1983).

Using isozyme markers, Rick and Fobes (1974) first found a tight genetic linkage between acid phosphatase locus, *Aps–1*, and the locus, *Mi*, controlling nematode resistance in tomato. This finding was significant in stimulating the use of isozymes as diagnostic markers for plant improvement. Linkages between isozyme loci and economically important traits show that isozyme markers provide a powerful tool for studying the inheritance of both qualitative and quantitative traits.

This chapter will demonstrate ways in which isozyme technologies have already provided the means for tagging important genes in plant breeding as well as suggesting possible future uses of these techniques in plant breeding programmes.

13.2 Advantages of Isozymes in Plant Genetics and Breeding

In general, isozyme markers are superior to morphological markers for genetic analysis. The alleles at most isozyme loci are co-dominant, allowing heterozygotes to be distinguished from homozygotes. In addition, isozyme markers only rarely exhibit epistatic interactions, thus allowing classification of any number of such markers segregating simultaneously. Screening the isozyme banding patterns (zymograms) of plants has several advantages over new DNA-based molecular markers, such as restriction fragment length polymorphisms (RFLPs) and random amplified polymorphic DNAs (RAPDs). The equipment is relatively inexpensive, and it is often possible to screen large numbers of individuals at the seedling stage, using only small amounts of plant tissue. As the process is non-destructive, the desirable genotypes can be grown and their seeds harvested.

Helentjaris (1992) has emphasized that limitations on the numbers of informative isozyme loci within many species restricts their use. However, in many cases sufficient variation has been found and has maintained an interest in these approaches.

13.3 Concepts for Using Isozyme Markers in Plant Breeding

Isozymes have been widely used to screen the variability present in plant populations and to select desirable genotypes for plant improvement. Linkages between specific isozyme alleles and agronomically important genes have contributed to tracing simply inherited agronomic traits such as disease resistance. Where genetic linkage exists, an isozyme is much easier to assay than testing susceptibility to a pest, since the screening can be conducted at seedling stage and is unambiguous. However, to be useful as a genetic marker, a tight linkage must exist between the desired locus and isozyme marker.

Backcrossing is an important and established technique by which the resistance gene of the donor parent is transferred into a susceptible cultivar in plant breeding, and the backcross lines are essentially near-isogenic to the recurrent parent. A potential linkage between an isozyme allele and the introgressed gene is suggested when the near-isogenic lines and their donor parent possess the same allele at an isozyme locus. The other technique to reveal desirable linkages is to compare isozymes between a resistant parent and its derived lines carrying the resistance gene. When almost all the lines possess the same isozyme allele as the resistant parent, the isozyme allele must be tightly linked with the resistance gene.

Most traits of interest to plant breeders are complex in inheritance, and quantitative traits such as yield and most of its components are normally assumed to be polygenic in nature mainly because the phenotypic expressions of these traits form continuous distributions. Genetic markers can be analysed for linkage with quantitative traits and the basic concepts for this procedure were first proposed by Sax (1923). He demonstrated an association of seed size (a quantitatively inherited character) with seedcoat pigmentation (a discrete monogenic character) in *Phaseolus vulgaris*. To use the isozyme alleles as genetic markers for identifying and localizing the quantitative trait loci (QTLs), a genetic linkage map comprised of isozyme loci uniformly distributed throughout the genome is required. In maize, for example, a total of 18 isozyme loci and two easily scorable morphological marker loci are located on nine of the ten maize chromosomes and are distributed in 40 to 45% of the regions of the genome within about 20 cM (Stuber, 1991). However, it is difficult to saturate the genome completely and uniformly by isozyme markers and for analysing QTLs isozyme markers need to be combined with other molecular markers such as RFLPs and random amplified polymorphic DNAs (RAPDs).

13.4 Isozyme Marker Associations with Qualitative Traits

13.4.1 Disease resistance

Useful resistance genes are often found in varieties of the same species, but when adequate resistance does not appear to exist in cultivated species, a resistance gene needs to be introgressed from related species or genera. Breeding programmes for disease resistance can make use of biological assays based on artificial inoculation at the seedling stage. However, biological assays can be ambiguous particularly if multiple disease resistance genes are involved. The use of linked isozyme markers to resistance genes can therefore assist in the classification of resistance genotypes for breeding programmes.

Resistance to fungus

Fusarium oxysperum f. sp. *lycopersici*, which causes Fusarium wilt of tomato, is widespread in most tomato production regions, differentiating into races 1, 2 and 3. For a biological assay seedlings at the two to four true leaf stage can be inoculated, and scored at 10–14 days and approximately 1 month after inoculation. Using this assay to examine segregating populations, resistances to the three races were either controlled by the disease resistance gene *I-3* alone or by gene(s) tightly linked to it. Furthermore, tight linkage was detected between the isozyme locus *Got-2* and resistances to the races, and the map distance was estimated to be 2.5 cM between *I-3* and *Got-2* on chromosome 7 (Bournival *et al.*, 1989, 1990).

Bacterial pustule of soyabean, caused by *Xanthomonus campestris* pv. *glycines*, is one of the most prevalent bacterial diseases in southern and midwestern regions of the USA. The resistant *rxp* allele was identified in the cultivar CNS, and the locus was linked to MDH (malate dehydrogenase) locus with a recombination level of 15.18±3.81% (Palmer *et al.*, 1992).

Wild species related to crops are a potential source of genes for resistance to a range of diseases afflicting those crops (Harlan, 1976). A good example is that of barley, whose ancestral wild species, *Hordeum spontaneum*, is known to contain diverse resistance to important diseases in barley. Resistance to leaf rust (*Puccinia hordei*) in barley is controlled by single partially dominant genes. One of the resistance genes (*Rph10*) has been linked to the isozyme locus *Est2* on chromosome 3 (recombination value, 15±5%), and another gene (*Rph11*) has been linked with two isozyme loci, *Acp3* (7±2%) and *Dip2* (11±2%) (Feuerstein *et al.*, 1990). *Hordeum spontaneum* also carries resistance genes to leaf scald, caused by *Rhynchosporium secalis*. Genetic studies demonstrated linkage of these resistance genes to the *Acp2* locus (recombination values 7±3%, 18±4%) on chromosome 4, *Est5* locus (recombination value 15±4%) on chromosome 1, and *Dip1* locus (recombination value 29±5%) on chromosome 6 (Abbott *et al.*, 1992).

Strawbreaker foot rot (*Pseudocercosporella herportrichoides*) is the most important soilborne disease of autumn-sown wheat in northwestern USA. Although the cultivar 'Cappelle-Desprez' from France shows some tolerance, the highest level of resistance occurs in an alien grass species, *Aegilops ventricosa*. For screening resistance genotypes, biological assays based on conidial suspension sprays are tedious and not always reliable. McMillin *et al.* (1986) found a close association between the resistance and the endopeptidase allele (*EP-V1*) derived from *Ae. ventricosa*, indicating that the endopeptidase allele should be a useful marker for breeding programmes designed to incorporate strawbreaker foot rot resistance into commercial wheat cultivars.

Resistance to virus

Pea enation mosaic virus (PEMV) is an important pathogen in the garden pea, affecting many areas in the USA and western Europe. Resistance to PEMV is controlled by a single dominant gene located at the *En* locus. A close linkage between *Adh-1* and *En* has been detected, and with a recombination frequency of 4% may make the isozyme marker *Adh-1* a more reliable predictor of PEMV phenotype than the direct screening process using inoculum (Weeden and Provvidenti, 1988). Another close linkage in the garden pea was investigated between *Pgm-p* encoding for the plastid-specific form of phosphoglucomutase and *Mo* controlling resistance to bean yellow mosaic virus (BYMV). A map distance of approximately two recombinant units was determined for the *Pgm-p* and *Mo* linkage, suggesting that the isozyme locus can be used as a genetic marker for resistance to BYMV (Weeden *et al.*, 1984).

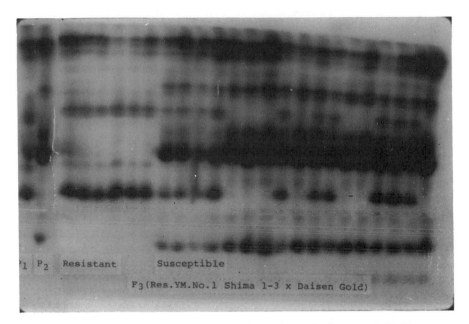

Fig. 13.1. Esterase zymogramme showing a linkage relationship between BaYMV resistance locus and esterase complex locus in barley (P_1: resistant RR parent, P_2: susceptible SS parent, and their F_3 progenies consisting of 6 RR, 10 RS and 7 SS).

Barley yellow mosaic virus (BaYMV) causes serious disease problems in East Asian malting barley and European winter barley. Genetically resistant varieties provide one of the few possibilities for preventing considerable yield losses caused by this soilborne virus. Therefore, extensive screening programmes have been conducted to find suitable genetic resources of resistance against BaYMV (e.g. Takahashi *et al.*, 1973; Friedt *et al.*, 1985; Proeseler and Lehmann, 1986; Kawada, 1991). Mokusekko 3, a Chinese landrace, is an important genetic resource for resistance to BaYMV, carrying a partial dominant resistance gene *Ym* and an additional gene with a slight resistance effect (Takahashi *et al.*, 1973). Using this genotype, large numbers of the resistant varieties and lines were developed in Japan. Almost all of them possessed the same esterase isozyme genotype as that of Mokusekko 3, suggesting that one of the resistance loci in Mokusekko 3 may be linked with the complex locus for esterase isozymes, *Est1*, *Est2* and *Est4*, on chromosome 3 (Konishi and Matsuura, 1987). Preliminary tests indicated that one of the resistance genes has been linked to the esterase complex locus with recombination values ranging from 1.26 to 5.0% (Konishi *et al.*, 1989). Further linkage analysis revealed that the resistance locus was closely linked to the esterase complex locus with 2.45±0.73% recombination (Konishi and

Kaiser, 1991) (Fig. 13.1). Because of difficulty of the artificial inoculation with BaYMV and irregular infection in the field, the isozyme marker provides a reasonable alternative for selecting resistant genotypes in barley breeding.

Resistance to barley mild mosaic virus (BaMMV) in German winter barley, a soilborne virus similar to BaYMV, is controlled by a recessive gene (*ym4*) on the long arm of chromosome 3 (Kaiser and Friedt, 1992). The resistance locus is closely linked with a DNA-based RFLP probe MWG10 (1.2 cM) near the distal end of the long arm of chromosome 3 (Graner and Bauer, 1993), as well as with the esterase complex locus (recombination value 4.7 cM). This indicates the esterase isozyme marker will provide a valuable screening for BaMMV resistance in barley.

13.4.2 Male sterility and self-incompatibility

Male sterility is often used for heterosis breeding programmes in self-pollinated crops, and is governed either by nuclear genes or by cytoplasmic genes. Since the nuclear male sterility is usually determined by a recessive gene, the male sterile stock is maintained by crossing male sterile plants with heterozygous fertile female plants. Identification of male sterile homozygotes is then best carried out by a closely linked gene that is easily assayed in segregating populations. In this case, isozyme markers are useful for the identification of the male sterile and heterozygous fertile genotypes in early growth stages.

In tomato, a genetic linkage between a nuclear male sterile locus, *ms-10*, was tested with two isozyme marker loci on the long arm of chromosome 2. The results indicated that the gene order was *Est-1*, *Prx-2* and *ms-10*, and that *ms-10* and *Prx-2* were approximately 1.5 cM apart, while the *Prx-2* and *Est-1* distance was 1.8 cM. The close linkage of *ms-10* and *Prx-2* suggested that *Prx-2* would provide a selectable marker for male sterility (Tanksley *et al.*, 1984). An association between *Pgm1* and *ms2* in soyabean was investigated in an F_2 population. The map distance of 18.73 cM was estimated and was not close enough for efficient, indirect, selection (Sneller *et al.*, 1992).

Another type of male sterility is that determined by the interaction of cytoplasmic and nuclear genes. The restoration of male fertility in cytoplasmic male sterile hybrids is accomplished by a restorer gene. In rapeseed (*Brassica napus*), a restorer gene for cytoplasmic male sterility was introduced from radish (*Raphanus sativus*) through intergeneric crosses, since no restorer was found in rapeseed. Co-segregation studies of isozyme markers and male fertility restoration showed that the restorer gene was linked with an isozyme marker, *Pgi-2* (recombination value was 0.25±0.02%). The tight genetic linkage relationship indicates that the isozyme marker is useful for distinguishing male fertile from male sterile plants, as well as for separating homozygous and heterozygous restored plants (Delourme and Eber, 1992).

Gametophytic self-incompatibility is a genetic mechanism that prevents individuals from effecting self-fertilization. *Lycopersicon peruvianum* is one of several self-incompatible species in the tomato genus. By interpreting differential segregation ratios from crosses between two different populations of *L. peruvianum*, the two isozymes *Idh-1* and *Prx-1* were found to be closely linked to a single self-incompatibility locus (*S*). The most likely gene order was *Idh-1* (2.2 cM), *Prx-1* (9.9 cM), *S* (Tanksley and Loaiza-Figueroa, 1985).

Rye (*Secale cereale*) is a normally self-incompatible cereal. Under controlled growth chamber conditions at 30°C during flowering time, self-pollinated seeds can be obtained in self-incompatible plants. In this way it is possible to carry out reciprocal crosses between homozygous and heterozygous individuals for *Prx7* alleles. Segregation at the *Prx7* locus in progenies of these crosses provides clear evidence of a close linkage between *Prx7* and one of the two incompatibility loci (0–2% recombination; Wricke and Wehling, 1985).

13.5 Isozyme Marker Associations with Quantitative Traits

13.5.1 Skewness of marker frequency in varieties and bred lines

Isozyme variation among varieties and hybrid populations provides useful evidence for suggesting that some associations may exist between isozyme markers and quantitative traits. In barley (*Hordeum vulgare*), for example, the frequency of genotypes consisting of allelic combinations at three esterase loci, *Est1*, *Est2* and *Est4*, forming a single complex locus, were monitored in the pedigree of Japanese two-rowed varieties. Two major genotypes for esterase isozymes were detected in the varieties derived from 'Golden Melon' (allelic combination designated *Ca*, *Dr* and *Nz* at the *Est1*, *Est2* and *Est 4* loci, abbreviated to *Ca-Dr-Nz*) and 'Prior' (*Pr-Fr-Su*). During the period 1956 to 1980, the frequency of the *Ca-Dr-Nz* varieties (which had been predominant in 1956) decreased rapidly. In contrast, varieties of the *Pr-Fr-Su* genotype, which were developed in different breeding stations, increased to occupy almost all the acreage of two-rowed barley cultivation in Japan, as illustrated in Fig. 13.2. This resulted from the indirect selection of *Pr-Fr-Su* genotypes in many crosses between *Ca-Dr-Nz* and *Pr-Fr-Su* genotypes (Konishi and Matsuura, 1987). A similar result was found for Japanese six-rowed naked barley. Crosses between Japanese six-rowed naked and foreign two-rowed covered barleys were made for breeding two-rowed naked varieties in Shikoku Agricultural Experiment Station of Japan. Forty-five of the 50 two-rowed naked lines examined possessed the same esterase genotype as that of the six-rowed naked parents adapted to barley cultivation in Japan (Konishi *et al.*, unpublished).

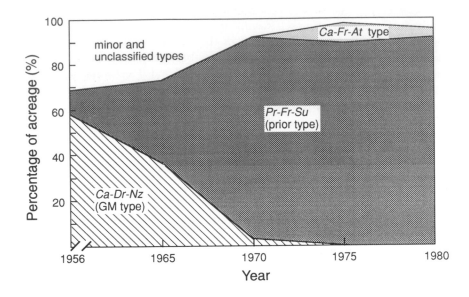

Fig. 13.2. Change in acreage occupied by two-rowed barley varieties with different esterase isozyme genotypes (expressed by percentage of total acreage of Japanese two-rowed barley).

Table 13.1. Esterase isozyme genotypes in European barley varieties derived from cross combinations of *Ca–Fr* × *Pr–Fr* parents.

Genotype Ca–Fr 11	Country (year of release)	Genotype Pr–Fr 2	Country (year of release)
Agio	DNK (1950)	Julia	NLD (1968)
Bine	DNK (1971)	Sultan	NLD (1966)
Carlsberg	DNK (1947)		
Flavina	SWE (1975)		
Foma	SWE (1961)		
Herta	SWE (1949)		
Proctor	GBR (1952)		
Rika	SWE (1951)		
Seger	SWE (1914)		
Tyra	DNK (1975)		
Zephyr	NLD (1965)		

$\chi^2 (1:1) = 6.23$.
DNK, Denmark; GBR, Great Britain; NLD, The Netherlands; SWE, Sweden.

Genotypic variation at two esterase loci (*Est1* and *Est2*) in European barley varieties was also investigated by Linde-Laursen *et al.* (1982). In this study the genotypic frequency of distinct esterase-complex alleles in 13 varieties derived from cross-combinations between parents carrying *Ca-Fr* and *Pr-Fr* genotypes was examined. It was found that 11 of the 13 varieties possessed the *Ca-Fr* genotype after breeder selections based on phenotype in different countries, as shown in Table 13.1 (Konishi *et al.*, unpublished). These results suggest that some QTLs for important traits in barley breeding may be linked with the esterase-complex locus located near the telomere of the long arm of chromosome 3.

13.5.2 *Genetic changes in populations from selection experiments*

In experimental barley populations subjected to natural selection, extensive studies by Allard and colleagues (summarized in Allard, 1988, 1992) documented continually increasing reproductive capacity over 50 generations. These changes were associated with highly significant changes in allele frequency at numerous marker loci. In barley populations of composite cross (CCII) grown at Davis, California, changes in frequencies of alleles at *Est1*, *Est2*, *Est3* and *Est4* loci and their allelic combinations were investigated. Although 13 of the 16 possible four-locus gametic types were present in the early generations, only one gametic type, 2112, increased steadily in frequency to 95% in F_{53}. On the other hand, when the same populations were grown at Bozeman, Montana, for 19 generations from F_{27} to F_{46}, the gametic type 2112 (the favoured type in Davis) declined rapidly, whereas gametic types 1221, 1112 and 1212 increased in frequency. In this connection, all commercial varieties of the Central Valley of California carried gametic type 2112 whereas 2121 was the most common gametic type of North Plains and Prairie Province varieties. The picture of genetic change that emerges is one in which increases in adaptation were correlated with the development of clusters of alleles of loci. These clusters are closely linked to many different morphological and physiological characteristics and gradual coalescence of clusters in a few large synergistically interacting complexes.

Regarding artificial selection, frequency changes of alleles at eight isozyme loci have been monitored in four long-term maize selection experiments. Allelic frequencies at these loci showed significant changes greater than would be expected with random drift alone, suggesting that the frequencies were due to selection for increased grain yield (Stuber *et al.*, 1980). Based on this result, it was hypothesized that manipulation of allelic frequencies at the appropriate isozyme loci should produce responses in the correlated quantitative traits. Experimental results indicated that selections based solely on manipulation of allelic frequencies at seven isozyme loci significantly increased grain yield and ear number (Stuber *et al.*, 1982). In a somewhat similar study, conducted in a population generated from a composite of elite inbred lines, isozyme frequency manipulations produced responses similar to those found for phenotypic selection in the same population (Frei *et al.*, 1986).

13.5.3 Marker associations in F_2 and backcross populations

F_2 populations are often used for identifying and locating quantitative trait loci (QTLs). Stuber and colleagues conducted extensive molecular-marker-facilitated investigations of quantitative trait loci in maize. Associations between 17–20 mapped isozyme markers and 82 quantitative traits including grain yield were investigated in each of two F_2 populations. Significant associations were found for all traits in each population. An average of 10.2 and 13.8 marker loci was significantly associated with factors influencing the expression of each trait in both populations, although numbers varied somewhat among traits. Yield and many yield-related traits, such as kernel number, kernel depth and ear length and circumference, were affected by factors associated with a large proportion of the marker loci. For example, members of tightly linked pairs of isozymes on chromosome 1 (*Adh1-Phi1*) and chromosome 6 (*Pdg1-Enp1* and *Idh2-Mdh2*) probably reflected associations with the same QTLs (Edwards *et al.*, 1987; Stuber *et al.*, 1987). Similar results were obtained using six F_2 populations derived from five elite inbred lines, indicating that a significant effect was found to be associated with at least one chromosomal region for each of 21 quantitative traits analysed. Comparisons made between populations showed that certain areas of the genome consistently affected specific traits, for example ear length and grain weight at *Amp3* on chromosome 5 and ear circumference at *Glu1* on chromosome 10 (Abler *et al.*, 1991). It should be noted, however, that effects detected by marker loci diminished relative to the true effect of QTL as the distance between marker locus and the QTL increased (Edwards *et al.*, 1987; Lander and Botstein, 1989).

Several investigations involving the associations of isozyme markers and quantitative traits in interspecific crosses of tomato have been reported. Tanksley *et al.* (1982) demonstrated that 27 of the 48 possible combinations of 12 mapped isozyme marker loci with four quantitative traits showed significant association in a backcross population of 400 plants derived from *Lycopersicon esculentum* and *Solanum pennellii*. In another interspecific backcross between the cultivated tomato, *L. esculentum*, and a high-altitude, cold-tolerant *L. hirsutum*, Vallejos and Tanksley (1983) examined linkages between segregating isozyme loci and genes responsible for cold tolerance. A minimum of three QTLs responsible for growth at low temperature were detected – two had positive effects and the other was negative. One marker locus, *Pgi-1*, gave a significant and positive effect only at low temperature. Another linkage study on quantitative traits was made in an interspecific F_2 population between tomato and a wild relative, *L. pimpinellifolium* by Weller *et al.* (1988). Out of 180 possible combinations of 10 markers, including 4 isozyme markers and 18 quantitative traits, 85 combinations showed significant associations. In soyabean, a total of 480 backcross lines derived from two crosses between *Glycine max* and *G. soya* were used by Suarez *et al.* (1991) to study the association between isozyme markers and agronomic seed composition traits. Significant associations were found between particular isozyme genotypes and every trait analysed. The estimated effect of genes linked to *Pgm1* locus was a delay in maturity of six days. The *Idh2* locus was associated with a QTL for linolenic acid content.

13.5.4 *Mapping quantitative trait loci in homozyous populations*

Quantitative traits usually show continuous variation and are influenced by environmental factors. Recombinant inbred (RI) lines are valuable for studying QTLs and can be readily produced by inbreeding the progenies of F_2 individuals derived from two well-established progenitor inbreds. Burr *et al.* (1988), for example, obtained RI families from approximately 50 plants in each of two F_2 populations by inbreeding for six generations by ear-to-row propagation. A genetic map based largely on isozymes and RFLPs has been produced to cover virtually the entire maize genome. Doubled haploid (DH) lines, produced through anther culture of hybrids and interspecific or intergeneric crosses, are also used in genetic mapping. Kleinhofs *et al.* (1993) constructed a map of the barley genome using 150 DH lines developed from the 'Steptoe' × 'Morex' cross by the *Hordeum bulbosum* method (Kasha and Kao, 1970). Ullrich *et al.* (1993) evaluated these DH lines grown in different locations, and demonstrated that QTL and QTL × environment interaction effects for agronomic and malting quality traits were found using a 123-point linkage map (containing RFLP, isozyme and morphological markers). These studies identified six QTLs for grain yield, six for lodging, ten for plant height, nine for heading date, six for grain protein, and nine for α- and β-amylase genes.

13.6 Conclusions and Prospects

Results from studies in plants have proved the effectiveness of isozyme marker techniques for identifying and mapping desired loci for qualitative and quantitative traits. Although the number of isozyme markers is limited, compared with DNA based markers, they have contributed to identifying appropriate genotypes for disease resistance and otherwise. The ease of assay of isozymes provides an attractive means for transferring a linkage block involving a resistance gene to produce a new variety. In the case of accumulating multiple resistance genes into one variety, the tight linkage of each gene to a suitable isozyme allele provides important diagnostic markers for achieving the desired new variety. It is clear that isozyme markers will continue to contribute to manipulating desired alleles in plant breeding, particularly when they are combined with newly developed molecular markers such as RFLP, RAPDs and microsatellites.

References

Abbott, D.C., Brown, A.D.H. and Burdon, J.J. (1992) Genes for scald resistance from wild barley (*Hordeum vulgare* ssp. *spontaneum*) and their linkage to isozyme markers. *Euphytica* 61, 225–231.

Abler, B.S.B., Edwards, M.D. and Stuber, C.W. (1991) Isoenzymatic identification of quantitative trait loci in crosses of elite maize inbreds. *Crop Science* 31, 267–274.

Allard, R.W. (1988) Genetic changes associated with the evolution of adaptedness in cultivated plants and their wild progenitors. *Journal of Heredity* 79, 225–238.

Allard, R.W. (1992) Predictive methods for germplasm identification. In: Stalker, H.T. and Murphy, J.P. (eds), *Plant Breeding in the 1990s*. CAB International, Wallingford, UK, pp. 119–146.

Bournival, B.L., Scott, J.W. and Vallejos, C.E. (1989) An isozyme marker for resistance to race 3 of *Fusarium oxysporum* f. sp. *lycopersici* in tomato. *Theoretical and Applied Genetics* 78, 489–494.

Bournival, B.L., Vallejos, C.E. and Scott, J.W. (1990) Genetic analysis of resistance to races 1 and 2 of *Fusarium oxysporum* f. sp. *lycopersici* from the wild tomato *Lycopersicon pennellii*. *Theoretical and Applied Genetics* 79, 641–645.

Burr, B., Burr, F.A., Thompson, K.H., Albertson, M.C. and Stuber, C.W. (1988) Gene mapping with recombinant inbreds in maize. *Genetics* 118, 519–526.

Delourme, R. and Eber, F. (1992) Linkage between an isozyme marker and a restorer gene in radish cytoplasmic male sterility of rapeseed (*Brassica napus* L.). *Theoretical and Applied Genetics* 85, 222–228.

Edwards, M.D., Stuber, C.W. and Wendel, J.F. (1987) Molecular-marked facilitated investigations of quantitative-trait loci in maize. I. Number, genomic distribution and types of gene action. *Genetics* 116, 113–125.

Feuerstein, U., Brown, A.H.D. and Burdon, J.J. (1990) Linkage of rust resistance genes from wild barley (*Hordeum spontaneum*) with isozyme markers. *Plant Breeding* 104, 318–324.

Frei, O.M., Stuber, C.W. and Goodman, M.M. (1986) Yield manipulation from selection on allozyme genotypes in a composite of elite corn lines. *Crop Science* 26, 917–921.

Friedt, W., Huth, H., Mielke und Zuchner, S. (1985) Resistanztrager gegen barley yellow mosaic virus. *Nachrichtenbl. Deutsch Pflanzenschutzd.* 37, 129–135.

Graner, A. and Bauer, E. (1993) RFLP mapping of the *ym4* virus resistance gene in barley. *Theoretical and Applied Genetics* 86, 689–691.

Hamrick, J.L. and Godt, M.J.W. (1990) Allozyme diversity in plant species. In: Brown, A.H.D., Clegg, M.T., Kehler, A.L. and Weir, B.S. (eds), *Plant Population Genetics, Breeding, and Genetic Resources*. Sinauer Associates, Sunderland, MA, USA, pp. 43–63.

Harlan, J.R. (1976) Genetic resources in wild relatives of crops. *Crop Science* 16, 329–333.

Helentjaris, T.G. (1992) RFLP analyses for manipulating agronomic traits in plants. In: Stalker, H.T. and Murphy, J.P. (eds), *Plant Breeding in the 1990s*. CAB International, Wallingford, UK, pp. 357–372.

Kaiser, R. and Friedt, W. (1992) A gene for resistance to barley mild mosaic virus in German winter-barley is located on chromosome 3L. *Plant Breeding* 108, 169–172.

Kasha, K.J. and Kao, K.N. (1970) High frequency haploid production in barley (*Hordeum vulgare* L.). *Nature (London)* 225, 874–876.

Kawada, N. (1991) Resistant cultivars and genetic ancestry of the resistance genes to barley yellow mosaic virus in barley (*Hordeum vulgare* L.). *The Bulletin of the Kyushu National Agricultural Experiment Station* 27, 65–79.

Kleinhofs, A., Kilian, A., Saghai Maroof, M.A., Biyashev, R.M., Hayes, P., Chen, F.Q., Lapitan, N., Fenwick, A., Blake, T.K., Kanazin, V., Ananiev, E., Dahleen, L., Kudrna, D., Bollinger, J., Knapp, S.J., Liu, B., Sorrells, M., Heum, M., Franckowiak, J.D., Hoffman, D., Skadsen, R. and Steffenson, B.J. (1993) A molecular, isozyme and morphological map of the barley (*Hordeum vulgare*) genome. *Theoretical and Applied Genetics* 86, 705–712.

Konishi, T. and Kaiser, R. (1991) Genetic difference in barley yellow mosaic virus resistance between Mokusekko 3 and Misato Golden. *Japanese Journal of Breeding* 41, 499–505.

Konishi, T. and Matsuura, S. (1987) Variation of esterase isozyme genotypes in a pedigree of Japanese two-rowed barley. *Japanese Journal of Breeding* 37, 412–420.

Konishi, T., Kawada, N., Yoshida, H. and Sohtome, K. (1989) Linkage relationship between two loci for the barley yellow mosaic resistance of Mokusekko 3 and esterase isozymes in barley (*Hordeum vulgare* L.). *Japanese Journal of Breeding* 39, 423–430.

Lander, E.S. and Botstein, D. (1989) Mapping Mendelian factors underlying quantitative traits using RFLP linkage maps. *Genetics* 121, 185–199.

Linde-Laursen, I., Doll, H. and Nielsen, G. (1982) Giemsa C-banding patterns and some biochemical markers in a pedigree of European barley. *Zeitschrift für Pflanzenzüchtung* 88, 191–219.

McMillin, D.E., Allan, R.E. and Roberts, D.E. (1986) Association of an isozyme locus and strawbreaker foot rot resistance derived from *Aegilops ventricosa* in wheat. *Theoretical and Applied Genetics* 72, 743–747.

Palmer, R.G., Lim, S.M. and Hedges, B.R. (1992) Testing for linkage between the *Rxp* locus and nine isozyme loci in soybean. *Crop Science* 32, 681–683.

Proeseler, G. and Lehmann, C.O. (1986) Resistenzeigenschaften im Gersten- und Weizensortiment Gatersleben. 25. Prufung von ihr Verhalten gegenüber Gerstengelmosiak-Virus (barley yellow mosaic virus). *Kulturpflanze* 34, 241–248.

Rick, C.M. and Fobes, J.F. (1974) Association of an allozyme with nematode resistance. *Report of Tomato Genetics Cooperation* 24, 25.

Sax, K. (1923) The association of size differences with seedcoat pattern and pigmentation in *Phaseolus vulgaris*. *Genetics* 8, 552–560.

Sneller, C.H., Isleib, T.G. and Carter, T.E., Jr (1992) Isozyme screening of near-isogenic male sterile soybean lines to uncover potential linkages: linkage of the *Pgm1* and *ms2* loci. *Journal of Heredity* 83, 457–459.

Stuber, C.W. (1991) Isozyme markers and their significance in crop improvement. In: Khanna, K.R. (ed.), *Biochemical Aspects of Crop Improvement*. CRC Press, Boca Raton, FL, USA, pp. 59–77.

Stuber, C.W., Moll, R.H., Goodman, M.M., Schaffer, H.E. and Weir, B.S. (1980) Allozyme frequency changes associated with selection for increased grain yield in maize (*Zea mays* L.). *Genetics* 95, 225–236.

Stuber, C.W., Goodman, M.M. and Moll, R.H. (1982) Improvement of yield and ear number resulting from selection at allozyme loci in maize population. *Crop Science* 22, 737–740.

Stuber, C.W., Edwards, M.D. and Wendel, J.F. (1987) Molecular marker-facilitated investigations of quantitative trait loci in maize. II. Factors influencing yield and its component traits. *Crop Science* 27, 639–648.

Suarez, J.C., Graef, G.L., Fehr, W.R. and Ciazio, S.R. (1991) Association of isozyme genotypes with agronomic and seed composition traits in soybean. *Euphytica* 52, 137–146.

Takahashi, R., Hayashi, J., Inouye, T., Moriya, I. and Hirao, C. (1973) Studies on resistance to yellow mosaic disease in barley. I. Tests for varietal reactions and genetic analysis of resistance to the disease. *Berichte des Ohara Instituts fur landwirtschafte Biologie, Okayama Universität* 16, 1–17.

Tanksley, S.D. and Loaiza-Figueroa, F. (1985) Gametophytic self-incompatibility is controlled by a single major locus on chromosome 1 in *Lycopersicon peruvianum*. *Proceedings of the National Academy of Sciences, USA* 82, 5093–5096.

Tanksley, S.D. and Rick, C.M. (1980) Isozymic gene linkage map of tomato: applications in genetics and breeding. *Theoretical and Applied Genetics* 57, 161–170.

Tanksley, S.D., Medina-Filho, H. and Rick, C.M. (1982) Use of naturally-occurring enzyme variation to detect and map genes controlling quantitative traits in an interspecific backcross of tomato. *Heredity* 49, 11–25.

Tanksley, S.D., Rick, C.M. and Vallejos, C.E. (1984) Tight linkage between a nuclear male-sterile locus and an enzyme marker in tomato. *Theoretical and Applied Genetics* 68, 109–113.

Ullrich, S.E., Hayes, P.M., Liu, B.H., Knapp, S.J., Chen, F., Jones, B.L., Blake, T.K., Franckowiak, J.D., Rasmusson, D.C., Sorrells, M.E., Wesenberg, D.M., Kleinhofs, A. and Nilan, R.A. (1993) Quantitative trait locus effects and environmental interaction in barley population. *Abstract of 15th North American Barley Research Workshop*, 27.

Vallejos, C.E. and Tanksley, S.D. (1983) Segregation of isozyme markers and cold tolerance in an interspecific backcross of tomato. *Theoretical and Applied Genetics* 66, 241–247.

Weeden, N.F. and Provvidenti, R. (1988) A marker locus, *Adh-1*, for resistance to pea enation mosaic virus in *Pisum sativum*. *Journal of Heredity* 79, 128–131.

Weeden, N.F., Provvidenti, R. and Marx, G.A. (1984) An isozyme marker for resistance to bean yellow mosaic virus in *Pisum sativum*. *Journal of Heredity* 75, 411–412.

Weller, J.I., Soller, M. and Brody, T. (1988) Linkage analysis of quantitative traits in an interspecific cross of tomato (*Lycopersicum esculentum* × *Lycospersicon pimpinellifolium*) by means of genetic markers. *Genetics* 118, 329–339.

Wricke, G. and Wehling, P. (1985) Linkage between an incompatibility locus and peroxidase isozyme locus (*Prx 7*) in rye. *Theoretical and Applied Genetics* 71, 289–291.

The Use of Carbon Isotope Discrimination Analysis in Plant Improvement

R.A. Richards and A.G. Condon

CSIRO, Division of Plant Industry, and Cooperative Research Centre for Plant Science, GPO Box 1600, Canberra, ACT 2601, Australia.

14.1 Introduction

Stable isotopes of many biologically important elements including carbon, oxygen, nitrogen, hydrogen and sulphur exist in nature, in addition to unstable, radioactive isotopes. The natural abundance of the most common stable isotope of these elements is generally far greater than that of the other forms. For example, 98.9% of atmospheric carbon is ^{12}C with the remaining 1.1% ^{13}C, while the abundance of ^{1}H, ^{14}N and ^{16}O all exceed 99.5%. In recent years improvements in measurement techniques and an understanding of the factors that lead to variation in isotopic composition have opened up new insights into biological systems. Accordingly, the measurement of isotopic composition has become a powerful tool in agricultural and ecological research.

In this chapter variation in carbon isotope composition in C_3 plant species, and how it can be used to identify plants that may be more productive in dry environments, will be examined. However, carbon isotope analysis also has other applications. As a diagnostic, carbon isotope composition was originally used to differentiate between species with either the C_3, C_4 or crassulacean acid metabolism (CAM) photosynthetic pathways, since these plant types have a distinct carbon isotope signature. In terms of crop science this application has had limited use, one notable exception being in assisting in the prosecution of wine makers suspected of 'doctoring' the juice of C_3 grapes with sugar from C_4 sugarcane! Apart from this basic physiological diagnostic, analysis of carbon isotope composition continues to provide more detail into other aspects of physiological activity associated with leaf gas exchange and biochemistry. There is also increasingly widespread use of measurements of N, O and H isotopic composition. The

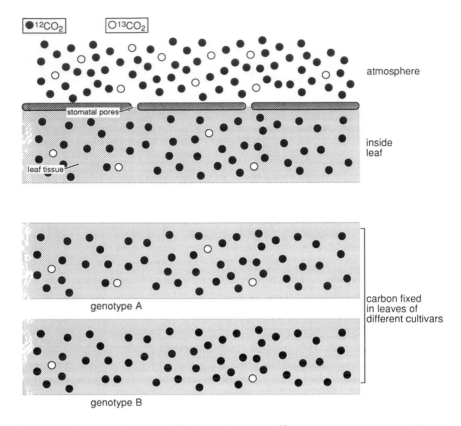

Fig. 14.1. Diagrammatic view of discrimination against $^{13}CO_2$, in a cross-section of the top of a leaf, leading to variation in carbon isotope discrimination. The partial pressure of CO_2 inside the leaf is less than in the air outside. CO_2 diffuses from the air into the leaf through the stomata and there is some discrimination against the rarer $^{13}CO_2$ during the diffusion process. There is further discrimination against $^{13}CO_2$ during carboxylation. Substantial genotypic variation in the amount of discrimination against $^{13}CO_2$ is commonly found in plant tissues of C_3 plants.

measurement of ^{15}N is now being extensively used to estimate symbiotic nitrogen fixation by plants. Hydrogen and oxygen isotope analysis are being used to trace water movement from the soil through plants to the atmosphere as well as providing tools to increase our understanding of the physiology of hydrogen and oxygen metabolism in plants. Recently, measurements of $^{18}O/^{16}O$ isotopic composition of atmospheric CO_2 have been used to assess vegetation effects on global CO_2 (Farquhar *et al.*, 1993).

Plant material generally has a different C, H, O or N isotopic composition to the raw materials from which these elements are supplied. This arises because small differences in the properties of the isotopes of these elements cause fractionations during the physical and chemical processes that occur as they are incorporated into plant dry matter. In the case of carbon fixation in C_3 photosynthesis, there is fractionation between $^{13}CO_2$ and $^{12}CO_2$ during diffusion of CO_2 through the stomatal pores at the leaf surface (Fig. 14.1). The diffusion of $^{12}CO_2$ is more rapid than that of $^{13}CO_2$. Further physical fractionation occurs as CO_2 enters solution and diffuses to the site of biochemical fixation (carboxylation) in the chloroplasts. Within the chloroplasts there is still greater fractionation at the primary carboxylating enzyme, Rubisco (Farquhar *et al.*, 1989). It is the recent development of our understanding of these processes leading to variation in isotopic composition that has expanded the potential to use carbon isotope analysis in crop science. This theme will be developed further in the following sections.

14.2 Isotope Analysis and Terminology

Isotope analysis of plant carbon is most commonly carried out using isotope ratio mass spectrometry. A recent review by Preston (1992) gives an excellent overview of the procedures involved for carbon and other elements. For carbon, the technique requires the production of a pure sample of CO_2 gas from plant material. This is generally achieved by combusting a small sample of the dried and finely ground plant material in the presence of pure oxygen. The CO_2 produced is purified chemically and/or cryogenically before being introduced into the mass spectrometer. Within the instrument the CO_2 molecules are ionized and the isotopes separated on the basis of mass-to-charge ratio using a strong magnetic field.

Because the less common stable isotopes are in such low amounts, it is technically difficult to measure the absolute isotopic composition of plant (or any other) material. The isotopic composition is therefore conventionally determined by comparing the molar abundance ratio (R_p) of the stable isotopes of interest in the plant sample, in this case $^{13}C/^{12}C$, to the value of the same ratio in a standard (R_s). Isotopic composition (δ_p) is calculated as $(R_p - R_s)/R_s$. For historical reasons, the conventional standard for carbon is CO_2 generated from Pee Dee belemnite (PDB), a carbonaceous rock. Values of carbon isotope composition are given in units of $\delta^{13}C_{PDB}$. Generally values of carbon isotope composition are expressed as parts per thousand or per mil (‰). With respect to PDB, the isotopic composition of current C_3 plant material generally has a value that is small but negative, near –28‰. The value of carbon isotope composition is negative because the $^{13}C/^{12}C$ ratio of plant material is lower than that of PDB. This is largely the result of two related factors. Firstly, there is a net discrimination against atmospheric ^{13}C during photosynthetic carbon fixation by C_3 plants, as outlined above. Secondly, the current atmospheric $^{13}C/^{12}C$ ratio is lower than when the Pee Dee belemnite was laid down. This is a result of the largely anthropogenic release of CO_2 from fossil fuels made up from plants that also discriminated against ^{13}C during their photosynthesis.

Carbon isotope composition ($\delta^{13}C$) is a useful empirical measure, but the development of an understanding of the processes causing variation in isotopic composition in plant material (outlined briefly above) led Farquhar and Richards (1984) to propose a new terminology to highlight the process of discrimination. They defined carbon isotope discrimination, Δ, as the molar ratio (R_a) of $^{13}C/^{12}C$ in atmospheric CO_2, the carbon source, divided by the same ratio in the plant product (R_p), but expressed as the deviation from unity, i.e. $\Delta = R_a/R_p - 1$. Carbon isotope discrimination is thus equivalent to a fractionation factor which accounts for the net positive discrimination against ^{13}C during carbon fixation in photosynthesis. The value of R_a/R_p is a number only slightly greater than one, so the deviation from unity is used for numerical convenience to obtain values that are small but positive, near 20‰ for C_3 plants. The definition of carbon isotope discrimination requires that values for both R_a and R_p are determined. For most field-grown material this is not a concern since the isotopic composition of the atmosphere is known and is essentially constant (currently –8‰ relative to PDB). As will be shown in the following section, the use of discrimination rather than composition means that the biological significance of variation in the $^{13}C/^{12}C$ ratio found in plant material can be more easily interpreted.

14.3 The Biological Significance of Variation in Carbon Isotope Discrimination

As noted earlier, the knowledge that there was widespread variation among plant species in their carbon isotope composition was initially used as a descriptive tool in ecological studies. The pioneering work of Farquhar *et al.* (1982) led to an understanding of the functional significance of variation in carbon isotope discrimination within and between C_3 plants and also a recognition of the potential value of carbon isotope analysis in plant breeding.

Farquhar *et al.* demonstrated that the value of Δ should be related to the ratio of the internal and external partial pressure of CO_2, p_i/p_a. In its simplest form their expression for discrimination by C_3 plants was:

$$\Delta = a + (b - a)p_i/p_a \times 10^{-3} \tag{14.1}$$

where a is the fractionation that occurs during diffusion of CO_2 in air (4.4‰) and b accounts for the subsequent fractionations dominated by carboxylation by Rubisco (27‰). Equation 14.1 represents an essentially instantaneous process. Importantly, when measured in plant dry matter, Δ gives an assimilation-weighted average value of p_i/p_a over the life of the plant material being analysed.

14.3.1 The significance of variation in p_i/p_a

The value of p_i/p_a (and of Δ over the long term) is determined by the balance between stomatal conductance and photosynthetic capacity, i.e. the balance between how freely CO_2 can diffuse into and out of the leaf through the stomata and the capacity for the photosynthetic machinery within the leaf to fix CO_2. If stomatal conductance is high or photosynthetic capacity is low (or both), then the values of p_i/p_a and Δ will be high, and vice versa. Clearly, it is not possible, simply from the value of p_i/p_a or Δ, to determine either conductance or photosynthetic rate. Only the balance between the two is assessed. Nevertheless, this balance is important since it is a major determinant of variation in the transpiration efficiency of leaf gas exchange, i.e. the instantaneous ratio of the CO_2 assimilation rate of a leaf, A, to its transpiration rate, T. Thus:

$$\frac{A}{T} = \frac{p_a(1 - p_i/p_a)}{1.6v} \qquad (14.2)$$

where v is the leaf-to-air vapour pressure difference and 1.6 is the ratio of the diffusivities of water vapour and CO_2 in air. Where there is adequate ventilation p_a is fixed and v is an independent variable. Transpiration efficiency is therefore negatively related to p_i/p_a and inversely related to v. For C_3 species growing under favourable conditions p_i/p_a is typically 0.7 ± 0.1 and the variability is $(1 - p_i/p_a)$, and hence A/T is of the order of 33%. This quite large variation is a potentially important component of variation in plant and crop transpiration efficiency. But identifying and exploiting variation in $(1 - p_i/p_a)$ using gas exchange techniques is difficult and laborious. Farquhar *et al.* (1982) also recognized that Δ should be *negatively* related to A/T because of the independent relationships of Δ with p_i/p_a and of A/T with $(1 - p_i/p_a)$. They proposed that the measurement of Δ could be a useful surrogate for the measurement of p_i/p_a in identifying variation in plant and crop transpiration efficiency.

14.3.2 Selecting for lower Δ to improve water use efficiency

The prospect of selecting for lower Δ in plant populations to improve water use efficiency was rapidly taken up by plant breeders and crop physiologists after confirmation that the theory relating Δ to A/T could be extended to whole-plant transpiration efficiency (W), i.e. the ratio of dry matter produced per unit of water transpired by the plant. Farquhar and Richards (1984) grew a small set of wheat genotypes under a range of watering regimes and found a negative relationship between Δ and *W*. The negative relationship between Δ and *W* was subsequently confirmed in pot studies for a range of C₃ species (Fig. 14.2) including peanuts (Hubick *et al.*, 1986), barley (Hubick and Farquhar, 1989), wheat (Condon *et al.*, 1990), common bean (Ehleringer *et al.*, 1991), cowpeas (Ismail and Hall, 1992), sunflower (Virgona *et al.*, 1990), tomato (Martin and Thorstenson, 1988), range grasses (Johnson *et al.*, 1990) and potato (Vos and Groenwold, 1989). In parallel with these studies, the positive relationship between Δ and p_i/p_a was confirmed in wheat (Evans *et al.*, 1986; Condon *et al.*, 1990), peanut (Hubick *et al.*, 1988) and bean (Ehleringer *et al.*, 1991) using gas exchange techniques.

The same relationship between Δ and p_i/p_a and hence to *W* found in C₃ species does not hold for C₄ species. There are two main reasons. Firstly, during the primary carboxylation in C₄ species (by PEP carboxylase) there is actually a slight discrimination in favour of ¹³C. Thus the relationship between Δ and p_i/p_a should be slightly negative. But Δ is also related to the proportion of CO_2 produced by decarboxylation in the bundle sheath cells which then leaks back into the mesophyll cells where it mixes with CO_2

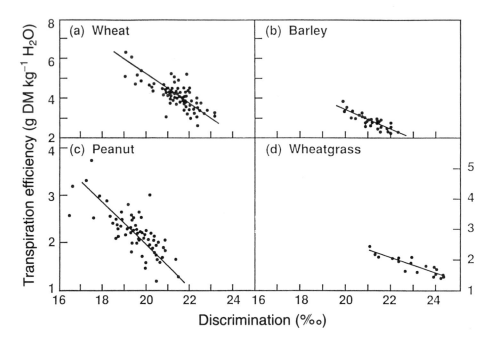

Fig. 14.2. Relationship, for pot-grown plants, between carbon isotope discrimination and transpiration efficiency in wheat, barley, peanut and wheatgrass. (Data adapted from Farquhar and Richards, 1984; Hubick *et al.*, 1988; Hubick and Farquhar, 1989; Johnson *et al.*, 1990; Turner, 1993; reprinted from Richards *et al.*, 1993.)

diffusing in through the stomata (Farquhar, 1983). Depending on the amount of leakage, which can be significant (Evans *et al.*, 1986), the slope of the relationship between p_i/p_a and Δ will be negative, positive or, more typically, zero. Accordingly, Δ cannot be related directly to p_i/p_a or to W for C_4 species. It is noteworthy, however, that variation in Δ in C_4 species may still be important as it should be related to light use efficiency (Farquhar, 1983).

14.4 The Importance of Transpiration Efficiency and Δ for Crop Improvement

The relationship between Δ and W in C_3 species was seized upon because previous attempts to improve yields in dry environments had not been very successful. The use of Δ provided a new, theoretically sound avenue to increase the yield obtained from a given supply of water. Some of the reasons for the lack of progress in yield improvement in rainfed environments are worth examining. Dry environments are characterized by greater seasonal variation in yield than more favourable environments. For this reason progress in traditional breeding approaches based on field evaluation is very slow and sometimes non-existent. Genotype × environment interactions are usually large and there is often poor expression of genetic variation. In fact there is little evidence that traditional breeding approaches have improved either total biomass or the efficiency of water use of our most intensively bred crops. Thus alternative approaches to selection for improved yields in dry environments may be of some consequence. Furthermore, selection for more efficient use of water is not easy. It requires the measurement of biomass as well as water use and both of these measurements can be difficult. In some crops leaves fall to the ground when they die, where they may then decompose or blow away, making biomass measurements imprecise. The determination of water use is more demanding. Transpiration can be determined with some accuracy in pots as they are easily weighed and soil evaporation can be prevented by mulching. But in the field, crop water use is very difficult to determine because water can be lost by drainage below the rooting depth or laterally as well as by evaporation into the atmosphere from the soil surface. Although there are ways to overcome these difficulties, it is not possible to determine either crop water use efficiency or transpiration efficiency on a large number of genotypes. Thus rapid plant measurements to determine variation in W are very appealing.

14.5 Some Complications in Using Δ to Improve Crop Yields

14.5.1 Translating greater W into greater yield

Despite the promise in using Δ to select for higher W, actual yield increases
do not automatically follow with an increase in W and/or a lower Δ because
there are a number of complications. The relationship between Δ and W
depends on there being little variation in respiration and the allocation of
carbon to the roots. There has been evidence for both of these in some of the
pot studies referred to earlier. For a crop where the economic product is the
above-ground biomass, the amount of biomass in a dry environment is a
function of the amount of water transpired, T, and the transpiration
efficiency, W, namely:

$$\text{Biomass} = T \times W \qquad (14.3)$$

This equation ignores the loss of water by evaporation from the soil surface.
Although this may be low in some environments, for example where rainfall
is very seasonal and crops are sown after this rainfall, in other environments
where crops are grown at a time when rainfall events are frequent the
evaporation of rainfall from the soil surface may be a large component of
total crop water use (ET). If ET represents the sum of crop transpiration (T)
and the evaporation of water from the soil surface then we can express
Equation 14.3 as:

$$\text{Biomass} = (ET \times T/ET) \times W \qquad (14.4)$$

This equation shows that not only is variation in W important but also
variation in the amount of ET and the proportion of ET evaporated from the
soil surface (T/ET). For a sparse crop growing where the soil is often wet,
T/ET can be less than 0.5 (Cooper *et al.*, 1983). There are further
complications, of course, where grain is the economic product. It does not
automatically follow that if biomass increases then grain yield also increases;
although this often is the case.

The relationship between yield and Δ may also be different to that
expected on the basis of differences in transpiration efficiency if water supply
is not a major limitation to crop growth. Positive relationships between yield
and Δ were obtained when wheat genotypes were grown under favourable
conditions (Condon *et al.*, 1987). This could occur for several reasons. An
obvious one is if the major source of variation in Δ is stomatal conductance.
The rate of photosynthesis (and Δ) will be higher in genotypes with high
conductance. In cases such as this, transpiration is likely to be greater for high
yielding genotypes but this carries no penalty for yield when water is
plentiful. Indeed, it should be associated with greater yield.

14.5.2 *Translating variation in* p_i/p_a *(or* Δ*) into crop transpiration efficiency*

An additional complication in achieving greater crop transpiration efficiency via Δ may arise if variation in Δ is the result of variation in stomatal conductance. An important simplification in Equation 14.2 relating A/T to $(1 - p_i/p_a)$ is that the leaf-to-air vapour pressure difference, v, is an independent variable. In most circumstances this is unlikely. Leaf temperature and therefore v will tend to be greater in genotypes with relatively low stomatal conductance unless the air around the leaves is very well stirred. Higher leaf temperature will drive transpiration faster than would be expected if differences in stomatal conductance had no effect on leaf temperature at all. Also, if the leaf is, say, one of many in a crop canopy all with a relatively low conductance then the air in the canopy around the leaves will become drier and hotter, causing a further increase in v and further offsetting the expected decrease in transpiration. The net result is that any difference in crop W will not be as great as might be expected from a difference in Δ. To evaluate the possible significance of these effects, two wheat genotypes differing in stomatal conductance were contrasted in a high yielding environment in large fields. These are conditions that should minimize the advantage of low Δ achieved via low conductance. From Δ values collected from plants over the entire season a maximum possible difference in crop transpiration efficiency was calculated to be 19%. The observed advantage was 15% (Condon and Richards, 1993).

14.6 The Use of Carbon Isotope Discrimination in Plant Breeding

Carbon isotope discrimination as a measure of W fulfils many of the measurement and genetic criteria that are required for use in a plant breeding programme.

14.6.1 *Measurement*

Measurements of Δ can be made on vegetative material early in the life of a plant and only a small sample of material is required. Thus sampling is relatively non-destructive and most of the plant remains and continues to grow. This means that as well as collecting seed from selected plants, these plants can also be used for hybridization. After sampling, the plant material must be dried in preparation for isotopic analysis and therefore it can be stored without concern over its degradation. An important feature of the measurement of the isotopic composition of the plant material is that it is an integrative measure of the p_i/p_a during the life of the sampled material and of the carbon used to form it. Thus it is not an instantaneous estimate like direct measurements of p_i/p_a by gas exchange. This latter point is very important and is probably the reason why the measurement of Δ is generally very repeatable.

The measurement of Δ by mass spectrometry is very precise and it can be automated. Although the degree of precision adds to the cost, high precision also means that fewer replications are required. Error in the measurement of any sample is very low and coefficients of variation are typically about 2%. This value contrasts with coefficients for W, biomass or grain yield which are typically between 10 and 20%.

A limitation to the use of Δ in breeding programmes is the cost of measurement. As a consequence of the technology required to measure isotopic ratios the cost of a single determination of Δ may vary between 5 and 20 US dollars depending on the precision required and the laboratory. Although this is high in contrast with simpler measurements such as flowering time or height, it is low compared with other costs for genotype evaluation such as sowing and harvesting field plots to determine grain yield, some grain quality components, or the extraction and characterization of DNA. Furthermore, as mentioned above, the degree of precision in the measurement of Δ is high and this can reduce the number of replications required. In efforts to reduce the cost a number of surrogates for Δ have emerged. These are specific leaf weight in peanuts (Wright et al., 1988) and plant mineral content which can be determined simply by ashing plant material (Masle et al., 1992). Although these measurements will not be as effective as the measurement of Δ itself, they may be useful in the initial screening of large populations.

14.6.2 Genetic variation

Genetic variation in Δ is substantial in all crops in which it has been studied. Typically there is a difference of about 2‰ between extreme genotypes, although differences of up to 3‰ have also been found in some surveys of widely divergent genotypes. In a relative sense and on the basis of the relationship between Δ and p_i/p_a given in Equation 14.1, the transpiration efficiency of a genotype with a Δ value of 19‰ could be 33% greater than that of a genotype with a Δ value of 21‰. In the field the actual difference achieved is not likely to be as large as this because of complications such as those discussed above.

There is no evidence that Δ is under simple genetic control, although certain major genes found in some species, such as those near-lethal mutants that keep stomata permanently open, will affect Δ. Because variation in Δ is a consequence of the balance between stomatal conductance and assimilation capacity, both of which are influenced by many genes each having a small effect, then Δ is also under the control of many genes. Nevertheless, genotypic variation in Δ is highly repeatable. Broad sense heritabilities, which indicate the degree of genetic variation as well as the repeatability, are substantial and can be as high as 95% (Hubick *et al.*, 1988; Ehdaie *et al.*, 1991; Condon and Richards, 1992). Narrow sense heritabilities, which represent the progress possible in selection, are smaller but still substantial (78% in wheat) (Z. Zhen, R.A. Richards and A.G. Condon, unpublished).

An additional genetic advantage for using Δ in selection is the low genotype $\times$ environment ($g \times e$) interaction when plants are grown during the normal field season. Typically, $g \times e$ values for Δ are a very small proportion of the total genetic variation (Hubick *et al.*, 1988; Hall *et al.*, 1990; Condon and Richards, 1992). An example of this low $g \times e$ was the consistency in ranking of a group of wheat genotypes grown in fields 3000 km apart (Condon *et al.*, 1987). An exception has been found in cowpea where, although genotypes ranked consistently in wet and dry environments in different years at the same location (Hall *et al.*, 1993), when grown over a broader range of environments from California to Senegal there was much less correspondence in the ranking of genotypes. Similarly, we have found inconsistencies in the values for Δ of wheat when grown out of season or in pots away from the typical field environment.

14.7 Concluding Remarks

The use of Δ in plant improvement is likely to have application where traditional methods of improvement have reached a plateau or where it is desirable to introduce new genetic variation for W into breeding programmes. If Δ is used in breeding it will be just one of many of the tests that plant breeders will rely on to select desirable genotypes in their target environments. It is most likely to have application where an improvement in transpiration efficiency is of greatest importance such as the more water-limited environments and where W is least influenced by other factors in the environment. Factors such as a rapidly changing vapour pressure deficit of the atmosphere and/or where a large proportion of water is evaporated from the soil surface are most likely to reduce the importance of a high leaf W and hence a low Δ in breeding. However, there are other important opportunities to use Δ in plant breeding that have yet to be fully explored. Selection for high Δ may be important to maximize stomatal conductance and hence assimilation rate of crops in irrigated or moist environments (Condon *et al.*, 1987; Morgan *et al.*, 1993). A high Δ may also result in faster growing canopies (Richards, 1992; Condon *et al.*, 1993) which may be important in short season environments or where water loss from the soil surface is substantial. On the other hand, selection for low Δ to improve light use efficiency in C_4 species also offers promise.

In conclusion, carbon isotope discrimination provides an important new approach to overcome the so far intractable and important problem of genetically improving water use efficiency and biomass production of crops. Its potential lies in its integrative power, rapid measurement, genetic variation and repeatability in breeding.

References

Condon, A.G. and Richards, R.A. (1992) Broad sense heritability and genotype × environment interaction for carbon isotope discrimination in field-grown wheat. *Australian Journal of Agricultural Research* 43, 921–934.

Condon, A.G. and Richards, R.A. (1993) Exploiting genetic variation in transpiration efficiency in wheat – an agronomic view. In: Ehleringer, J.R., Hall, A.E. and Farquhar, G.D. (eds), *Stable Isotopes and Plant Carbon–Water Relations.* Academic Press, San Diego, CA, pp. 435–450.

Condon, A.G., Richards, R.A. and Farquhar, G.D. (1987) Carbon isotope discrimination is positively correlated with grain yield and dry matter production in field grown wheat. *Crop Science* 27, 996–1001.

Condon, A.G., Farquhar, G.D. and Richards, R.A. (1990) Genotypic variation in carbon isotope discrimination and transpiration efficiency in wheat: leaf gas exchange and whole plant studies. *Australian Journal of Plant Physiology* 17, 9–22.

Condon, A.G., Richards, R.A. and Farquhar, G.D. (1993) Relationships between carbon isotope discrimination, water use efficiency and transpiration efficiency for dryland wheat. *Australian Journal of Agricultural Research* 44, 1693–1711.

Cooper, P.J.M., Keatinge, J.D.H. and Hughes, G. (1983) Crop evapotranspiration – a technique for calculation of its components by field measurements. *Field Crops Research* 7, 299–312.

Ehdaie, B.A., Hall, A.E., Farquhar, G.D., Nguyen, H.T. and Waines, J.G. (1991) Water use efficiency and carbon isotope discrimination in wheat. *Crop Science* 31, 1282–1288.

Ehleringer, J.R., Klassen, S., Clayton, C., Sherrill, D., Fuller-Holbrook, M., Fu, Q.A. and Cooper, T.A. (1991) Carbon isotope discrimination and transpiration efficiency in common bean. *Crop Science* 31, 1611–1615.

Evans, J.R., Sharkey, T.D., Berry, J.A. and Farquhar, G.D. (1986) Carbon isotope discrimination measured concurrently with gas exchange to investigate CO_2 diffusion in leaves of higher plants. *Australian Journal of Plant Physiology* 13, 281–292.

Farquhar, G.D. (1983) On the nature of carbon isotope discrimination in C_4 species. *Australian Journal of Plant Physiology* 10, 205–226.

Farquhar, G.D. and Richards, R.A. (1984) Isotopic composition of plant carbon correlates with water-use efficiency of wheat genotypes. *Australian Journal of Plant Physiology* 11, 539–552.

Farquhar, G.D., O'Leary, M.H. and Berry, J.A. (1982) On the relationship between carbon isotope discrimination and intercellular carbon dioxide concentration in leaves. *Australian Journal of Plant Physiology* 9, 121–137.

Farquhar, G.D., Ehleringer, J.R. and Hubick, K.T. (1989) Carbon isotope discrimination and photosynthesis. *Annual Review of Plant Physiology and Plant Molecular Biology* 40, 503–537.

Farquhar, G.D., Lloyd, J., Taylor, J.A., Flanagan, L.B., Syvertson, J.P., Hubick, K.T., Wong, S.C. and Ehleringer, J.R. (1993) Vegetation effects on the isotope composition of oxygen in atmospheric CO_2. *Nature* 363, 439–443.

Hall, A.E., Mutters, R.G., Hubick, K.T. and Farquhar, G.D. (1990) Genotypic differences in carbon isotope discrimination by cowpeas under wet and dry field conditions. *Crop Science* 30, 300–305.

Hall, A.E., Ismail, A.M. and Menendez, C.M. (1993) Implications for plant breeding of genotypic and drought-induced differences in water-use efficiency, carbon isotope discrimination, and gas exchange. In: Ehleringer, J.R., Hall, A.E. and Farquhar, G.D. (eds), *Stable Isotopes and Plant Carbon–Water Relations*. Academic Press, San Diego, CA, pp. 349–369.

Hubick, K.T. and Farquhar, G.D. (1989) Genetic variation in carbon isotope discrimination and the ratio of carbon gained to water lost in barley. *Plant, Cell and Environment* 12, 795–804.

Hubick, K.T., Farquhar, G.D. and Shorter, R. (1986) Correlation between water-use efficiency and carbon discrimination in diverse peanut (*Arachis*) germplasm. *Australian Journal of Plant Physiology* 13, 803–816.

Hubick, K.T., Shorter, R. and Farquhar, G.D. (1988) Heritability and genotype × environment interactions of carbon isotope discrimination and transpiration efficiency in peanut. *Australian Journal of Plant Physiology* 15, 799–813.

Ismail, A.M and Hall, A.E. (1992) Correlation between water-use efficiency and carbon isotope discrimination in diverse cowpea genotypes and isogenic lines. *Crop Science* 32, 7–12.

Johnson, D.A., Asay, K.H., Tieszen, L.T., Ehleringer, J.R. and Jefferson, P.G. (1990) Carbon isotope discrimination: potential in screening cool-season grasses for water limited environments. *Crop Science* 30, 338–343.

Martin, B. and Thorstenson, Y.R. (1988) Stable carbon isotope composition ($\delta^{13}C$), water use efficiency, and biomass productivity of *Lycopersicum esculentum*, *Lycopersicum pennellii*, and the F_1 hybrid. *Plant Physiology* 88, 213–217.

Masle, J., Farquhar, G.D. and Wong, S.C. (1992) Transpiration ratio and plant mineral content are related among genotypes of a range of species. *Australian Journal of Plant Physiology* 19, 709–721.

Morgan, J.A., LeCain, D.R., McCaig, T.N and Quick, J.S. (1993) Gas exchange, carbon isotope discrimination, and productivity in winter wheat. *Crop Science* 33, 178–186.

Preston, T. (1992) The measurement of stable isotope natural abundance variations. *Plant Cell and Environment* 15, 1091–1097.

Richards, R.A. (1992) The effect of dwarfing genes in spring wheat in dry environments. II. Growth, water use and water use efficiency. *Australian Journal of Agricultural Research* 43, 529–539.

Richards, R.A., Lopez-Castaneda, C., Gomez-Macpherson, H. and Condon, A.G. (1993) Improving the efficiency of water use by plant breeding and molecular biology. *Irrigation Science* 14, 93–104.

Turner, N.C. (1993) Water use efficiency of crop plants: potential for improvement. In: Buxton, D.R., Shibles, R., Foisberg, R.A., Blad, B.L., Asay, K.H., Paulsen, G.M. and Wilson, R.F. (eds), *International Crop Science I*.

Crop Science Society of America, Madison, pp. 75–82.

Virgona, J.M., Hubick, K.T., Rawson, H.M., Farquhar, G.D. and Downes, R.W. (1990) Genotypic variation in transpiration efficiency, carbon isotope discrimination and carbon allocation during early growth in sunflower. *Australian Journal of Plant Physiology* 17, 207–214.

Vos, J. and Groenwold, J. (1989) Genetic differences in water-use efficiency, stomatal conductance and carbon isotope fractionation in potato. *Potato Research* 32, 113–121.

Wright, G.C., Hubick, K.T. and Farquhar, G.D. (1988) Discrimination in carbon isotopes of leaves correlates with water-use efficiency of field grown peanut cultivars. *Australian Journal of Plant Physiology* 15, 815–825.

Index